Developing Robust Date and Time Oriented Applications in Oracle Cloud

A comprehensive guide to efficient date and time management in Oracle Cloud

Michal Kvet

BIRMINGHAM—MUMBAI

Developing Robust Date and Time Oriented Applications in Oracle Cloud

Group Product Manager: Rahul Nair

Publishing Product Manager: Preet Ahuja

Senior Editor: Tanya D'cruz

Technical Editor: Rajat Sharma

Copy Editor: Safis Editing

Language Support Editor: Safis Editing

Project Coordinator: Ashwin Kharwa

Proofreader: Safis Editing

Indexer: Sejal Dsilva

Production Designer: Vijay Kamble

Marketing Coordinator: Nimisha Dua

First published: May 2023

Production reference: 1190423

Published by Packt Publishing Ltd.

Livery Place

35 Livery Street

Birmingham

B3 2PB, UK.

ISBN 978-1-80461-186-9

www.packtpub.com

This book has been in progress for a long time, and I would like to thank my family for their continuous support and understanding, as well as Professor Karol Matiaško for the opportunity to grow professionally and become a database expert. I want to thank all the members of Packt Publishing for guiding me and helping to make the book top-quality. Finally, I want to thank God for the gift of wisdom and the people who trusted me and stood by me in any way.

– Michal Kvet

Contributors

About the author

Michal Kvet is a researcher, educator, and database expert at the University of Žilina in Slovakia. His primary focus areas are databases, analytics, performance, and cloud computing. He works closely with Oracle and Oracle Academy. He is the co-author of multiple textbooks (a SQL and PL/SQL cookbook, a book on APEX application development, a book on temporal databases, and a MySQL cookbook), coordinates multiple Erasmus+ projects (`https://code-in.org/`, `https://beeapex.eu/`, and `https://evergreen.uniza.sk/`), and co-organizes several research conferences and database workshops. Besides this, he supervises engineering projects and bachelor's, master's, and doctoral theses. Over the years, his research has been associated with date and time management and temporal databases. He has Oracle's SQL, PL/SQL, Cloud, Analytics, and Administration certifications. His core knowledge of temporality is provided to you in this book.

About the reviewers

Heli Helskyaho is the CEO for Miracle Finland Oy. Heli holds a master's degree (computer science) from the University of Helsinki and specializes in databases. At the moment, she is working on her doctoral studies at the University of Helsinki. Heli has been working in IT since 1990. She is an Oracle ACE Director and a frequent speaker at many conferences. She is the author of several Oracle technology-related books.

I want to thank Michal for inviting me to work on this project. It is always a pleasure to read his writing and learn from his knowledge. Thank you, Ashwin Dinesh Kharwa, and the rest of the Packt team for the support. This book was a very fun project, and I am very glad I was able to be part of it.

Dr. Frane Urem is a highly skilled vice dean and professor at the Polytechnic of Šibenik and the University of Zadar. With a Ph.D. in computer engineering and professional experience in the IT industry, Dr. Urem is skilled in different areas of computer science. He advocates for project-based and non-formal learning, encouraging students to explore and develop their skills through real-world projects. As a tech reviewer, Dr. Urem provides valuable insights into and analysis on various tech products and services. His background in software engineering, databases, and information system design allows him to evaluate different technologies comprehensively.

I dedicate this review to my wife and kids, who inspire me daily to be my best self. Thank you for your love, patience, and support. I could not have done this without you.

Table of Contents

Part 1: Discovering Oracle Cloud

1

Oracle Cloud Fundamentals   3

2

Part 2: Understanding the Roots of Date and Time

3

4

Concepts of Temporality   107

Part 3: Modeling, Storing, and Managing Date and Time

5

Modeling and Storage Principles   127

6

Conversion Functions and Element Extraction 157

7

Date and Time Management Functions 183

8

Delving into National Language Support Parameters   219

Part 4: Modeling Validity Intervals

9

Duration Modeling and Calculations   243

10

Interval Representation and Type Relationships   277

11

Temporal Database Concepts   303

12

Building Month Calendars Using SQL and PL/SQL   331

Part 5: Building Robust and Secure Temporal Solutions

13

Flashback Management for Reconstructing the Database Image   351

14

Building Reliable Solutions to Avoid SQL Injection   363

Part 6: Expanding a Business Worldwide Using Oracle Cloud

15

16

Assessments   419

Index   425

Other Books You May Enjoy   436

Preface

Hi, developers, database managers, administrators, and anyone interested in expanding the knowledge of proper database date and time modeling.

Data processing is an inseparable part of information technology. The relational paradigm is currently widespread, based on the entities and relationships powered by relational algebra. In this environment, two significant streams and structures can be identified – operational data, defined by transactions, and data analytics, defined by data warehouses. Thus, it is necessary to store and evaluate not only current valid data but also states that were valid in the past and future plans, right? Proper decision-making is based on complex data, reflecting the history and evolution of the object states, and future prognoses of the object values. Evidently, database object states must be timeline-referenced, and this is exactly the area that I aim at with this book.

In this book, I offer my many years of experience with Oracle Database, from the on-premises world to current cloud solutions. Oracle Database provides sophisticated solutions for large data tuple management, ensuring performance, optimization, security, and reliable data storage.

This book is a practical guide on how to define, manage, and operate date and time-delimited data, operated by the Oracle Database. I will drive you through a definition of temporality, the evolution of date and time management, and its specifics. All the chapters are practically oriented, focusing on the functions and conversions available, as well as limitations.

By understanding the principles of date and time modeling, definitions, and processing, extended by regional rules and habits, as well as time-zone references, you will get the opportunity to grow your business faster and spread your applications and information systems worldwide.

Enjoy reading and get inspired!

Who this book is for

There are five main roles and groups that are the target audience of this content:

- **Students** who need to understand the complexity of data processing, its evolution, date and time definition techniques and aspects, and all the consequences of improper definitions and management.

- **Application developers and testers** who need to get insights into individual date and time data types, functionalities, conversion and extraction principles, as well as the complexity of region and time-zone references. Beyond that, they will understand the concepts, limitations, and techniques involved in building reliable solutions to be spread to any region or time zone by referring to time-zone synchronization.

- **Cloud managers and integrators** who need to focus on time-zone shifts with a proper reference for integrating and moving on-premises solutions to the Oracle Cloud environment by gaining the advantages of using autonomous databases and their automated optimization and administration processes.

- **Database administrators** who need to accept the Oracle Cloud fundamentals, migration perspectives, and overall date and time standardization by helping developers optimize their solutions and the performance of systems. Beyond that, they can cooperate with developers to define solutions that do not allow SQL injection, even for date and time processing.

- **Testers** who need to grasp the complexity of proper date and time processing, respecting the regional habits and national language parameters to offer universal solutions that are deployable worldwide.

- Naturally, this book is well suited to anyone interested in getting deeper knowledge about date and time processing in Oracle Database to build robust date- and time-oriented applications in Oracle Cloud.

What this book covers

Chapter 1, Oracle Cloud Fundamentals, provides a description of the Oracle Cloud concepts, deployment models, and the terminology used in the Oracle Cloud environment. It will drive you through the Oracle Cloud registration process and how to provision an autonomous database. It will also describe the core elements of the database systems by focusing on the core memory structures and background processes, as well as the database system architectures on offer and mapping principles between the instance and database.

Chapter 2, Data Loading and Migration Perspectives, navigates you through the techniques of moving data to the database repositories using SQL Loader and server- and client-side imports and exports. Moreover, it provides various enhancements to optimize the migration process with practical examples of techniques to use.

Chapter 3, Date and Time Standardization Principles, deals with the roots of date and time management and the ISO 8601 standard reference. It describes the principles of date value management, time element modeling, and composite value definition. It also summarizes period-of-time modeling and representation.

Chapter 4, Concepts of Temporality, is mostly theoretical. It describes the term temporality and associates it with multiple temporal aspects, such as Daylight Saving Time or summer and winter time, by emphasizing time-zone shifts and leap years. It explains the differences between calendar models and introduces the concept of a leap second.

Chapter 5, Modeling and Storage Principles, provides you with a summary of existing data types for modeling date and time values in Oracle Database, as well as applied arithmetic. Equally, it describes constructor functions and the limitation of the concept of storing date values in a numerical format.

Chapter 6, Conversion Functions and Element Extraction, continues directly from the previous chapter by discussing the methods for converting values across data types. It also deals with element extraction techniques, focusing on reliability and integrity issues. Finally, it provides a discussion regarding conversion validation.

Chapter 7, Date and Time Management Functions, provides a general description of the methods for dealing with date and time values, such as identifying next Sunday or adding a specified number of months consistently. Understanding the existing methods significantly simplifies the development process, explicit data management, and definitions. Equally, the concept of identifying people using a date of birth is discussed.

Chapter 8, Delving into National Language Support Parameters, focuses on the regional principles, rules, and habits related to date and time management. By reading this chapter, you will gain knowledge of the overall parameters that influence the processing and outputs. Moreover, Oracle Database offers on-the-fly implicit data conversions. You will become familiar with the techniques of specifying default formats and date and time element mapping.

Chapter 9, Duration Modeling and Calculations, is responsible for teaching you the overall techniques of duration modeling and calculation. However, the validity of states is often unlimited in time. We simply do not know in advance when the state will change or even, if ever. Therefore, we will show you ways to model unlimited validity.

Chapter 10, Interval Representation and Type Relationships, leads on from the previous chapter's introduction to duration modeling. In this chapter, the focus is on the interaction of states and the relationships between them. This chapter will also discuss how temporal validity is powered by defining periods. After reading the chapter, you will have learned how to identify period usage and some related limitations.

Chapter 11, Temporal Database Concepts, introduces temporal databases. First, it deals with temporal dimensions and categorization. Secondly, we define the temporal architectures and granularity used for processing, dealing with the object, attribute, and synchronization group levels.

Chapter 12, Building Month Calendars Using SQL and PL/SQL, analyzes how to build monthly calendars, either using SQL or PL/SQL. It aims at identifying the first day of the month, the weekday reference, and the number of days, taking leap years into account. Moreover, name day management is discussed.

Chapter 13, Flashback Management for Reconstructing the Database Image, focuses on getting historical data. Relational databases are transaction oriented, right? So, why not use those transaction logs to reconstruct the database as it existed in the past? Flashback technology allows you to use operating logs efficiently by introducing multiple techniques for this purpose.

Chapter 14, Building Reliable Solutions to Avoid SQL Injection, deals with security. I assume you don't want anyone to have access to your data, do you? We will show how the date and time can be misused to attack the database through SQL injection. Naturally, the techniques to prevent this are proposed as well.

Chapter 15, Timestamp Enhancements, recognizes extensions of the TIMESTAMP data type. The focus of this chapter is on time-zone management, shift calculation, and overall date and time synchronization across multiple regions. Beyond that, it discusses TIMESTAMP value normalization techniques. Alongside the entire concept of time-zone management, we will emphasize the importance of UTC references.

Chapter 16, Oracle Cloud Time-Zone Reflection, is the last chapter. It deals with the process of moving local systems and spreading them across regions. It requires using client time-zone references instead of server references given that clients can be geographically distributed across multiple regions. We will show you how to dynamically reflect the time zone of a server's location to the client.

To get the most out of this book

To get the most from this book, it is recommended to have basic knowledge of SQL and PL/SQL, as well as the architecture of Oracle Database. It is also recommended to have an Oracle Cloud account and an autonomous database provisioned, at least using the Always Free version to test and implement solutions practically.

Software/hardware covered in the book
Oracle SQL Developer Desktop or any other developer tool for data management
An Oracle Cloud account

Please note that the process of provisioning and installation is described in *Chapter 1, Oracle Cloud Fundamentals*.

If you are using the digital version of this book, we advise you to type the code yourself or access the code from the book's GitHub repository (a link is available in the next section). Doing so will help you avoid any potential errors related to the copying and pasting of code.

Download the example code files

You can download the example code files for this book from GitHub at https://github.com/PacktPublishing/Developing-Robust-Date-and-Time-Oriented-Applications-in-Oracle-Cloud. If there's an update to the code, it will be updated in the GitHub repository.

We also have other code bundles from our rich catalog of books and videos available at https://github.com/PacktPublishing/. Check them out!

Download the color images

We also provide a PDF file that has color images of the screenshots and diagrams used in this book. You can download it here: https://packt.link/s20k8.

Conventions used

There are a number of text conventions used throughout this book.

Code in text: Indicates code words in text, database table names, folder names, filenames, file extensions, pathnames, dummy URLs, user input, and Twitter handles. Here is an example: "The ADD_MONTHS function has two parameters – date value (date_val) and the number of months (number_months) to be added or subtracted."

A block of code is set as follows:

```
select ADD_MONTHS(TO_DATE('15.02.2022', 'DD.MM.YYYY'), 7)
  from dual;
--> 15.09.2022 00:00:00
```

Note, that statement results are prefixed by the --> notation.

When we wish to draw your attention to a particular part of a code block, the relevant lines or items are set in bold:

```
select to_date('30.february 2022'
                default null on conversion error,
             'DD.MM.YYYY')
  from dual;
```

Any command-line input or output, as well as syntax, is written as follows:

```
alter database set time_zone = '+09:00';
```

Bold: Indicates a new term, an important word, or words that you see onscreen. For instance, words in menus or dialog boxes appear in **bold**. Here is an example: "Database provisioning can be done by locating the menu on the home screen (**Launch Resources**) or by clicking on the hamburger menu and navigating to **Oracle Database | Autonomous Database**."

> **Tips or important notes**
> Appear like this.

Get in touch

Feedback from our readers is always welcome.

General feedback: If you have questions about any aspect of this book, email us at customercare@packtpub.com and mention the book title in the subject of your message.

Errata: Although we have taken every care to ensure the accuracy of our content, mistakes do happen. If you have found a mistake in this book, we would be grateful if you would report this to us. Please visit www.packtpub.com/support/errata and fill in the form.

Piracy: If you come across any illegal copies of our works in any form on the internet, we would be grateful if you would provide us with the location address or website name. Please contact us at copyright@packt.com with a link to the material.

If you are interested in becoming an author: If there is a topic that you have expertise in and you are interested in either writing or contributing to a book, please visit authors.packtpub.com.

Share Your Thoughts

Once you've read *Developing Robust Date and Time Oriented Applications in Oracle Cloud,* we'd love to hear your thoughts! Scan the QR code below to go straight to the Amazon review page for this book and share your feedback.

https://packt.link/r/1804611867

Your review is important to us and the tech community and will help us make sure we're delivering excellent quality content.

Download a free PDF copy of this book

Thanks for purchasing this book!

Do you like to read on the go but are unable to carry your print books everywhere?

Is your eBook purchase not compatible with the device of your choice?

Don't worry, now with every Packt book you get a DRM-free PDF version of that book at no cost.

Read anywhere, any place, on any device. Search, copy, and paste code from your favorite technical books directly into your application.

The perks don't stop there, you can get exclusive access to discounts, newsletters, and great free content in your inbox daily

Follow these simple steps to get the benefits:

1. Scan the QR code or visit the link below

https://packt.link/free-ebook/9781804611869

2. Submit your proof of purchase
3. That's it! We'll send your free PDF and other benefits to your email directly

Part 1: Discovering Oracle Cloud

It is evident that local data centers are relics of the past. Let this part assist you in understanding the core concepts of Oracle Cloud, deployment models, and their properties, specifics, and limitations, as well as the main terminology associated with cloud technology.

Soon, you will become familiar with Oracle Cloud registration, database definition, and provisioning. That's great. But what happens next? Data must be migrated to a cloud environment. Therefore, this part introduces `SQL Loader`, `server`, and the client-side `import` and `export` tools, and explores multiple migration techniques.

Are you are unfamiliar with the Oracle Cloud environment and do not know precisely where and how to start? This part is for you, with the whole process outlined step by step.

This part includes the following chapters:

- *Chapter 1, Oracle Cloud Fundamentals*
- *Chapter 2, Data Loading and Migration Perspectives*

1

Oracle Cloud Fundamentals

In the past, data was stored in a local server room with limited expansion opportunities. Each server had a specific hardware capability. Upgrades were often costly and technically demanding, resulting in the need to buy new equipment. Later, distributed architectures were created to ensure robustness and resilience, but one way or another, the solution was not so complex and robust. Scalability can partially be achieved by dynamically reacting to the current and expected workload; however, *cloud storage* and *databases* provide the technical foundation needed for easy scalability. In terms of *Oracle Cloud databases*, autonomous management and technologies are a significant milestone.

Automation is now present almost everywhere, whether in smart devices and smartphones, modern cars full of sensors that are partially operated autonomously, or smart homes and cities, including advanced functions associated with **Machine Learning (ML)**, **Artificial Intelligence (AI)**, or the **Internet of Things (IoT)**. **Autonomous Databases (ADBs)** go even further by providing a complex environment for your data handling, apps, and services to produce effective outcomes, reducing the costs and time required to set parameters, optimize the configuration, and so on.

As a business expands, the amount of data to be handled grows exponentially. It will no longer be sufficient to only cover current valid data. Historical data needs to be stored, manipulated, and evaluated, either in an original form or analytically aggregated in data warehouses, marts, or other analytical structures. As the data quantity grows, it cannot be managed manually by one local machine. It is necessary to ensure availability in an error-prone environment. Thus, additional servers must be employed to serve as backups, standby, and so on. The whole environment needs to be secured and properly interconnected over networks.

Managers and administrators have also realized that putting whole structures in one building is neither suitable nor secure, resulting in the need to rent other server rooms, usually geographically distributed rooms. That's exactly where the cloud comes into play. The entire administration, securing, distribution, and backup strategies are moved to the cloud environment, so we arrive at the concept of autonomous processing here again.

In this chapter, we're going to cover the following main topics:

- Oracle Cloud core concepts and the Always Free option

- Defining Oracle ADBs and their types and principles

- Deployment models and database architectures

- Process of database provisioning and connecting

- Database system architecture overview – database and instance levels

Note that the source code can be found in the GitHub repository accessible via this web address: `https://github.com/PacktPublishing/Developing-Robust-Date-and-Time-Oriented-Applications-in-Oracle-Cloud/tree/main/chapter%2001`.

Alternatively, you can scan the following QR code as well:

Oracle Cloud concepts

Oracle ADBs provide you with complexity, robustness, availability, and security with the following enhancements:

- **Self-driving**: ADBs reduce the human activity required to provision, secure, monitor, back up and recover, as well as troubleshoot and perform maintenance activities to optimize and tune the database and its overall performance. They strongly reduce the work required by administrators so that they can focus on other tasks, apps, and database optimization strategies. Moreover, ADBs are converged. They can serve any data structures and types – relational, graph, temporal, spatial, streams or object structures, XML, JSON, and so on. Therefore, provisioned databases do not need to be oriented to a single purpose. The Oracle Cloud architecture is based on the Exadata platform, covering dynamic **Online Transaction Processing** (**OLTP**) in **Autonomous Transaction Processing** (**ATP**) or an analytical support layer, defined in **Autonomous Data Warehouses** (**ADW**). A specific type is covered by Oracle **Autonomous JSON Database** (**AJD**), which is specialized for NoSQL-style applications that use **JavaScript Object Notation** (**JSON**) **documents**. It is a feature-scoped service for storing, manipulating, and retrieving JSON documents using SQL and Document APIs. JSON is very flexible, allowing us to process schemaless data by offering dynamic reactions to application changes. There's no need to normalize the content into relational tables. Oracle AJD typically uses **Simple Oracle Document Access (SODA)** APIs.

- **Self-securing**: Specific services ensure system protection via firewalls and threat detection. Individual updates and patches are applied without requiring user or administrator intervention, and even with zero downtime. Data always goes through end-to-end encryption. The cloud ensures security at all levels.

- **Self-repairing**: ADBs are more powerful, robust, and reliable than manually administered local databases. Data images are automatically mirrored and spread across different regions. This automatically protects the system from any physical failure (at the server or data center level) by shifting the workload to different standby databases. The fact that systems are multiplicated allows the system to be upgraded with no downtime.

Oracle Cloud Infrastructure (**OCI**) extends the original on-premises systems with high-performance computing power running in a cloud environment. The main advantage is elasticity, allowing the system to dynamically reflect the current workload, processing demands, and user activity. In addition, it uses Oracle autonomous services, an integrated security layer, robust functionality, and optimization techniques. OCI improves your performance and processing through autonomous services, easy migration, cost reduction, and performance enhancements.

The following list summarizes the available product categories in OCI:

- **Oracle Analytics** uses built-in ML and AI to propose a robust solution for the company and offer better decision-making opportunities. It covers Oracle Analytics Cloud, Oracle Big Data Service, Oracle Big Data SQL Cloud Service, Oracle Data Science, OCI Data Flow, and many more.

- The **application development environment** handles data-driven application development by simplifying the whole development process. It covers the API Gateway service, Blockchain Platform, OCI Data Science, Oracle Digital Assistant, Java functionality, OCI Events Service, Mobile Hub, Oracle MySQL Database Service, and more. Two solutions should be emphasized – **Oracle Application Express** (**APEX**) and Visual Builder. These tools provide a complex environment to create web- or mobile-based applications based on SQL, PL/SQL, or JavaScript functionality. Thus, using these tools makes implementation far easier, aided by rapid development. The solution can be created overnight.

- **Applied software technologies** include AI, blockchain, ML, data science, and digital assistants.

- **Compute nodes** ensure scalability and overall performance.

- **Databases** such as ATP, ADW, AJD, Oracle Base Database Service (formerly known as Database Cloud Service) (bare-metal/**Virtual Machine** (**VM**)), Exadata Cloud Service, and so on are offered.

- **Integration** is performed using API Gateway, Oracle GoldenGate, Oracle Data Integrator, OCI Data Integration, and Oracle SOA Cloud Service.

- **Observability and management** are offered by logging, monitoring, notifications, and other techniques, along with the OCI Resource Manager service.

- **Networking and connectivity** are managed by the DNS, email delivery, FastConnect, health checks, load balancing, **Virtual Cloud Networks** (**VCNs**), and so on.

- **Security, identity, and compliance** reduce the constant threat risk with security-first design principles, utilizing built-in tenant isolation and least privilege access. There are several defense layers that can be implemented, including **Identity and Access Management** (**IAM**), OCI Vault key management, Security Zones, Cloud Guard, Web Application Firewall, Bastion, and Vulnerability Scanning Service. The core element of overall security, however, is the always-on data encryption. Automated security is responsible for reducing complexity, the number of human errors, and costs with automated patching.

- **Storage** includes Archive Storage, Block Volume, Data Transfer, File Storage, Local NVMe SSD storage, Object Storage, and Storage Gateway.

Oracle Cloud technology is in widespread use across the whole world by both commercial and government entities. In Europe, clouds are located in multiple cities, including Amsterdam, London, Frankfurt, Zürich, and Newport. New locations are still being opened over time, and availability is widespread.

Now that you understand the core properties of Oracle Cloud, it's time to provision the database and get started with practicing. Oracle offers you the **Always Free** option with time-unlimited resources for testing and studying purposes. Although it is resource limited, as the name suggests, it's free. Later on, you can apply for the paid option to extend the functionality and resources if required.

The Always Free option

OCI has launched a significant project to offer cloud services to students and developers for testing and evaluation. The Oracle Cloud Always Free tier is provided to students to test the suitability of their environments. Services are time-unlimited with the following resource limitations:

- 2 ADBs, each limited to 1 OCPU and 20 GB of disk storage

- Compute VMs

- 2 Block Volume storage instances – 100 GB in total

- 10 GB object storage

- 10 GB archive storage

If you are unfamiliar with OCI, the Oracle Cloud Always Free option makes sense. It is implemented inside the Oracle Cloud Free Tier, which comes with a 30-day free trial with $300 of free credits and access to a wide range of Oracle Cloud services during the trial period (containing the Database, Analytics, Compute, and Container Engine for Kubernetes services). This is limited to no more than 8 instances across these services, and up to 5 **terabytes** (**TB**) of storage. After the free 30 days are up, it shifts to the Always Free tier.

Scan the following QR code or use the link to get Oracle Cloud Always Free option:

https://www.oracle.com/cloud/free/

The registration is straightforward – a wizard will walk you through it. First, you must specify some parameters: a username, which defines the tenancy, and a password. Note that a credit card must be provided, but it will not be charged. This allows you to dynamically make your account a paid account to take advantage of the multiple features offered. It is also used to make sure that a real person is registering and not a bot.

If you are a student, ask your teachers and representatives to provision a cloud account for you. If the school is part of the Oracle Academy project, the whole registration process is simplified and significantly faster. Moreover, there is no requirement to provide a credit card at all. Oracle Academy (https://academy.oracle.com) is free for schools and universities, offering you many resources, learning materials, and other teaching resources.

Figure 1.1 – Oracle Cloud home screen

Among that, Oracle offers OCI tutorials for free, such as LiveLabs, the OCI Architecture Center, or GitHub repositories:

LiveLabs	OCI Architecture Center	GitHub repositories

Figure 1.1 shows the home screen of the Oracle Cloud environment and highlights its main functionality. As you can see, multiple database types can be provisioned. In the next section, you will get an overview of ATP databases and analytical model covered by ADW.

Types of ADBs

ADB processing is available only in OCI and Exadata Cloud@Customer environments. There is no on-premises version of an ADB. There are three types of ADBs, which are distinguished by their formats and workload types:

- **ATP** is used for operational data defined by the short transactions changing the data.
- **ADW** is analytically oriented by focusing on storing long-term data. It involves a complex data retrieval process.
- **AJD** provides specialized document-oriented storage and management.

Besides these types, there is also an APEX type built for Oracle APEX application development, which provides the ability to define data-driven, low-code solutions. However, generally, APEX applications can be developed on any type of database.

We will now describe and compare these three main types. Selecting the right type for your workload is performance critical.

ATP handles **online transactional** data changes. The data structure should be normalized with no data redundancy or anomalies. Values, which can be calculated, are not stored in the system. ATP requires data consistency and integrity, ensured by the transactions that shift the database from one consistent state to another. The aim is to ensure the performance of data manipulation operations – data modifications (*insert*, *update*, and *delete*), as well as data retrieval (*select*). Thus, the index set must be balanced to serve all these operations.

ADW is a specific database repository that deals with data analysis. The essential operation is the data retrieval process, so overall optimization is focused on the *select* statement performance, ensured by a huge amount of indexes (*B+trees* and *bitmaps*). Data is often denormalized. Precalculated outputs are stored in a column format, compared to the row format used in ATP. There is no focus on data modification, which can, in principle, last any amount of time. The process of analytical processing of large data amounts through complex queries is essential and is the goal of performance optimization.

The main differences between ATP and ADW are summarized in the following table:

Category	ADW	ATP
Memory configuration	Parallel joins and complex aggregation processing in memory (focus on the **Private Global Area (PGA)**)	Transaction data processing in the **System Global Area (SGA)** – this limits I/O operations
Optimization	Complex SQL	Response times
Format	Column	Row
Data structure	Pre-calculation and indexes	Normalization and indexes
Statistics collection	Bulk operations	DML operations

Table 1.1 – ATP and ADW summary

The following table shows the resource service priority types. Although, generally, any type can be used for both workload types but in the environment operated in parallel, it is always important to set the operation priorities properly to ensure efficient performance of the online transaction operations. This is because the analytical reports can be too demanding and resource-consuming.

Resource service priority	ADW			ATP		
	Low, Medium, and High			Low, Medium, High, Tp, and Tpurgent		
	Type	SQL parallelism	Concurrency	Type	SQL parallelism	Concurrency
	Low	1	300 x CPUs	Tp	1	300 x CPUs
	Medium	4	1.25 x CPUs	Tpurgent	User specified	300 x CPUs
	High	All CPUs	3 x CPUs			

Table 1.2 – Resource service priority

ADW is determined by three priority levels – *Low*, *Medium*, and *High* (as seen in *Table 1.1*). The Medium and High options support parallelism, while the Low type uses a serial approach. These levels are used for priority definition in reporting and batch processing. ATP can use all priority levels, but the *Tp* and *Tpurgent* types are preferred for **Transaction Processing** (**TP**). The highest priority is covered by the *Tpurgent* option.

Finally, AJD is a non-relational database used to store, manage, and obtain data in JSON document format. The structure is not fixed, allowing a variety of data to be stored in each row. There is no normalization strategy.

Considering the historical evolution of cloud migration, it is clear that administrators and managers tended to store all the data locally in their companies. They simply believed that data was secure if it was stored and administered by them and the storage and all the hardware capabilities were physical and visible. However, sooner or later, the hardware could collapse, and data would be lost if the backups were not managed properly. Even if backups were available, a company's credibility was corrupted if any amount of online data was lost.

On the other hand, local on-premises data management was sometimes required due to various challenging limitations – the cost of service, interoperability, laws (meaning that data could not be stored *outside* the company), regional availability (cloud repositories were now uniformly distributed across regions), security, integration techniques (some applications were not able to run in the cloud), and so on.

Most of these challenges were solved consecutively using mechanisms that can handle operations and their complexity. By using cloud solutions, almost all the activities related to the infrastructure, hardware, and security are covered by the cloud vendors. You do not need to patch and update your system. It is always available and backups are handled automatically, creating reliable and secure solutions. In a nutshell, cloud companies generally offer these service models:

- **Infrastructure as a Service** (**IaaS**) provides a lower abstraction level and supplies the machines (both physical and virtual) with storage capacity, firewalls, network gateways, and workload balancers. Thus, the cloud provides you with the storage, server, network, OS, and overall software tools to run Oracle. You are allowed to bring your own software image and are responsible for application software, maintaining the OS, and installing patches.

- **Platform as a Service** (**PaaS**) provides a higher level of abstraction. You, as the customer, are not responsible for administering infrastructure and other cloud resources, such as the OS, database, and so on. PaaS serves you the database with no necessity to install Oracle software or configure its environment.

- **Software as a Service** (**SaaS**) offers the highest abstraction level. The user is not responsible for the platform or infrastructure. Applications are located in the cloud, and the whole responsibility is shifted to the vendor. You just pay for the usage.

Now that you have been familiarized with the types of cloud services available, which offer several different levels of abstraction, it is important to remember that from the physical access point of view, either individual cloud resources can be shared or private separate hardware can be provisioned just for you. A database is always private, but the instance resources can be shared by multiple users to optimize resources, as well as costs. To help you understand resource sharing, Oracle has provided four deployment models. The next section will walk you through these deployment models.

Understanding the deployment models

Many users and companies still require their data to be stored in their local data center but would like to enjoy the benefits of the cloud's robustness, stability, and power. To serve the varied requirements of different businesses, four deployment models have been introduced – public, private, hybrid, and community cloud.

A public cloud

The general solution is covered by a public cloud, in which all resources are part of the cloud provider data center, shared by the users. Users do not need to invest in the hardware. They just rent the resources available. Moreover, this ensures the dynamic scalability of individual resources, which can be provisioned at any time on demand, reflecting the workload. The disadvantage is that you do not have local data under your control. Thus, if laws and contracts do not allow you to store data outside of your company, you cannot use this option.

Note that the Oracle database can be run on various cloud providers. It is not strictly limited to Oracle Cloud – for example, **Microsoft Azure** or **Amazon Web Service** (**AWS**) can also be used. In 2022, Oracle and Microsoft announced the general availability of Oracle Database Service for Microsoft Azure. Microsoft Azure services can be directly provisioned, managed, and accessed in OCI. Thanks to this cooperation, users can build new applications (or migrate existing ones) on Azure and connect them to the high-performance, high-availability, managed Oracle Database services on OCI. This is done via the **Oracle Azure Interconnect** services.

A private cloud

This model provides you with full control over the resources. Data is kept in your local data center, placed on-premises, but you can still use the power of the cloud. Resources are not shared by multiple customers, making data access separation highly scalable and integrated. This is used for mission-critical enterprise systems that require especially high performance. It allows portability between public and private cloud systems.

The hybrid cloud

The hybrid cloud provides an intermediary between private and public clouds by providing a universal solution. Namely, some applications run in the public cloud, but some systems cannot be migrated there. Therefore, they are operated by private cloud systems. A typical example is an application that needs to be run exclusively on an older version of a database system. The Oracle Cloud environment does not support all versions, just the newest ones available.

The community cloud

The community cloud is the fourth type of deployment model, filling the gap between the other categories already covered. Although it is mostly only used in theory, Oracle supports it and it is therefore worth referencing. A community is characterized by a set of companies sharing the same objectives. Cloud infrastructure is provisioned for the whole community and supervised by the manager responsible for the cloud system.

We will take a different view of the data itself, along with the availability and storage of resources, in the following section. Individual resources can be shared, but the benefits of cloud access can also be used in your own data center using a dedicated type of architecture.

Shared versus dedicated architecture

Each database system comprises the physical data files forming the database and the instance itself, delimited by the memory structures and background processes. During the session creation process, the client contacts the database listener for interconnection and direct access. The client cannot access the database directly. It is operated by the background processes, stored in the memory, forming the result set or processing the data that has been modified.

There are two types of infrastructure options:

- With a *shared* deployment, all resources of Exadata Cloud Infrastructure are shared by users. This allows environments to be set up very quickly by provisioning resources and databases. Thus, the storage and instance are shared. Do not be afraid – naturally, data is not shared across users and applications.

- By contrast, a *dedicated* deployment allows you to separate your applications in a cloud environment in your own dedicated Exadata Cloud Infrastructure. This option is available in customers' data centers (Exadata Cloud@Customer) or a dedicated public cloud can be used.

To provide an overview of the principles and complexity involved in the Oracle Cloud technology, it is necessary to reference the basic terms related to the Oracle Cloud environment. In the next section, we will introduce the main terms related to Oracle Cloud, especially concerning geographical location, resources, and storage management.

Oracle Cloud terminology

This section will introduce you to the core terms of Oracle Cloud, focusing on regions, availability domains, realms, consoles, tenancies, and compartments. VCNs, instances, and images are also covered. We will look at Object Storage as its form of file storage. The complexity of Oracle Cloud and all its properties are very well summarized in the books listed in the *Further reading* section at the end of this chapter.

Region

A **region** is a geographical location from which resources are provided (for example, a VCN).

Availability domains

Each region has at least one availability domain. Each **availability domain** is independent, isolated from other domains, and fault-tolerant. Thus, configuring multiple availability domains can ensure high availability and failure resistance. Each *availability domain* contains three fault domains.

Realms

A **realm** is a logical collection of regions. Each *realm* is isolated and does not share any data with other realms. A *tenancy* (which will be explained next) is associated with just one realm and has access to the region belonging to the realm.

Tenancies

A **tenancy** is a specific cloud repository, usually devoted to an organization or company, and provides secure and isolated storage and processing partitions. You can manage, create, and associate cloud resources and services across a tenancy.

Consoles

The **cloud console** is a web application providing access and management functions for OCI.

Compartments

A **compartment** comprises a cloud resource (instances, VCNs, and so on) with specific privileges and quotas. It is a logical unit rather than a physical container. Note that Oracle provides you with a tenancy after registration, which is a root compartment holding and managing all cloud resources provided. Then, you can create a resource categorization tree. Each resource is associated with a compartment by definition. The core principle is based on granting only users the resources necessary for their work and nothing more.

VCNs

A **VCN** is a virtualized version of a conventional network, including subnets, routers, gateways, and so on. It is located within one region and can spread across multiple availability domains.

Instances

An **instance** is a compute host running in the cloud. Its main advantage is flexibility. You can utilize sources (physical hardware) on demand to ensure performance, high availability, and robustness and comply with the security rules you have set.

Images

An **image** is a specific template covering the operating system and other software installed. In addition, Oracle provides you with several virtual hard drives, which can be used in the cloud, such as Oracle Linux, CentOS, Ubuntu, or Windows Server.

Storage management

Storage management is an inevitable part of data processing. Storage is where external database files are present and logs and backups are accessible. A **block volume** is a virtual hard drive that provides persistent data storage space. It works following similar principles to hard drives in ordinary computers. It is possible to attach or detach them on demand, even to another instance, without any data or application loss.

Object Storage is a storage repository architecture available and accessible from anywhere via a web interface. Physical data files can have any structure and type. Their size is limited to 50 GB per file. Object Storage is a standard repository for backups or large data objects, neither of which are usually changed very often. A **bucket** is a lower architectural definition. It denotes a logical container within Object Storage. Several buckets can be present in any Object Storage instance. The amount of data (in terms of both size and count) is unlimited.

Before provisioning a database, let's reflect on the core element of OCI – IAM. This service allows you to create users, groups, and policies to control access to resources. All these resources are managed and set by the created users. During the provisioning, one user is automatically created, followed by granting them administrator privileges. Individual users can be part of a specific **group** with the same privileges, access options, and permissions. A **policy** specifies the user's access to a particular resource. It is typically set for the whole group using a tenancy or compartment. Individual resources are grouped into compartments, forming the fundamental element of OCI, and ensuring segregation, isolation, and proper organization.

The components of technology managed in the cloud, such as compute instances, database instances, block volumes, load balancers, and so on, are called **resources**.

To ensure the accessibility of individual components and resources, basic knowledge of networking principles is required.

Networking

A **network** is formed by a set of computer or device nodes, where individual nodes can communicate with each other. Each **node** is uniquely identifiable by an IP address. The **router** is the component used for traffic routing within a network. **Firewalls** are used to secure the resource by blocking packets that break security rules. A VCN is a private network running in one OCI region. A VCN is an important element for the definition and configuration of application servers, databases, load balancers, and other cloud services. The overall aim is to ensure the high availability of a robust and reliable solution by applying scalability and security rules. VCNs can be divided into several subnets. The **route table** consists of the rules for the traffic out of a VCN. **Security lists** commonly act as a regular firewall for the subnets. Similarly, **network security groups** act as firewalls for groups of instances across subnets.

If you have created a VCN, you can define **compute instances**. The next section will discuss this.

Compute instances

OCI allows you to define and provision compute hosts called **compute instances**. Each compute instance can be operated and administered independently. OCI offers **bare-metal** (with dedicated physical server access for the highest performance and strong isolation) and **VM instances** (sometimes shortened and expressed as VMs) only.

Compute instances are used to run individual applications or installations, such as Apache, Docker, and so on. For compute instances, various operating systems and versions can be used. They can be installed from the available images already present in the cloud repository, or your own images can be used.

Each instance is delimited by the number of CPUs and network resources and the amount of memory. The list of available platform images can be found in the OCI documentation. Oracle provides images for Oracle Autonomous Linux, CentOS, Ubuntu, Windows Server, and so on, so any system can generally be supported and migrated.

We covered a basic overview of the OCI principles, technology, and available resources in this section. As is evident, Oracle Cloud provides you with a robust solution for storing data and application references, ensuring availability, backup strategies, and patching. Oracle technology is the most relevant for complex information system support. Thus, there is no question of why to migrate to Oracle Cloud since the advantages are unambiguous. The only question is how to do it.

Database provisioning

After connecting to the cloud, you will arrive on the main dashboard screen. There is a hamburger menu in the top-left corner. Clicking on the **Oracle Cloud** logo navigates you to the home screen. Individual resources and configuration options are here. The top panel consists of the cloud location and profile information, containing your identification details, tenancy, user settings, and more, as shown in the following screenshot:

Figure 1.2 – Oracle Cloud home screen main menu

There are several arrows marked in *Figure 1.2*. The yellow arrow (**1**) points to the hamburger menu, while the Oracle Cloud logo (the blue arrow (**2**)) is used for home screen navigation. The current cloud location is present in the menu as well (indicated by the red arrow (**3**)). The green arrow (**4**) points to the user menu, language, notifications, and preference definition.

Database provisioning can be done by locating the menu on the home screen (**Launch Resources**) or by clicking on the hamburger menu and navigating to **Oracle Database | Autonomous Database**. The following screenshot shows the sub-elements for provisioning specific database types:

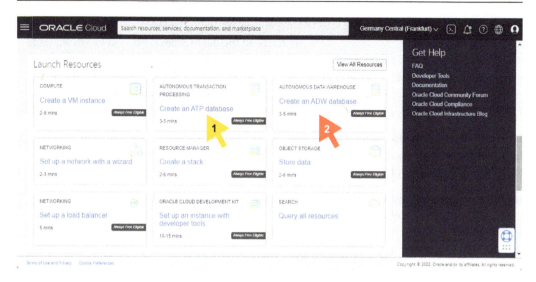

Figure 1.3 – Database provisioning

As can be seen in *Figure 1.3*, resource types are presented in panels, defined by their categories, along with indications on the estimated time required for the creation and whether the resource is available under the Always Free option or whether specific licensing is necessary. Clicking on **Create an ATP database** (the yellow arrow (**1**) in *Figure 1.3*) will take you to a new database parameter specification window, in which you can enter the **Compartment** information, **Database name**, as well as its user-friendly representation, **Display name**, as shown in the following screenshot. The given **Database name** must contain only letters and numbers. The first one should be a letter. The maximum length is 30 characters. The red arrow (**2**) in *Figure 1.3* points to the link to **Create an ADW database**.

Provide basic information for the Autonomous Database

Compartment

kvetmichal (root)

Display name

Database_for_library

A user-friendly name to help you easily identify the resource.

Database name

libraryDB

The name must contain only letters and numbers, starting with a letter. Maximum of 30 characters.

Figure 1.4 – Database parameter specification during database provisioning

Then, the workload type needs to be selected, depending on the intended usage. ADW is suitable for analytics and complex evaluation, with emphasis on the data retrieval process involving large datasets and pre-calculated values. By contrast, ATP is used for a high-concurrency environment with high transactional workloads. The third option is the JSON option, AJD, mainly associated with the document API and management of storage in JSON format. The APEX database type is optimized for building low-code (or even no-code), data-driven applications.

Then, you select the **Deployment type** option, choosing from **Shared** or **Dedicated** architecture, followed by the database configuration – **database system version**, **OCPU count**, and **storage capacity** (the value is expressed in TB).

Finally, the **administrator credentials** need to be defined. Oracle applies a strong password strategy. Currently, it must consist of at least 12 characters with at least 1 uppercase letter, 1 lowercase letter, and 1 number. It cannot contain `admin`, double quotes (`"`), or your username. The requirements for the credentials might evolve, but will always be summarized in the OCI documentation and outlined when entering the details in a pop-up window.

Optionally, network gateways and accessibility rules can be specified. For the licensing, it is possible to **Bring Your Own License (BYOL)**.

New database resources are provisioned by clicking on the **Create Autonomous Database** button. The status of the process is visible below the ATP logo. The orange color represents any maintenance or processing currently taking place, while the green color expresses the current availability of the resources. Individual parameters and properties are listed there as well.

Several buttons are present in the upper part of the database home screen, as shown in the following screenshot:

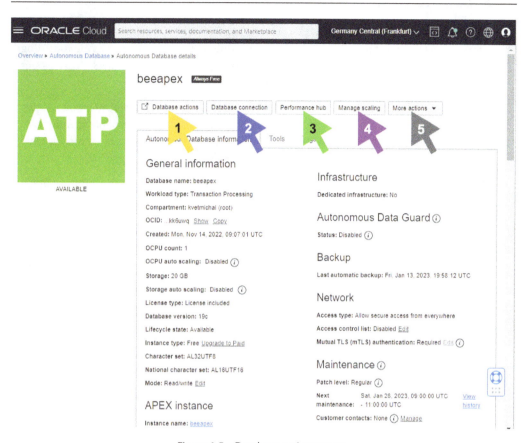

Figure 1.5 – Database actions menu

Referring to the colored arrows in the preceding screenshot, let's look at what each of these buttons does:

- **Database actions** (the yellow arrow (**1**)) launches the **SQL Developer Web** tool. It offers a graphical SQL interface, data modeler, data visualization packages, ML, or REST functionality in the **Development** section. There is also an **Administration** subsection for data import operation management, downloading **Client Credentials (Wallet)**, and user administration and APEX management; a **Monitoring** subsection for performance monitoring and evaluation; a **Downloads** subsection for downloading **Oracle Instant Client** or **SODA** drivers; and **Related Services**, dealing with RESTful, SODA, and ML services.

- **Database connection** (the blue arrow (**2**)) provides you with the client credentials and connection information to connect to the cloud database. In addition, it offers you a zipped file consisting of the encrypted **Client Credentials (Wallet)**. These are used to connect the SQL Developer Desktop environment launched locally on the client computer.

- **Performance hub** (the green arrow (**3**)) consists of extended statistics and services for monitoring activity, such as average active sessions, workloads, blocking sessions, SQL monitoring, **Automatic Database Diagnostic Monitor (ADDM)**, and more. You can monitor within a specific time zone and for a specified time range (database (server), client (browser), or UTC).

- **Manage scaling** (the purple arrow (**4**)) provides you with the ability to extend the storage capacity or shrink it, as well as optimize the OCPU count for the required workload.

- Finally, **More Actions** (the gray arrow (**5**)) allows you to scale, start, stop, or restart the database. In addition, there are options for license management and changing the administrator password.

Among the management buttons, three tabs consist of **Autonomous Database Information**, **Tools**, and **Tags**, as shown in *Figure 1.6*. The **Tools** tab presents **Oracle Application Express (APEX)** for creating low-code database applications, **Oracle ML User Administration**, and **SODA Drivers** for JSON document processing via a set of APIs. SODA drivers are available for REST, Java, Node.js, Python, PL/SQL, and C.

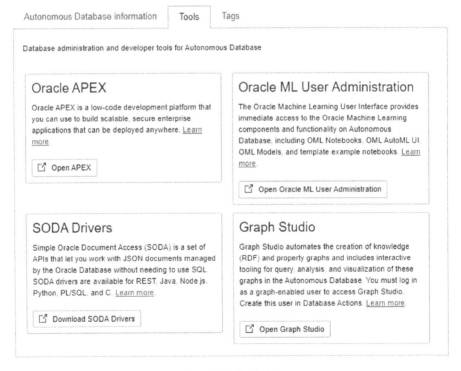

Figure 1.6 – Tools list

Graph Studio, however, allows you to create and manipulate graph databases by automating models and in-memory graphs. **Graph Studio** is a self-service graph database providing a lot of complex data management functionality, visualization tools, and an analytical environment. It is part of the ADB Free Tier and can be applied to ADW and ATP on shared infrastructure. Currently, it is not available for AJD. **Property Graph Query Language** (**PGQL**) is commonly used as an SQL variant, focusing on property graph structures formed by the vertices and edges. *Figure 1.7* shows an example query visualization.

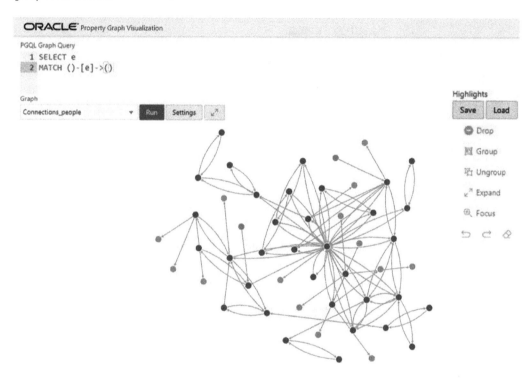

Figure 1.7 – Graph representation example (source: https://docs.oracle.com/en/
database/oracle/property-graph/20.4/spgdg/graph-visualization-application1.
html#GUID-6DDB37F7-C78E-49B7-B062-1240B5D38A5C)

By scrolling down on the Oracle Cloud dashboard, some usage metrics (**CPU Utilization**, **Storage Utilization**, **Sessions**, **Execute Count**, **Running Statements**, and **Queued Statements**) are revealed. Charts can be filtered by time.

You have successfully provisioned the database and are now familiar with the Oracle Cloud console. It's time to connect to the created database using the SQL Developer tool. We will examine both the desktop and cloud versions.

Connecting to the ADB

The easiest way to connect to the database is using **SQL Developer Web**, which is part of the cloud database menu (**Database Actions**).

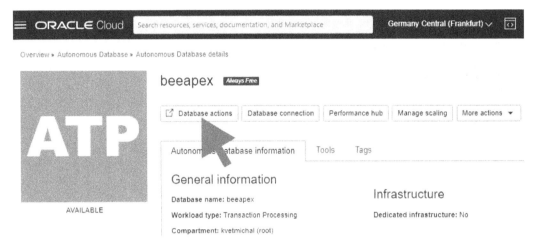

Figure 1.8 – Launching SQL Developer Web

A new browser tab will open, requesting the **Username** and **Password** details. In our case, we will provide the details of the administrator user that were set up when the database was defined and that were applied in the provisioning process.

Figure 1.9 – Launching SQL Developer Web – username definition

Figure 1.9 shows the main screen of SQL Developer Web. It provides the interface for defining SQL scripts, along with the environment for data modeling, APEX application building, as well as managing AJD databases and defining and deploying REST APIs. There are also other categories for easy data loading and exporting using wizards, administration, and monitoring interfaces.

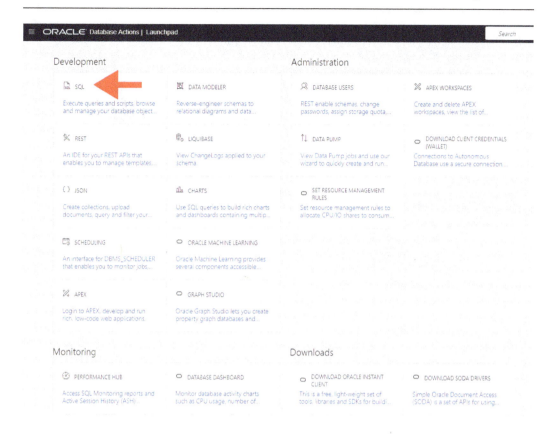

Figure 1.10 – SQL Developer Web – main screen

By clicking on the **SQL** button (the red arrow in the preceding figure), a new window is launched, consisting of three parts, as shown in the following screenshot:

Figure 1.11 – SQL Developer Web – SQL section

The left part consists of the data dictionary reference, highlighting the created objects (the yellow arrow (**1**) in the preceding figure). The upper part is used for SQL statement definitions and forms the core of the entire environment (the blue arrow (**2**)). The bottom part provides results and information summaries (the green arrow (**3**)).

Each ADMIN user automatically gets the privilege to use such a tool. They can also grant that privilege using the enable_schema procedure of ords_admin:

```
begin
  ords_admin.enable_schema
    (p_enabled => TRUE,
     p_schema => 'MICHAL', -- username for the grant
     p_url_mapping_type => 'BASE_PATH',
     p_url_mapping_pattern => 'michal',
     p_auto_rest_auth => NULL
    );
  commit;
end;
/
```

Besides this, REST services can be enabled in the **Administration | Database Users** section.

SQL Developer can also be launched locally in the desktop environment. It is downloadable from the official site:

```
https://www.oracle.com/tools/downloads/sqldev-downloads.html
```

You just need to choose the appropriate platform you are running (if you are using Windows, it is recommended to select the version, including the **Java Development Kit** (**JDK**), if it has not been installed manually before) and its version (the most up-to-date is preferred; new versions are released periodically).

Oracle SQL Developer Desktop does not need to be installed; just unzip the provided archive file. It is powered by Java and can be launched immediately. Before dealing with the database, the new connection must be defined. Click on the green plus symbol (✤) in the **Connections** section and specify the connection details. We will walk you through setting up the parameters and how to obtain them in the first phase. The filled-in dialog window is shown later in *Figure 1.15*.

The **Name** field of the connection is left to your preferences. Whatever you choose will then be listed in **Connection List**. **Database Type** is `Oracle`. SQL Developer Desktop can be used for managing different database system types if the particular drivers are installed. Leave **Authentication Type** set to `Default`. **Username** is `ADMIN` or any other user created in the ADB by you. The **Password** details for the `ADMIN` user were specified during the provisioning and can be changed at any time in the database section's main menu (navigate to **More Actions**). Let **Role** be the default value for ordinary users. If the user belongs to a particular privilege group, such as `SYSDBA`, `SYSOPER`, and so on, choose the appropriate one. For example, `ADMIN` is the database administrator with `SYSDBA` privileges granted.

Connection Type must be changed to **Cloud Wallet**, prompting you to specify the **Configuration File** information, consisting of the connection details. The question now is how to get the Cloud Wallet configuration file. Return to the cloud console, navigate to the database, and click on the **Database connection** button to obtain your **Wallet**.

Figure 1.12 – Getting the Oracle Wallet

By clicking on the button, a new pop-up window opens. **Client Credentials (Wallet)** are typically downloaded for the instance; however, there is also an option to download a specific **Regional Wallet** (consisting of all instance wallets used for administration purposes). For development, the **Instance Wallet** type should be selected:

Download client credentials (Wallet)

To download your client credentials, select the wallet type, and click **Download wallet**. You then enter a password for the wallet. This client credential download only contains information for mTLS connections. **You do not need a wallet for TLS connections.**

Figure 1.13 – Prompting Oracle Wallet

Cloud database connections are always secure, so you can provide the password for the wallet to be generated and downloaded. The downloaded wallet is a ZIP archive containing the following files:

- `ewallet.sso`: This consists of the encryption wallet details.

- `sqlnet.ora`: This specifies the general wallet location and encryption types.

- `tnsnames.ora`: This provides connection details – protocols, hosts, ports, and other parameters. The downloaded file consists of file connect strings delimited by the name, as well as parameters – *Low*, *Medium*, and *High* are preferred for the analytical interface, with *Tp* and *Tpurgent* for transactional processing.

Note that the connection strings can be listed when the wallet is generated in the **Connection Strings** subsection. The following screenshot shows an example of a connection string.

Database Connection

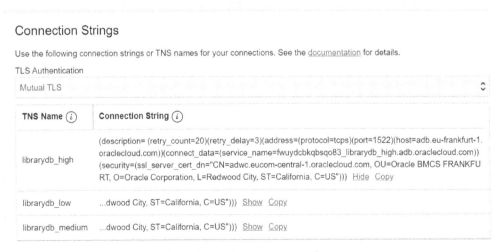

Figure 1.14 – Connection string list

The downloaded wallet can then be referenced in SQL Developer Desktop by specifying the **Configuration File** path (the yellow arrow (**2**) in the following screenshot), followed by the used service type (*Low*, *Medium*, *High*, *Tp*, or *Tpurgent*) (the red arrow (**1**)):

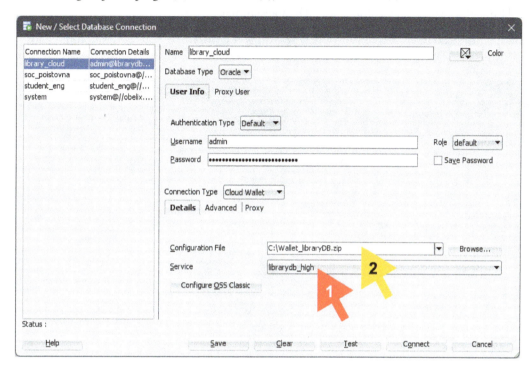

Figure 1.15 – Creating a new connection in SQL Developer Desktop

Passwords can be optionally stored in an encrypted format. The list of stored connections and encrypted passwords can be found in the application data. For Windows, the location is as follows:

```
%APPDATA%\SQL Developer\system<VERSION>\o.jdeveloper.
db.connection.<VERSION>\connections.json
```

For Linux, the analogous path is the following:

```
~/.sqldeveloper/system<VERSION>/o.jdeveloper.
db.connection.<VERSION>/connections.json
```

This file contains all the parameters specified during the connection definition.

The downloaded wallet consists of encryption keys, as well as connection details. This wallet is used by the users, developers, managers, administrators, and all other IT staff to access the Oracle Cloud databases. However, what if someone leaves the company? How can we ensure that some particular data cannot be used later on? The solution is to use *wallet rotation*, which invalidates existing client keys for the database instance owned by the cloud account in a region. It can be done immediately or after a grace period (from 1 to 24 hours). Even besides instances where people leave a company, it is also generally beneficial to rotate wallets regularly based on the organization's policies. Wallet rotation can be done by clicking on the **Rotate wallet** button shown in *Figure 1.13*.

Now, the connection is specified and a new session is created, where you can write commands or statements:

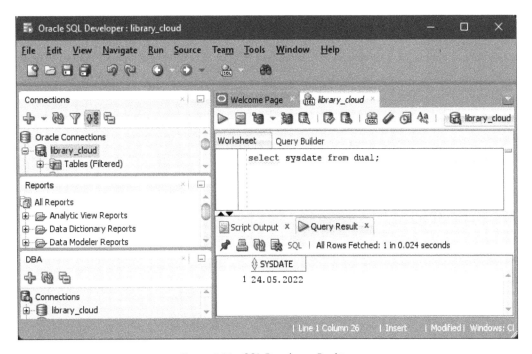

Figure 1.16 – SQL Developer Desktop

The preceding screenshot shows a `select` statement, providing you with the current date and time value. What about the output format? How can we reference individual elements, time elements, and particular time zones? We can already perceive the complexity of the whole problem here. A complete description of date and time management in the Oracle database environment will be covered in later chapters. Enjoy!

Throughout this chapter, we have often referenced the term *resource sharing*. However, what this means in practice is explained in the following section. We will focus on the database system instance itself and summarize the processes and memory structures. To ensure the best performance and optimize the access strategy, it is beneficial to understand the data flow, core elements, memory structures, and database and instance interconnection.

Database system architecture overview

Database systems (DBSs) are made up of databases and **data management systems (DBMSs)**. A database comprises the physical files holding a collection of data. It consists of data files, log files, parameter files, and so on. A database is commonly controlled by a DBMS. A database instance is a set of background processes manipulating the data and memory structures used for data processing, evaluation, and retrieval. Background processes are primarily responsible for asynchronous I/O data operations but also manage memory structures and general maintenance tasks. There are many background process categories. The most relevant for this book are the following:

- **Database Writer (DBWn)** is responsible for writing a modified data block from the *buffer cache* memory structure to the particular blocks of the files that constitute the database.

- **Log Writer (LGWR)** is responsible for the transaction logging by capturing data and storing it in the online redo log file.

- **Process Monitor (PMON)** performs recovery if a user process fails by cleaning the caches and deallocating the assigned session resources.

- **System Monitor (SMON)** performs instance recovery in case of its failure.

- **Archiver (ARCn)** is responsible for copying the online redo log file if a log switch occurs. Log files are consecutively rewritten. However, by copying them, it is possible to reconstruct the database into a historical image by identifying any changes made since a certain point in the archived logs.

- **Checkpoint (CKPT)** is responsible for the checkpoint event management by taking all modified buffers and writing their content to data files. Control files are also modified.

- **Manageability Monitor (MMON)** performs maintenance tasks, such as issuing alerts, taking snapshots, and capturing statistics.

- **Job Queue Processes (CJQ0** and **Jnnn)** are responsible for launching and executing the jobs (scheduled user-defined tasks that are executed once or planned to be executed periodically).

The general architecture of the database server is shown in *Figure 1.17*. To connect the client to the server, first, the user process is invoked on the client site, contacting the database server *listener*. By cooperating with the other background processes (mostly PMON), a new server process is created. It takes the small memory structure called the **PGA** that is private to the server process. It stores the current state, cursor variables, local variables, parameters, and so on. The server process is the manager of the user process on the server side, serving user requests for processing. After creating a server process, a client can communicate with the server process directly, without the listener.

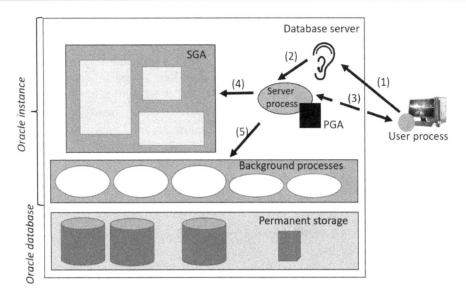

Figure 1.17 – Oracle database server architecture

Each instance is formed by the background processes already discussed, along with memory structures, which we will learn about in the following section.

Memory structures

You are now adequately familiar with the background processes. However, what about the memory structures they operate? Dozens of structures are part of the memory that ensures performance and maintenance activities. In this section, we'll mainly look at memory structures applicable to the database data. These structures are shared among the sessions forming the SGA. Let's look at them in some detail:

- The **SGA** is allocated when an instance is started up and released on shutdown. It consists of various memory structures. Most of them are dynamic in size. The available structures can be divided into two types – required structures (such as buffer caches, log buffers, and shared pools) or optional structures (such as large pools, Java pools, or Streams pools).

- The **Database buffer cache** is a work area for executing SQL as an interlayer between databases. All data changes are done in the memory and then saved to the database. The buffer cache is a block-oriented matrix. The block itself can be *clean* (no changes have been made to the data held there), *dirty* (data with changes), or *empty* (not yet used).

- The **Log buffer** is a small, short-term staging area holding data change vectors (modifications applied to data) before they are written to the *redo log file* on the disk. *Redo* logging ensures that no data can be lost. Due to performance issues, redo data is not directly written to the redo log file, but a near-real-time approach is used, processing in batches. There are many triggers for copying redo logs from the memory to the storage:

  - Every 3 seconds

  - By reaching a *commit* or *checkpoint*

  - If it is 1/3 full

  The log writer background process operates the log buffer.

- The most complex structure is a **Shared pool**, consisting of dozens of substructures. It is managed internally. The most important structures are the *Library cache* (storing recently executed code in parsed form), the *Data dictionary cache* (holding metadata such as object definitions: tables descriptions, indexes, users, and so on), the *PL/SQL area* (holding cached data to prevent repeated reading), and the *Result cache* (storing result sets temporarily for SQL and PL/SQL).

- The optional **Large pool** extends the shared pool and covers large objects such as backups. The **Java pool** is a heap space for *Java-stored procedures* to be run by the database system. The **Streams pool** is used by Oracle Streams architecture. The principle is to extract change vectors from the redo log to reconstruct statements to be executed, which requires additional memory.

So, now we know the core elements that make up the database systems, as well as the sub-components of each element. However, what about the interconnection between the instance and the database? Is one instance devoted to only one database? Well, in the following sections, we will highlight individual strategies, focusing on database containerization in the cloud environment. We will list all the key features and principles of database management related to system architecture and approaches.

Database system architecture

Each database system is formed by the instance characterized by the background processes and memory structures and the database itself. In general, various mapping strategies are available to operate the database, representing the ratio between them. Namely, one database can be accessed from multiple instances. However, one instance is dedicated only to one database (container) at a time.

Single-tenant databases

A single-tenant database (also known as a non-**Container Database** (CDB)) consists of a set of data files, control files, transaction logs (redo log files), parameter files, and metadata files. For the database, one instance is created. Before version 12c, this architecture was the only one used. Instance memory was solidly interconnected with the database (commonly named *ORCL* by default). The database listener was responsible for connection creation at the session level. Data was interchanged between the processes on the client and server part directly without listener interaction. Oracle has now deprecated this architecture. However, despite being unsupported in version 21c, it is still widely used.

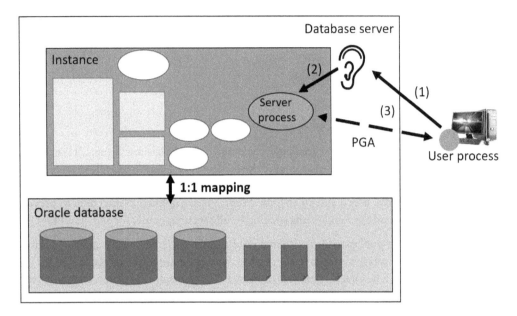

Figure 1.18 – Single-tenant database architecture

The preceding diagram shows the single-tenant database architecture. The mapping between the instance and database is 1:1. The extension of this architecture just involves a single-tenant RAC environment consisting of one database, operated by multiple instances.

Single-tenant RAC databases

Single-tenant (non-container) database can be run on multiple computers (nodes, servers, and hosts) by managing multiple instances operating (mounting, opening, and managing) one database. The main advantages are high performance, availability, fault resistance, and scalability. Thus, new instances with additional memory and CPUs can be added dynamically to serve the workload. The client is navigated to a **Single-Client Access Name** (**SCAN**) RAC listener. The listener connects the client to the most suitable instance based on balancing the current workload.

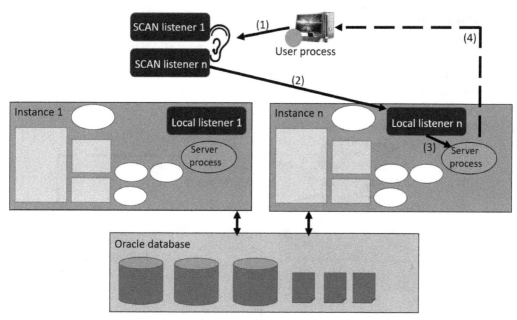

Figure 1.19 – Single-tenant RAC database architecture

The preceding diagram illustrates the RAC architecture of the single-tenant database, meaning that one database is operated by multiple instances. The workload is balanced by SCAN listeners navigating the client to a particular instance listener. By generalizing this architecture, containerization is created. It allows the databases to be attached and detached dynamically from the root container.

Multi-tenant CDBs

Multi-tenant CDBs were introduced in Oracle Database 12c. Also known as root CDBs, these contain a limited set of data files, control files, redo log files, parameter files, and metadata. However, there are no application objects or code in the data files. This kind of database is self-contained and can be mounted and opened without any other physical structures.

A **Pluggable Database** (**PDB**) is only made up of data files. They contain application data, objects, and the code itself. No metadata is present, so each PDB needs a container to be plugged into it. This type of database inherits its data repository from the container (such as redo log files, control, files, and parameter files).

The accessibility and mapping must apply certain rules. The root container can manage multiple PDBs simultaneously, but each database is associated with just one container at a time. Thus, one instance is associated just with one (root) container; however, it can reference multiple PDBs. One PDB is part of one container at a time.

Multi-tenant RAC databases

Multi-tenant RAC databases provide a general solution consisting of SCAN listeners.

Each instance has a separate local listener; however, the overall workload is balanced using the SCAN listeners. This architecture is shown in *Figure 1.20*. The user process is routed to the SCAN listener layer, followed by the transition to the specific instance. The database layer is enclosed by the container with dynamic database attachment and detachment functionalities.

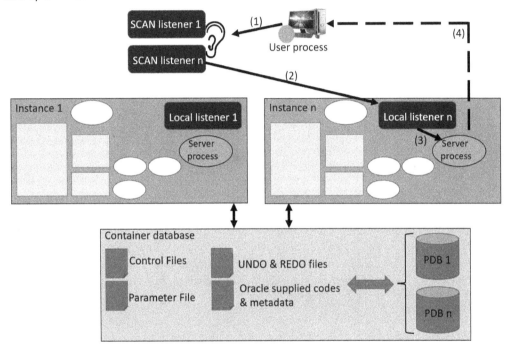

Figure 1.20 – A multi-tenant CDB

The preceding figure illustrates this architecture. The architecture and user management are similar to *Figure 1.18*, with the difference being that the database view is composed of configuration and parameter files and PDBs that can be managed dynamically. It provides a robust solution by only managing active databases.

Finally, a sharded database architecture will be discussed next, which divides the structure of the database into several fragments.

Sharded databases

Oracle Database version 12.2 introduced sharded databases – a logical database horizontally partitioned across a pool of physical databases. Each database (shard) has its own dedicated server and instance. However, from the user's point of view, it looks like one single database. Data is distributed across shards using the sharding key of the table. Following the system architecture, there is a connection pool to a sharded database, operated by the shard directors (listeners routing the processing based on the sharding key) and shard catalog. It provides robust linear scalability, fault tolerance (as shards exist on separate independent hardware and do not influence other shards), and data independence (updates can be applied to each shard independently). Moreover, it allows distribution across any number of regions. Each shard can be also be configured with different hardware or storage systems, ensuring compliance with laws and regulations related to the specific data positioning restrictions.

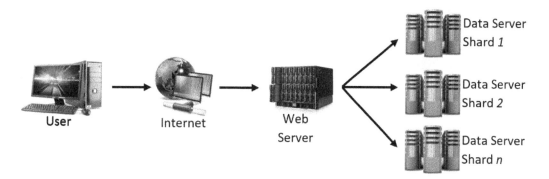

Figure 1.21 – Sharded database

The practical usage of sharding is associated with the elimination of manual data preparation as a necessity, leading to time savings by emphasizing scalability, high availability, response time, and write bandwidth. Sharded tables are distributed and duplicated across servers, which are then associated with the particular application and usage by limiting the amount of data covered by the server. Moreover, each shard can be optimized for specific usage and applications via the parameters, index set, partitioning, and other optimization techniques. A significant advantage is achieved by this parallelism and elasticity – data can be resharded without any downtime.

Summary

In this chapter, you learned about the core concepts of the Oracle Cloud technology based on self-driving, self-securing, and self-repairing properties, followed by an examination of the types of ADBs that can be provisioned in any deployment model. Besides this, you were also introduced to a selection of important terms related to the Oracle Cloud environment.

As databases and applications are continuously being migrated from the on-premises world to the cloud, we summarized the various database architectures in use. It is important to know and understand the individual existing architectures and be able to adapt the existing on-premises solutions to the architecture used in the cloud environment.

At this point, you know how to create cloud databases, what the individual types are, and how to connect to them using Oracle Wallet. The next chapter deals with data management, focusing on data loading, import, and export techniques.

Questions

1. What are the main attributes of OCI?

    A. Self-repairing only

    B. Self-securing only

    C. Self-driving only

    D. Self-repairing, self-securing, and self-driving

2. Which database type is commonly associated with bitmap indexes? Reflect on the performance aspects of the index construction.

    A. ATP

    B. ADW

    C. JSON

    D. PGA

3. Which priority levels are preferred for ATP?

    A. Low

    B. Medium

    C. Medium and High

    D. Tp and Tpurgent

4. Which of these services provides you with the highest level of abstraction?

    A. IaaS

    B. PaaS

    C. SaaS

    D. All of them provide the same level

5. Which background process is responsible for instance recovery?

    A. ARCn

    B. PMON

    C. LGWR

    D. SMON

6. Which database architecture involves a one-to-one mapping between the database and instance?

    A. Single-tenant architecture

    B. A container database

    C. RACs generally

    D. A multi-tenant container database

7. A cloud repository, usually devoted to the organization or company to provide secure and isolated storage and processing partitions, is called which of the following?

    A. A compartment

    B. A tenancy

    C. An image

    D. A sub-domain

8. Oracle Cloud Wallet consists of which of the following?

    A. An Ewallet.sso file only

    B. Sqlnet.ora and the control file only

    C. An embedded connect string in the Sqlnet.ora file

    D. The Ewallet.sso, Sqlnet.ora, and Tnsnames.ora files

Further reading

- *Oracle Cloud Infrastructure for Solutions Architects: A practical guide to effectively designing enterprise-grade solutions with OCI services* by Prasenjit Sakrar and Guillermo Ruiz. This provides practical tips and tricks to create robust cloud-based solutions: https://www.packtpub.com/product/oracle-cloud-infrastructure-for-solutions-architects/9781800566460.

- *Practical SQL for Oracle Cloud* by Michal Kvet, Karol Matiaško, and Štefan Toth. It emphasizes the provisioning process with the SQL and PL/SQL language used in ATP. It can be downloaded free of charge using the following web link: `https://oraclecloud.uniza.sk/`.

You can also scan this QR code:

- *Getting Started with Oracle Cloud Free Tier: Create Modern Web Applications Using Always Free Resources* by Andrian Png and Luc Demanche. This outlines the best practices for using the Oracle Always Free tier, covered by the Free Tier option.

2
Data Loading and Migration Perspectives

The data loading process is essential for feeding databases with external data sources. It can be used to load new objects or to update existing objects and their content in the database. Appropriate knowledge is required; otherwise, everything would need to be done on the fly without any robust support, which would raise many problems. Data loading is a complex process of copying and loading data while maintaining data integrity. Commonly, huge datasets need to be loaded, whether from the Oracle database, a different database system, or any system generally.

Three solutions are discussed in this chapter – **SQL Loader**, **client-side import**, and **server-side import** using **dump files**. Finally, **Oracle Cloud Infrastructure (OCI) Database Migration** is discussed, which is based on the dump files as a source or direct connection using a database link. The whole process emphasizes date and time management and provides a reference for proper element mapping. The second part of this chapter provides a summary of the migration perspective. It deals with the most significant methods for data loading and unloading. It does not attempt to provide a full description; it just aims to provide you with an overview of the method and technique, which can be used and studied in more detail.

In this chapter, we will cover the following main topics:

- SQL Loader for loading Oracle external data

- Client and server-side import and export

- Data migration techniques

The source code for this chapter can be found in the GitHub repository accessible via this link: `https://github.com/PacktPublishing/Developing-Robust-Date-and-Time-Oriented-Applications-in-Oracle-Cloud/tree/main/chapter%2002`. You could also try using the following QR code:

Understanding SQL Loader

SQL Loader is a tool for inserting third-party system data in various format types, such as CSV, TXT, or any delimited format. The structure of the data is defined by the associated *control file* controlling the mapping. SQL Loader is a user (client) process defined by the extraction of data rows in the first phase, followed by loading tuples. The loading process can be done conventionally (the `insert` statements are generated with regard to the `UNDO` and `REDO` data logs) or via a direct path (in this case, the *buffer cache* is bypassed, and data is loaded directly into the data files in a block format). No UNDO data log content is generated. The **High Water Mark** (**HWM**) pointer is shifted to point to another location after processing, making new blocks with data accessible to the table structure. The HWM pointer references the last associated block of the particular table. Thus, the direct path type is significantly faster and requires less storage (even REDO logging can be disabled). There are, however, some negative aspects and limitations that need to be mentioned.

Firstly, the process is not enclosed by the transaction, so integrity cannot be ensured. Consequently, referential integrity control mechanisms need to be dropped or disabled to allow the direct path method; otherwise, it cannot be used. Similarly, insert triggers are not fired, whereas the processing does not use ordinary `insert` statements. And finally, the whole processed table is protected by the data *locks*. These data locks prevent **Data Manipulation Language** (**DML**) statements (the `insert`, `update`, `delete`, and `select` statements) from being executed during loading.

The user can set the method of loading (conventional/direct) to be used during the SQL Loader command definition (`SQLLDR`).

SQL Loader requires three file types to operate:

- A SQL file consisting of the structure definition (tables, relationships, indexes, triggers, and so on)
- **Control** (**CTL**) files containing instructions for the loading (how to map data files into the defined table)
- **Data** (**UNL**) files consisting of the values to be loaded into the database

The loading process starts with the database schema definition (executing code inside the SQL file), typically obtained by the existing system (reverse engineering), or our own definition can be used generally. The DBMS_METADATA package can provide you with the script of the existing object definition in a large textual format (CLOB):

```
DBMS_METADATA.GET_DDL
(
    object_type      IN VARCHAR2,
    name             IN VARCHAR2,
    schema           IN VARCHAR2 DEFAULT NULL,
    version          IN VARCHAR2 DEFAULT 'COMPATIBLE',
    model            IN VARCHAR2 DEFAULT 'ORACLE',
    transform        IN VARCHAR2 DEFAULT 'DDL'
)
RETURN CLOB;
```

The script can be launched by SQL Developer or executed in any other tool (SQL Client, SQL*Plus, and so on).

The data files can vary based on your preference and the output of the existing solution. Generally, one line in an input file typically references one tuple, and individual values are separated by the defined delimiter.

The control file looks like the following code snippet. The first part deals with the data location (INFILE) and the destination table name (INTO TABLE). Next, individual attribute values must be delimited in some way. In this case, the delimiter is a pipe (|) – FIELDS TERMINATED BY '|'. Afterward, the data structure is defined by the order of the data represented in the *.unl file:

```
LOAD DATA
INFILE 'book.unl'
INTO TABLE book
FIELDS TERMINATED BY '|'
(
  BOOK_ID,
  TITLE_ID,
  PRICE,
  REGISTRATION_DATE       DATE 'MM/DD/YYYY',
  DISPOSAL_DATE           DATE 'MM/DD/YYYY',
  LOST_DATE               DATE 'MM/DD/YYYY'
)
```

It is necessary to focus on date and time processing. In the UNL data file, all values are treated as character strings. When mapping the values into the database, it is strongly recommended not to rely on implicit conversion. Be aware that the amount of data to be loaded is typically high. The bulk loading process can require less time and fewer resources. Therefore, appropriate preparation of the data and instructions is crucial.

The referenced data file can look like the following given the preceding format, where MM represents the month, DD represents the day, and YYYY represents the year:

```
279|17|9|09/23/2002|08/02/2014|12/29/2014
```

In the preceding example, the date value constructor was used, as defined by the format and mapping elements. However, using user-defined functions for the operation can sometimes be more useful and powerful. It is done using the bind variable reference. To apply the function reference, encapsulate its call in the CTL file in double quote marks. The column reference is then generated by a bind variable, denoted by the colon at the beginning of the reference (for example, `"replace(:telephone,' ','')"`).

The following code snippet uses the `replace` function to remove any spaces from the telephone number. The `replace` function has three parameters – the source, the search string, and the value to replace the identified string:

```
LOAD DATA
INFILE 'person_contact.unl'
INTO TABLE person
FIELDS TERMINATED BY '|'
(
  PERSON_ID,
  NAME,
  SURNAME,
  VALID_FROM        DATE 'MM/DD/YYYY',
  TELEPHONE         "replace(:telephone,' ','')",
  EMAIL             "replace(:email,'(at)','@')"
)
```

The next example is a bit more complex. The date is modeled by individual elements representing day, month, year, and time. However, it can happen that the format is not the same across the whole dataset, and therefore direct date construction is not applicable. Let's consider the VALID_FROM attribute. Employment contracts are commonly associated with the first day of the month. Thus, the day element value does not need to be stated in the UNL file. However, for employees hired on another day within a month, this day element is crucial. Therefore, the loading operation needs to be dynamic and should route the processing based on the identified data. In the following snippet, the length of the input string value (before processing, the original string is trimmed, meaning empty symbols, such as spaces, are removed from the left as well as the right part of the string) is obtained. Based on its value, the original day element is either extracted from the input file or the 01 value is used if it is not present. The whole operation is done using case:

```
LOAD DATA
INFILE 'person_contact.unl'
INTO TABLE person
FIELDS TERMINATED BY '|'
(
  PERSON_ID,
  NAME,
  SURNAME,
  VALID_FROM "case length(trim(:valid_from)<=7
              then to_date(:valid_from, 'MM/YYYY')
              else to_date(:valid_from, 'DD/MM/YYYY')
            end",
  TELEPHONE   "replace(:telephone, ' ', '')",
  EMAIL       "replace(:email, '(at)', '@')"
)
```

The loading process is supervised by the instruction file (the control file) referencing the data to be operated on in the data files. The correct order of the process is critical (a table reflecting another table's primary key must be loaded later, whereas the foreign key value must refer to existing data). It can be done using the following command (on Linux):

```
$ sqlldr login@connect_string
                    control='control_file_name.ctl'
```

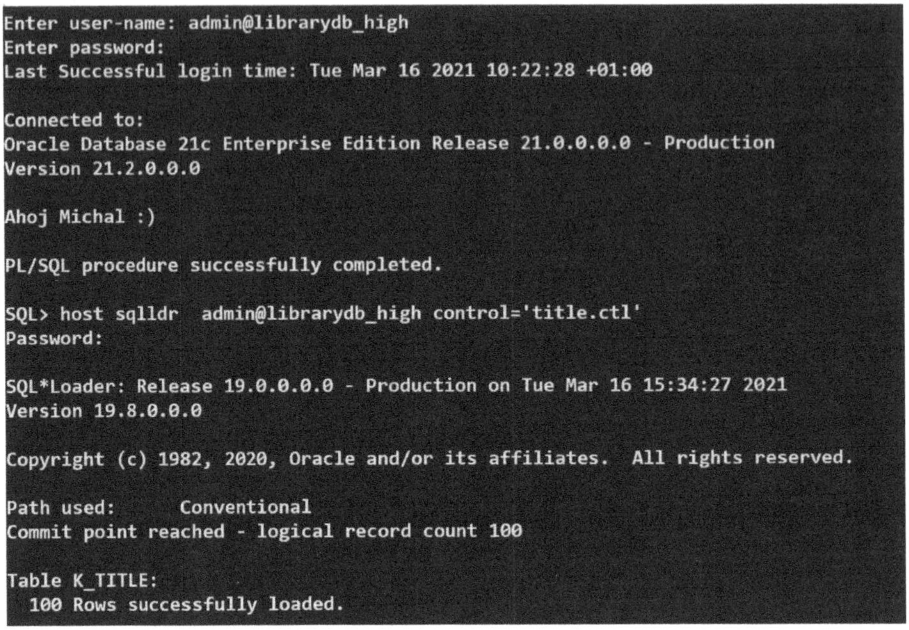

```
Enter user-name: admin@librarydb_high
Enter password:
Last Successful login time: Tue Mar 16 2021 10:22:28 +01:00

Connected to:
Oracle Database 21c Enterprise Edition Release 21.0.0.0.0 - Production
Version 21.2.0.0.0

Ahoj Michal :)

PL/SQL procedure successfully completed.

SQL> host sqlldr  admin@librarydb_high control='title.ctl'
Password:

SQL*Loader: Release 19.0.0.0.0 - Production on Tue Mar 16 15:34:27 2021
Version 19.8.0.0.0

Copyright (c) 1982, 2020, Oracle and/or its affiliates.  All rights reserved.

Path used:      Conventional
Commit point reached - logical record count 100

Table K_TITLE:
  100 Rows successfully loaded.
```

Figure 2.1 – SQL Loader process output

Check the correctness of the data immediately after it is loaded. The results shown on the screen do not reflect the amount of inserted data, but only the number of rows read from the UNL data file. If there is any problem, solve it before continuing (errors are mainly based on integrity constraint violations).

Appropriate information about the execution process can be found in the log file. This file has the same name as the control file, but the extension is * . log:

```
person.ctl --> person.log
```

The log file will consist of a header denoting the version of SQL Loader:

```
SQL*Loader: Release 21.0.0.0.0 - Production on Mon Mar 21
09:54:28 2022
Version 21.3.0.0.0
```

Then, there is a summary of the used file references, namely Control File, Data File, Bad File, and Discard File. Rows that do not pass the user-defined conditions for loading are copied to Discard File. Bad File consists of the rows that have been refused because of errors, such as incorrect data types or referential integrity:

```
Control File:      person.ctl
Data File:         person.unl
```

```
Bad File:              person.bad
Discard File:          none specified
```

The third part of the log file deals with the loading parameters, pointing to the number of rows to be loaded, the maximum number of refused rows, and so on. There is also the name of the method used for loading (`Path used`), which can be `Conventional` or `Direct`:

- The `Conventional` path is the default option. It is based on creating ordinary `insert` statements by using a bind array buffer to load data into the database tables, which limits the performance, whereas it competes with other processes for buffer resources. To load the data, the database manager must find suitable blocks to serve new tuples. In principle, some blocks can be empty or contain some existing rows. One way or another, it is always ensured that the row to be loaded fits the block in terms of the size demands. Moreover, the database optimizer tries to find the most suitable data block to limit block fragmentation. Thanks to that, the number of blocks is optimized. On the other hand, this can slow down the bulk loading process rapidly.

- The `Direct` path's load is optimized for maximum data loading capability. Instead of using conventional `insert` statements, the `Direct` path creates new data blocks and associates them with the table. These blocks are directly written to the database. During the `Direct` path loading, the whole table is locked. After the process, the pointer to the last associated block HWM is extended to cover newly created blocks.

There are several advantages, properties, and limitations of both methods that should be highlighted:

- The `Direct` path:

  - Partial blocks are not used; no read operation is necessary

  - Fewer writes are performed

  - Bind arrays are not used; the `insert` statements are not generated

  - The `Direct` path does not use the buffer cache, thus dirty blocks are not created

  - Particular tables and associated indexes are locked during the whole operation

  - The operation (formed by multiple `insert` statements) is asynchronous, allowing you to execute `insert` statements in parallel

  - No transaction logs are generated because the operation is not part of the transaction

  - It can be optimally used for large datasets to be loaded using parallel operations to maximize performance, either to empty or non-empty tables

  - When using the `Direct` path, some prerequisites must be applied:

    - Tables cannot be clustered.

    - No pending transaction is active for the table.

- Integrity constraints based on the processed table, such as NOT NULL, are checked. Constraints depending on other tables, such as *referential integrity*, must be disabled before and re-enabled after the processing (this can be done using the REENABLE clause of SQL Loader) by checking those constraints. Rows that do not apply the rules are logged.

- The Conventional path:

  - It can be applied to a clustered table.

  - The operation processing is associated with log data generation (common insert statements are executed). The log then protects the data by providing integrity, rules, and transaction support.

  - It is useful for loading small datasets.

  - Data storage is optimized by locating free space available through the block set.

  - It is used if the function calls need to be applied to the data fields.

The SQL Loader log report can be divided into several parts. The first part deals with the method used and parameters for the load, such as the number of rows to be loaded or skipped:

```
Number to load:     ALL
Number to skip:     0
Errors allowed:     50
Bind array:         64 rows, maximum of 256000 bytes
Continuation:       none specified
Path used:          Conventional
```

The next part deals with the table structure and mapping. It consists of the destination table and selected options:

```
Table K_PERSON, loaded from every logical record.
Insert option in effect for this table: INSERT
```

Then, there is a list of columns with the defined order to be mapped. The definition consists of the column name, its table position, length, termination, enclosure, and data type category:

```
Column Name        Position      Len   Term      Encl Datatype
------------------------------------------------------------

NAME               FIRST           *   |              CHARACTER
SURNAME            NEXT            *   |              CHARACTER
PERSON_ID          NEXT            *   |              CHARACTER
STREET             NEXT            *   |              CHARACTER
```

ZIP	NEXT	*		CHARACTER
TOWN	NEXT	*		CHARACTER
DISTRICT	NEXT	*		CHARACTER
REGION	NEXT	*		CHARACTER
STATE	NEXT	*		CHARACTER

The next part deals with the data loading summary by listing the number of processed and categorized rows. The first section deals with the table content, which is processed by summarizing the number of successfully loaded and rejected rows. Rejection can be due to data errors, the WHEN clauses, or the NULL values for all fields. Next, there is a summary of the amount of space used, followed by the number of logical records skipped, read, rejected, and discarded from the input file:

```
Table K_PERSON:
  100 Rows successfully loaded.
  0 Rows not loaded due to data errors.
  0 Rows not loaded because all WHEN clauses were failed.
  0 Rows not loaded because all fields were null.


Space allocated for bind array:      148608 bytes(64 rows)
Read buffer bytes:                    1048576


Total logical records skipped:             0
Total logical records read:              100
Total logical records rejected:            0
Total logical records discarded:           0
```

Error information about the rejected rows can also be present if there are any refused rows. Namely, for each row, extra information about the raised exception is present, as follows:

```
Record 29: Rejected - Error on table B_PARKING, column
POSITION.
ORA-01400: cannot insert NULL into ("BIKESHARING"."B_
PARKING"."POSITION ")
```

Moreover, a file with a .bad extension is created. Its content consists of copies of the rejected rows. Therefore, rows that were not loaded can be easily identified. The reason for the rejection is included as part of the log, defined by the raised exception.

The content of the BAD file can look like the following snippet. The values are delimited by a semicolon; however, the first value is missing – the row starts with a semicolon:

```
;265;01.03.2020 15:32:11;03.03.2020 12:28:20;
```

Finally, the log refers to the time by getting the `Run` `began` and `Run` `end` attributes, and `Elapsed` `time`:

```
Run began on Ne Sep 17 18:26:39 2021
Run ended on Ne Sep 17 18:26:42 2021


Elapsed time was:          00:00:02.50
```

The UNL file can be composed dynamically of the existing data. The following code gets the data using the SQL command. The used `chr(10)` function expression represents the *line feed* (end-of-line representation) issued explicitly:

```
set echo off newpage 0 space 0 pagesize 0 feed off
spool author.unl
   select trim(name) || '|' ||
          trim(surname) || '|' ||
          trim(author_id) || '|' ||
          to_char(registration_date,'MM/DD/YYYY') || '|' ||
          trim(note) || '|' || chr(10) from author;
spool off
```

The `spool` command tells the command line SQL*Plus to direct the output of any query to a flat server-side file.

SQL Data Loader can also be invoked from the SQL Developer Desktop and Web versions. In the following section, we will show you how to do that. The application wizard covers the whole process step by step, so it is more user friendly and easy.

Importing data using SQL Data Loader in SQL Developer Desktop

Connect to the cloud instance of the database. Expand the list of tables in the left panel. Then, select the **Import Data...** option by right-clicking on the particular table name that is to be used as the data destination:

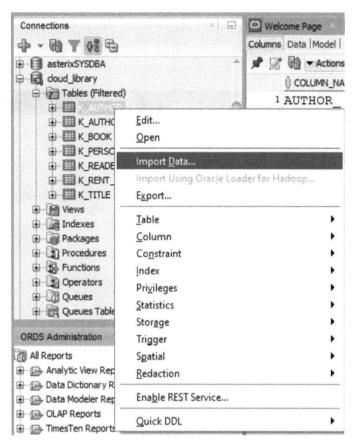

Figure 2.2 – SQL Developer Desktop – importing data with SQL Loader

The navigation wizard consists of multiple self-explanatory steps that do not need to be described in detail.

An analogous data loading operation can be found in the SQL Developer Web version, part of the cloud instance, by navigating to **SQL**.

Importing data using SQL Data Loader in the cloud interface

Let's look at how to load data from the textual source file using SQL Developer Web. For starters, navigate to the **SQL** section, as shown in the following screenshot:

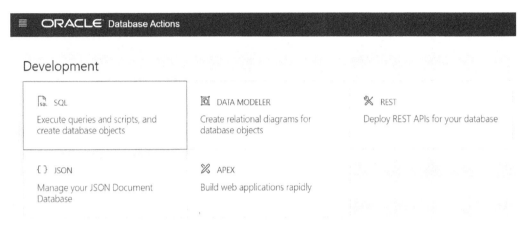

Figure 2.3 – The SQL interface of SQL Developer Web

Next, click on the **Data Loading** tab. Data input can be in a delimited format (CSV, DSV, TXT, and so on), an Excel sheet (XLS, XLSX), XML, JSON, or AVRO:

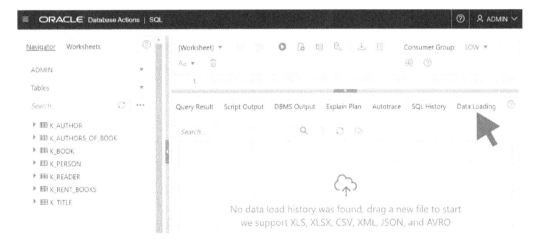

Figure 2.4 – SQL Loader in the SQL Developer Web tool

Data rows are automatically extracted from the source file:

Figure 2.5 – Uploading data into the table (SQL Loader in SQL Developer Web tool)

It is, however, necessary to specify the format, data enclosure, and delimiters. The first row can be optionally used as a column name list for a table specification:

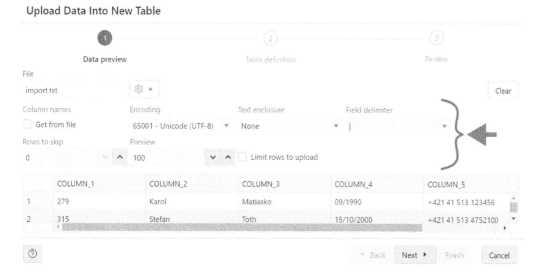

Figure 2.6 – Mapping data into the table (SQL Loader in the SQL Developer Web tool)

Then, the destination table will be specified, covering column names and data types with precision, the primary key set, and the NULL constraint. As is evident, the format needs to be specified for the date values. The limitation of that wizard is the impossibility of using functions for the data value input conversion. The `Format mask` column is similar to the mapping rule used in conversion functions (`TO_DATE`):

Figure 2.7 – Date and time mapping (SQL Loader in the SQL Developer Web tool)

Finally, the script for the structure definition and the mapping summary are present. Thus, the loading operation can be launched:

```
CREATE TABLE ADMIN.PERSON
    (
    PERSON_ID   NUMBER(38),
    NAME        VARCHAR2(50)    NOT NULL,
    SURNAME     VARCHAR2(50)    NOT NULL,
    VALID_FROM  DATE            NOT NULL,
    TELEPHONE   VARCHAR2(30),
    EMAIL       VARCHAR2(50)
    )
    LOGGING;


ALTER TABLE ADMIN.PERSON
    ADD CONSTRAINT PERSON_PK PRIMARY KEY ( PERSON_ID ) ;
```

A summary of the loading process is present after the execution. In our case, the first row cannot be imported because the date format does not match the input data:

Failed rows

	Error Message		person_id	name	surname	valid_from	telephone	email
1	ORA-01843: not a valid month	🖉	279	Karol	Matiasko	09/1990	+421 41 513 12345	karol.matiasko(at)uniza.sk

Figure 2.8 – Failed data rows (SQL Loader in the SQL Developer Web tool)

So far, we have explained how to process data in the SQL Loader interface with SQL Developer Desktop and Web, as well as a command-line environment. The data loading process can be handled directly from the Oracle Cloud interface in the database management section. Navigate to your database and click on the **Database Actions** button in the main panel, as shown in the following screenshot:

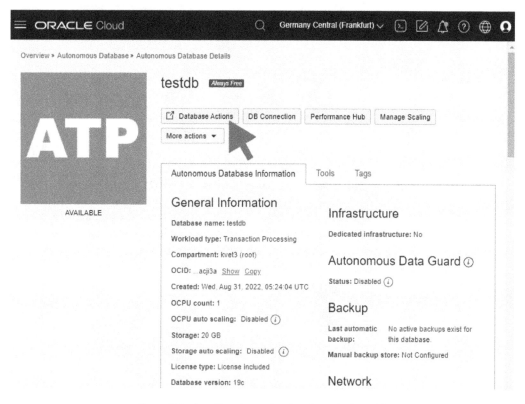

Figure 2.9 – Loading data using Database Actions (1)

A new window will open, listing the available functionalities in the **Launchpad** (refer to *Figure 2.10*). You will find three categories of data interaction:

- **LOAD DATA**: To load data from local files into Oracle Autonomous Database, **LOAD DATA** and **LOCAL FILE** should be selected, followed by clicking on the **Next** button. Then, select the files to be loaded by specifying the path or by using the drag-and-drop option. The supported data formats are AVRO, CSV, JSON, TSV, delimited TXT, XML, XLSX, and XML.

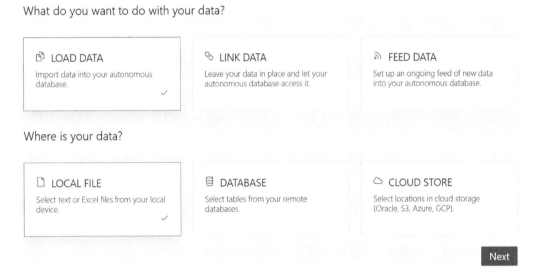

Figure 2.10 – Loading data using Database Actions (2)

Each source file is visualized by a rectangle on a summary screen. The loading process is launched by clicking on the play icon in the top-left corner while individual parameters and processing options can be set by clicking on edit (the button with a pencil symbol), as shown in the following screenshot:

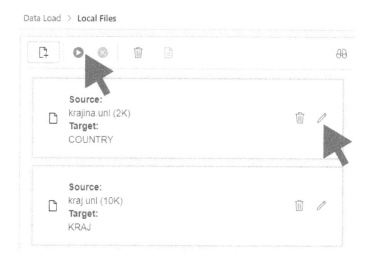

Figure 2.11 – Loading data using Database Actions – local file list

- The **Name** field specifies the destination table. In the **Option** section, the following actions can be selected: **Create Table**, **Insert into Table**, **Replace Data**, **Drop Table and Create New Table**, or **Merge into Table**. The **Properties** section allows you to set the input style of the rows, such as **Encoding**, **Text enclosure**, and **Field delimiters**. The values are set using drop-down lists. The **Mapping** section focuses on the target column names and data types:

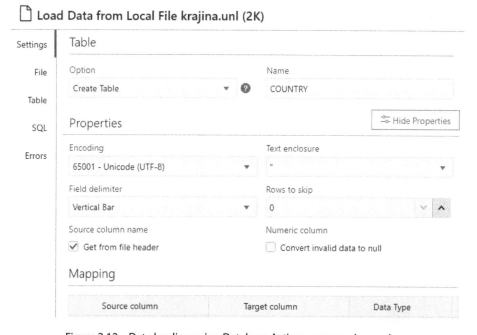

Figure 2.12 – Data loading using Database Actions – processing options

If the process is done successfully, the source file is marked green, showing the status and the elapsed processing time:

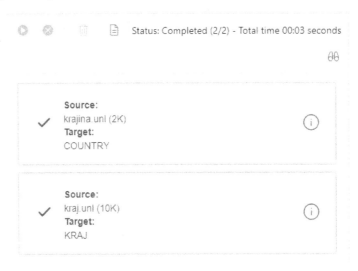

Figure 2.13 – Loading data using Database Actions – completed

- The **LINK DATA** option is used if you want to access the cloud object store bucket from your Oracle Autonomous Database. The access is done using the *external tables*. The supported formats are AVRO, CSV, JSON, Parquet, ORC, and delimited TXT. Linking data from the database source requires a database link object definition.

  The **FEED DATA** option is for feeding streams of data into your Oracle Autonomous Database. The data source can be a local file, a database, or a cloud store. Tutorials and step-by-step instructions are available in the cloud repository for your reference. The principles of date and time management are the same as described previously. **Data Tools** on the **Launchpad** also contains options for defining the Data Pump operation, data insights (to discover anomalies, patterns, and outliers in data), and data analysis.

Tracking the external source and mapping data to the database can also be done using an *external table*. This provides the same access to the data as if it were physically stored in the database. In reality, however, the data is only mapped from an external source. The next section deals with the *external table* definition and individual mapping parameters.

Accessing data stored outside the database using an external table

External table management can be considered complementary to the SQL Loader functionality by treating the external sources similarly to conventional internal database tables. Since Oracle 12c Release 2 (12.2.0.1), external table data can be partitioned to achieve additional performance improvements, similar to traditional table partitioning. External tables are created using the `organization external` keyword extension of the `CREATE TABLE` command.

There are several techniques for specifying the external table and access driver. `ORACLE_LOADER` is the default option. The source data must be a textual type used to load data into internal tables. The `ORACLE_DATAPUMP` option can perform load and unload operations. The external data is in binary dump files. Loading is done by fetching data from the binary dump files. The method's access driver can write dump files only as part of creating an external table. It is done with the `Create Table as Select` command. `ORACLE_HDFS` extracts data from the **Hadoop Distributed File System** (**HDFS**), while `ORACLE_HIVE` extracts data from Apache Hive.

Among the types of external table definitions discussed here, there are other significant parameters in the external table definition. `DEFAULT DIRECTORY` refers to the external data source, using an Oracle directory as an interlayer between the database and physical directory path. `ACCESS PARAMETERS` and `LOCATION` will be explained shortly.

During the definition of `ORACLE DIRECTORY`, the database system does not check the existence of the source file, which points to the Oracle directory (used as a mapper between the database system and physical storage).

The following code snippet shows the external table definition. It consists of the object name, followed by the list of attributes and data types. Then, there is `ORGANIZATION EXTERNAL`, followed by the `TYPE`, `DEFAULT DIRECTORY`, `ACCESS PARAMETERS`, and `LOCATION` specifications. In the `ACCESS PARAMETERS` section, there is a definition of the data delimiter and source mapping. From a date and time perspective, the input value is treated as a character string, pointing to the `DATE` data type in a table. Thus, the date element's format must be specified:

- The Oracle directory definition:

```
create or replace directory ext_tab_dir
                        as 'X:\source_data';
```

- The external table definition:

```
CREATE TABLE person_tab
    (person_id          INTEGER,
     title              VARCHAR(20),
     first_name         VARCHAR(20),
     last_name          VARCHAR(15),
     birth_date         DATE,
     retirement         DATE
     )
ORGANIZATION EXTERNAL
```

```
(TYPE ORACLE_LOADER
 DEFAULT DIRECTORY ext_tab_dir
 ACCESS PARAMETERS
   (RECORDS DELIMITED BY NEWLINE
    FIELDS TERMINATED BY ','
    missing field values are null
      (person_id, title, first_name, last_name,
       birth_date DATE 'DD.MM.YYYY HH:MI:SS',
       retirement DATE 'MM.YYYY')
   ) LOCATION ('person_file.dat')
);
```

A solution with partitioning using ORACLE_LOADER is shown in the following code snippet. It creates a `student_tab` table consisting of a student identifier (`student_id`), `class`, the beginning of the study period (`start_date`), and the date of completion (`final_date`). In this approach, range partitioning is used based on `class`. Bachelor study data is stored in three files. The source files are defined by the `location` clause, while the master study data consists of only one source file:

```
create table student_tab
 (student_id integer, class integer,
  start_date date, final_date date)
   organization external
    (type oracle_loader default directory ext_tab_dir)
     partition by range(class)
       (partition bachelor values less than (4)
         location('class1.dat', 'class2.dat', 'class3.dat'),
        partition master values less than (6)
         location('all_classes.dat')
    );
```

As you can see, source data does not need to be physically stored in the database by using load techniques. Instead, external tables can be used. In that case, data is stored in the external file, but the user data access experience is similar to that of ordinary tables. This approach can provide fast data insights but is limited to data retrieval (new data rows cannot be added or existing rows updated).

SQL Loader is designed for processing data from external sources. The next sections deal with Oracle database import/export done either on the client or server side by migrating from the on-premises world to the cloud or by moving the environment to a new server. It can be enhanced by transitioning to a newer version.

Getting to know client-side import/export

In the past, data import and export were done on the *client side*. Before processing, a user session must be created, composing the data tunnel between the database server and the client. The `select` statement did the export, and the `insert` operation did the import. Regular communication was via the internet, which limited the approach due to the fidelity, network failures, and packets loss. Firstly, the processing time was strongly dependent on the network's speed. Secondly, if the interconnection was not stable, the whole process would fail. After losing the connection, it was impossible to restore it and continue with the import or export process. The operation was aborted and needed to start once again from the beginning. Therefore, over time, client-side **export** (**EXP**) became obsolete and was replaced by server-side management. However, client-side import and export management principles still need to be covered and emphasized because many existing systems still use them. If the export was created by a client-side tool, it must be imported in the same manner – on the client side. The same approach must be used if the server-side option is used.

Client-side export is invoked by the EXP tool by defining the `connect` string followed by the list of structures (tables) to be exported and the destination file. If the list of tables is omitted, the whole structure is exported:

```
$ exp <login>@<connect_string>
      tables='<list_of_tables>' file='<file_name>.exp'
$ exp <login>/<password>@<connect_string>
        tables='k_person k_reader' file='library.exp'
```

The opposite operation is invoked by the client-side **import** (**IMP**) command, followed by the connection definition and at least the source file. It can be extended by the list of tables to be imported and an optional clause telling the system how to handle exceptions in a specific manner. An `ignore` clause can ensure that raised exceptions are ignored, which mostly occur when the destination table already exists. The source file created by the export tool contains not only the data but also the table definition:

```
$ imp <login>@<connect_string>
       tables='<list_of_tables>' ignore=Y
              file='<filename.exp>'
```

If the source and destination users are not the same, it is necessary to remap the structures using the `fromuser` and `touser` options. The point is, how do we figure out the username of the person who performed the export if it's not provided directly? Simply, such information is part of the exported file and can be obtained during the import attempt:

```
$ imp <login>@<connect_string>
       fromuser=<old_login> touser=<new_login>
          file='<filename.exp>'
```

Ordinary users can import data only to their schemas. However, a user with DBA privileges can import data to any user schema using the previous command.

Server-side Data Pump import/export (**impdp/expdp**) was introduced in the Oracle version 10g, and the original imp and exp methods have been marked as deprecated. Although decades have passed, we still come across systems that use the original processes on the client side.

Examining server-side import/export using dump files

The Oracle Cloud environment promotes Data Pump operations, through which all activities are done in the cloud infrastructure. A client just supervises the process. Therefore, it is far more effective and reliable compared to conventional solutions, such as client-side import and export definitions. To carry out this process using the cloud, in the first phase, export data is copied to the cloud data repository or Object Storage, followed by ensuring its accessibility from the database layer. The user of the database system does not have the right to access the filesystem where the export resides. Therefore, an additional database object must be created, called the *Oracle directory*. It accepts the security rules of the database, as well as the operating system and storage. Thus, appropriate **Read** and **Write** privileges must be granted via it. The Oracle directory is associated with the storage directory, through which the directory content can be mapped to the database. It simply provides access to the filesystem.

The overall loading process is managed by the Data Pump master process and workers. The following sections deal with the prerequisites for Data Pump management. The Data Pump import is executed on the server side, and it is necessary to copy the export file to the cloud Object Storage and make it available for the database and management processes. Then, credentials for the database instance and an authentication token object must be created. Finally, the process can be launched. We will take you through the whole process. The implementation will be done using the SQL Developer tool.

Object Storage and buckets

Oracle Cloud Object Storage is a high-performance cloud storage solution – a reliable, resistant, and cost-efficient data repository. Object Storage can cover an unlimited number of files with any structure. The *Always Free* version is limited to 20 GB of capacity. Oracle Cloud Object Storage data can be easily, safely, and securely managed and retrieved over the internet or by using a cloud platform. It is not associated with a specific *compute instance*. The core element is the region itself.

Object Storage can be accessed from the left panel of the cloud management tool by selecting **Storage | Object Storage & Archive Storage**.

Figure 2.14 – The Storage option

A **bucket** is a logical container for storing data files. Each bucket is created and associated with the *compartment* relating to the policies limiting the actions that can be executed in that compartment. We will use the general term *object* as a file of any data structure and format for cloud storage. The object is defined by its representation and metadata. Each object is stored within the bucket.

To create a new bucket, the following parameters are defined:

- **Bucket Name**: The name of the bucket used for the consecutive reference.

- **Bucket Storage Tier**: This can be either **Standard** or **Archive**.

  - The **Standard** tier is a default option for Object Storage service data. It is primarily intended for hot data that needs to be addressed and accessed quickly and frequently. A core use case is a content repository for data, images, music, video, or logs. Another usage is related to easily accessible backups. The tier selection cannot be changed later, and any uploaded object automatically inherits these parameters.

  - The **Archive** tier is used for archive storage service data operations. Its definition is related to cold storage in that data is accessed rarely but must be preserved for a long time. The minimum retention period is 90 days; thus, even if the data is deleted (or overwritten), the smallest processed (and paid) interval is 90 days.

- The **Object Auto-Tiering** option (*enabled* or *disabled*) allows the system to move infrequently accessed objects to the less expensive storage repository (if the paid option is used). It can be enabled during the bucket creation or any time later. Bucket-level granularity is used for auto-tiering. If the object size is larger than 1 MiB and the object is not accessed very often, it can be dynamically moved to the **Infrequent Access Tier**. Once the frequency of access is increased, the object file is automatically moved to the original repository.

- The **Object Versioning** option preserves all versions of the data object when creating and uploading a new object version or by deleting and overwriting an older object version. It is enabled at the bucket level. There is always the current version, marked as the latest, and zero or more previous versions. There are three states that object versioning can hold:

  - **Disabled**: Object versioning has never been enabled for the bucket; thus, if you attempt to upload a file with a name that has already been used, that particular original file is overwritten. Similarly, file deletion is permanent.

  - **Enabled**: All versions are covered, and a unique version identifier is assigned for each object. Object versioning cannot be disabled if it was enabled before. In that case, the **Suspended** option is used.

  - **Suspended**: The **Upload** and **Delete** operations are not versioned (similar to the **Disabled** options). Versions created before the suspension remain in the system unless the user explicitly requests them. You can re-enable the versioning at any time.

- **Emit Object Events**: This option automates the state changes (for the object or the whole bucket) using pre-defined events (such as user notifications) for the **create, read, update, and delete (CRUD)** operations.

- The **Encryption** type: Using either **Oracle-managed keys** or **customer-managed keys**.

- **Tags**: These are metadata by which the resources can be categorized and tracked inside the *tenancy*. **Tags** consists of the **Tag Key** and **Tag Value** pairs.

Create a new bucket for the dump file repository. It will hold relevant import, export, and log files for Data Pump operations. Set the **Default Storage Tier** bucket type to **Standard**. Object changes do not need to be monitored or versioned. Let **Encryption** be done and managed by Oracle:

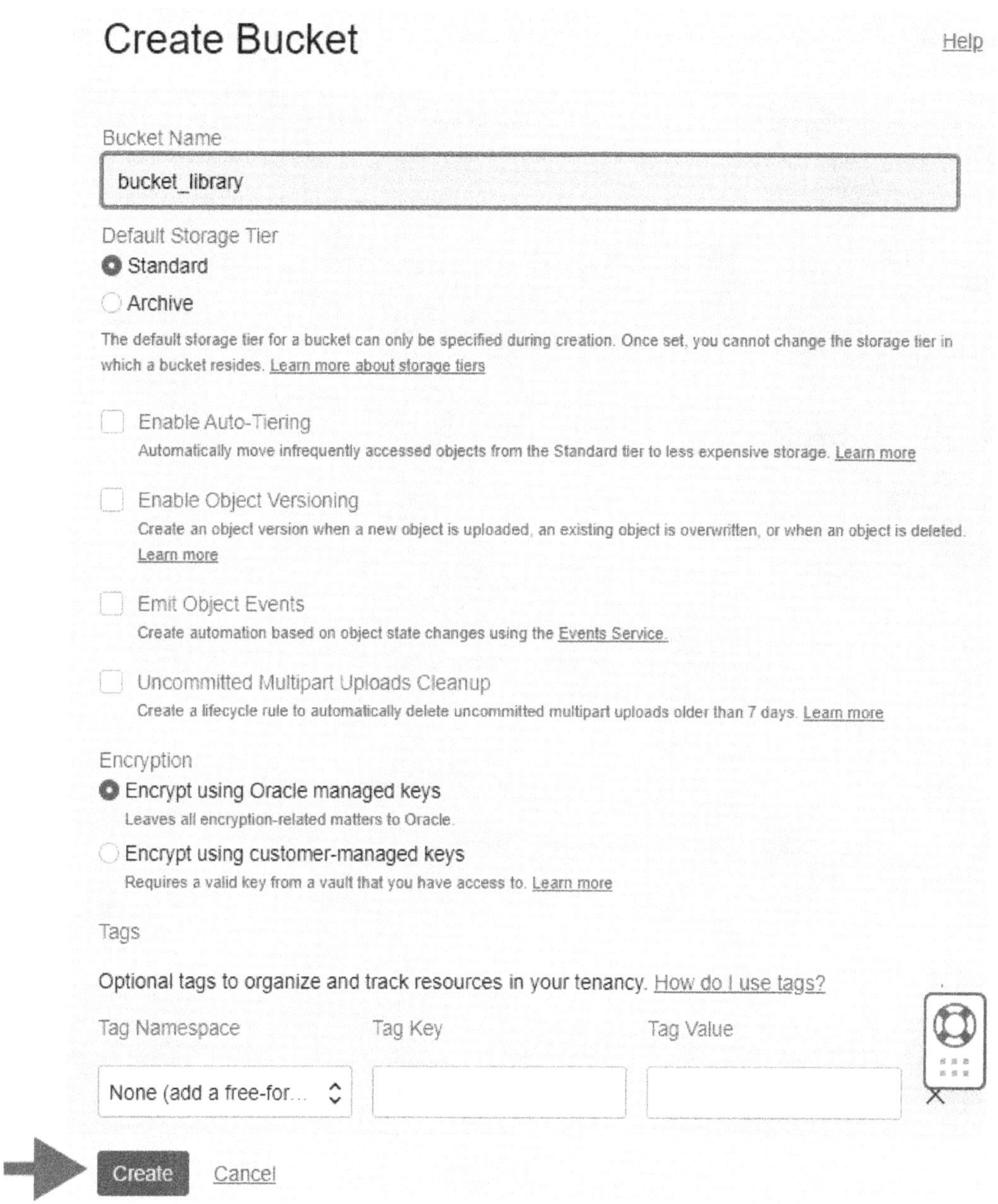

Create Bucket

Bucket Name

bucket_library

Default Storage Tier

● Standard

○ Archive

The default storage tier for a bucket can only be specified during creation. Once set, you cannot change the storage tier in which a bucket resides. Learn more about storage tiers

☐ Enable Auto-Tiering

Automatically move infrequently accessed objects from the Standard tier to less expensive storage. Learn more

☐ Enable Object Versioning

Create an object version when a new object is uploaded, an existing object is overwritten, or when an object is deleted. Learn more

☐ Emit Object Events

Create automation based on object state changes using the Events Service.

☐ Uncommitted Multipart Uploads Cleanup

Create a lifecycle rule to automatically delete uncommitted multipart uploads older than 7 days. Learn more

Encryption

● Encrypt using Oracle managed keys

Leaves all encryption-related matters to Oracle.

○ Encrypt using customer-managed keys

Requires a valid key from a vault that you have access to. Learn more

Tags

Optional tags to organize and track resources in your tenancy. How do I use tags?

Tag Namespace	Tag Key	Tag Value
None (add a free-for... ⌄		

Create   Cancel

Figure 2.15 – Creating a bucket

When you press the **Create** button, the system will create the bucket. After it has been created, the new bucket will be visible in the list.

If you click on the **Bucket Name** field, you will be able to see its definition, followed by the parameters, availability, and list of objects stored there. For now, the bucket does not hold any data, as shown in the following screenshot:

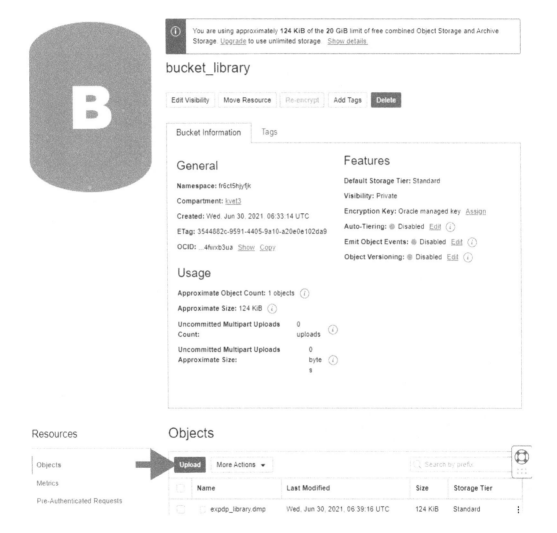

Figure 2.16 – Bucket list

Upload the exported dump file to the bucket and click on the **Upload** button under the object definition.

> **Important note**
> **Object Prefix Name** is optional and will be treated as the left-most part of the original object name extension. The original name is self-explanatory, so the input field can remain empty.

After the upload operation, the file we are currently uploading is available in the object list of the bucket, meaning that the file is accessible via the internet or through various interfaces.

At this stage, the file has been successfully uploaded into Cloud Object storage. However, it is not accessible from the database. It is not even accessible through the local SQL Developer instance. The solution is to create **credentials** by invoking the `create_credentials` procedure of the `dbms_cloud` package.

Creating credentials

Credentials are database objects holding the `username` and `password` pairs used for authentication, impersonating the `EXTPROC` functions, external jobs, remote jobs, and the `SCHEDULER` file watchers. Each credential object is identifiable by its unique name. To create a new credential, the `CREATE_CREDENTIAL` procedure is used. In order to do that, the user must have the `CREATE CREDENTIAL` or `CREATE ANY CREDENTIAL` privileges.

The syntax of the method consists of seven parameters. The first three are mandatory. The others have the `DEFAULT` clauses:

```
DBMS_CREDENTIAL.CREATE_CREDENTIAL (
    credential_name    IN    VARCHAR2,
    username           IN    VARCHAR2,
    password           IN    VARCHAR2,
    database_role      IN    VARCHAR2    DEFAULT NULL
    windows_domain     IN    VARCHAR2    DEFAULT NULL,
    comments           IN    VARCHAR2    DEFAULT NULL,
    enabled            IN    BOOLEAN     DEFAULT TRUE);
```

Let's take a closer look at these parameters:

- `credential_name`: This is a unique name used for reference. It cannot be undefined (`NULL`) and is automatically converted to uppercase unless the value is enclosed in quotation marks.

- `username`: This is a definition of the connection (the *tenancy*) to the cloud database.

- `password`: This is provided by the authentication token.

- The `database_role` parameter: This specifies the administration privileges (`SYSDBA`, `SYSDG`, `SYSADMIN`, or `SYSBACKUP`). By default, the connection is made via standard user privileges, as defined by the `NULL` value.

- `windows_domain`: This is applicable for the operating system Windows and the specific domain that the referenced user belongs to.

- The `comments` parameter: This defines a text string for easier user consecutive reference. It does not impact the execution.

- The `enabled` parameter: This defaults to `TRUE`, specifying the current usability of the credentials. The following snippet shows an example:

```
BEGIN
DBMS_CLOUD.CREATE_CREDENTIAL(
    credential_name => 'CREDENTIAL_NAME',
    username => 'tenancy_name',
    password => 'authentication_token');
END;
/
```

The `credential_name` parameter is the unique identifier. This value is then referenced in the code alongside `tenancy`. Naturally, `password` is not specified in plain text. Instead, the **authentication token** is provided.

Authentication token

To define new credentials, it is necessary to obtain an authentication token. There are, in principle, two ways to obtain one. The first approach is by clicking on **Profile** and selecting a username.

The second approach is to access it using the left navigation panel menu: **Identity & Security | Users**. Then, by clicking on the user, the available resources are listed on the left part of the screen. Navigate to **Auth Tokens** (in the **Resources** section), click on **Generate Token,** and provide an explanatory description:

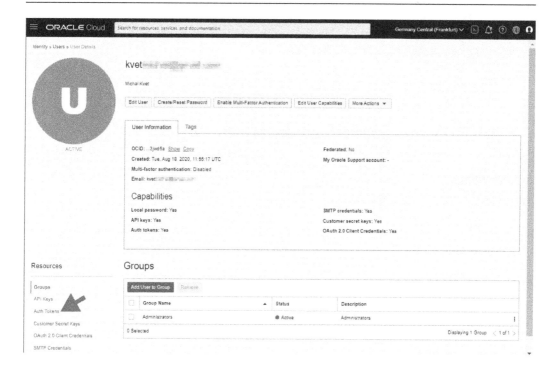

Figure 2.17 – Authentication token definition

Copy the generated token to the clipboard. Note that it is impossible to get it later. You would have to remove the token and create a new one.

Now you have all the required values to create credentials, so let's return to the defined syntax. First, connect to the cloud instance as the administrator user (admin) via SQL Developer Desktop or Web or Instant Client. Then, execute the code in the following code snippet to create credentials. The process of getting a username and authentication tokens has already been specified in the previous paragraph.

An example of the usage is placed in the following code snippet:

```
BEGIN
DBMS_CLOUD.CREATE_CREDENTIAL(
    credential_name => 'ATP_CREDENTIAL_MK',
    username => 'kvetmichal',
    password => 'H26oW:]:q]xb2KuWRmvQ');
END;
/
```

The list of the created credentials can be obtained by retrieving data from the `all_credentials` data dictionary view. It consists of the following attributes:

- `owner`
- `credential_name`
- `username` (the name of the user that will be used to log in)
- `windows_domain` (the attribute to use when logging in)
- `comments`
- `enabled` (the status indicating whether the credential is enabled (`true`) or disabled (`false`))

The following `select` statement provides a pair of `credential_name` and `username` from the `all_credentials` data dictionary view:

```
select credential_name, username
  from all_credentials
    order by credential_name;
```

The last step is to get access to the Data Pump file to be imported. It is necessary to get a link (URI address) for the file in Object Storage, which is done by a **pre-authenticated request**.

Pre-authenticated request

Pre-authenticated requests establish a way for users to access a structure (object or bucket) in Object Storage without having their own credentials. In the **Object Storage** section of **Cloud Management**, select the particular bucket.

Click on **Pre-Authenticated Requests** in the **Resources** section, and a new pop-up window appears, prompting you to specify the name and target, access type, and expiration date:

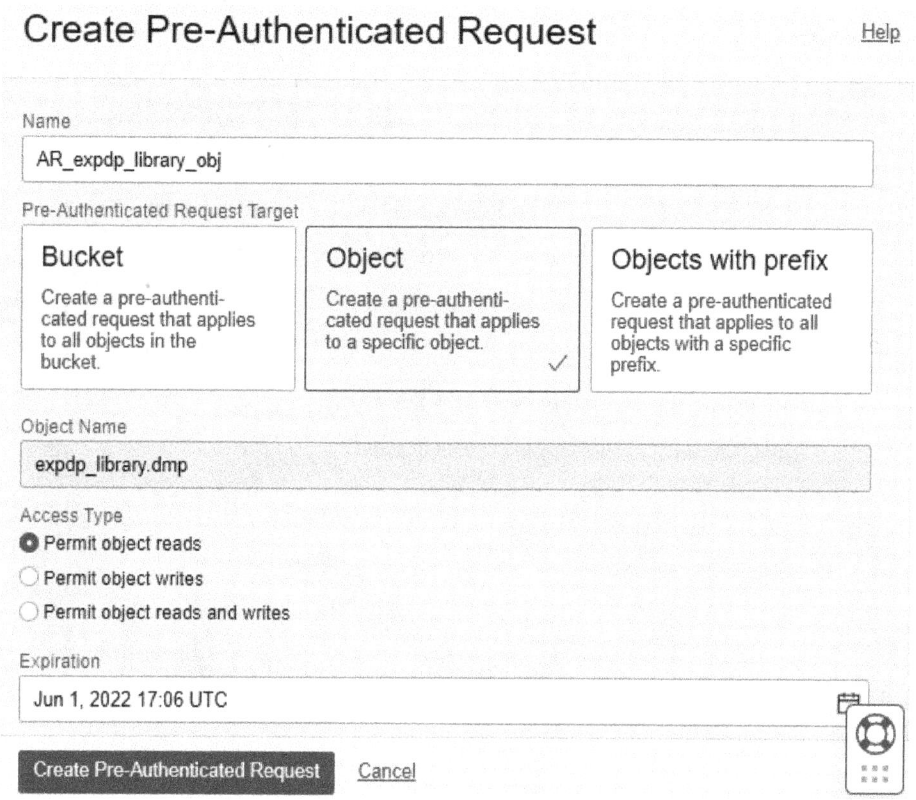

Create Pre-Authenticated Request

Name

AR_expdp_library_obj

Pre-Authenticated Request Target

Bucket	Object	Objects with prefix
Create a pre-authenticated request that applies to all objects in the bucket.	Create a pre-authenticated request that applies to a specific object. ✓	Create a pre-authenticated request that applies to all objects with a specific prefix.

Object Name

expdp_library.dmp

Access Type

● Permit object reads
○ Permit object writes
○ Permit object reads and writes

Expiration

Jun 1, 2022 17:06 UTC

Create Pre-Authenticated Request   Cancel

Figure 2.18 – Pre-authenticated request definition

A pre-authenticated request can be created for two precision levels, **Object** (the name of the file has to be specified) or **Bucket** granularity can be used. Object level can then be further divided into a specific object (**Object**) or all objects with the defined prefix (**Object with prefix**).

File pre-authentication requests can also be created by clicking on an individual object and selecting **Create Pre-Authenticated Request**. The window that opens is similar to the object configuration in *Figure 2.18*, but in this case, the object target is directly pre-selected, and the object name reference is already filled:

Objects

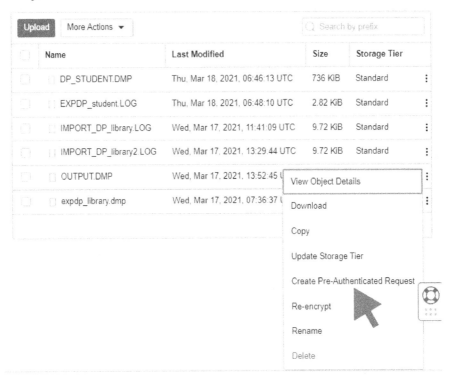

Figure 2.19 – Creating a pre-authenticated request for the object

Create a new pre-authenticated request for the particular export file to be imported. In this case, since the file content will not be changed, the **Read** privilege is suitable. The pre-authenticated request URL looks like this:

```
https://objectstorage.eu-frankfurt-1.oraclecloud.com
/p/o2q3BsQ-lhHKZ80nO2eTZdzWcqW1U1hvaJuoxstSdV3UrNmI4BB2tucxXTw
D4ft1
/n/frrt85axbzme
/b/bucket_library
/o/expdp_library.dmp
```

Let's evaluate its structure:

- p is the unique identifier of the pre-authenticated request reference.

- n covers the namespace.

- b defines the bucket.

- o expresses the object. If the bucket is to be referenced, the o section is empty.

The prerequisites and necessary object definitions are passed, so let's do the Data Pump import and export. The first part deals with SQL Developer Desktop, then we look at the Web version.

Import process using dump files

That's all regarding pre-processing and references. Now, it's time to use the wizard for the Data Pump importing. To do so, a destination user is commonly created. The appropriate privileges must be granted to the created user:

```
create user library_user identified by *******;
grant connect, resource, unlimited tablespace
        to library_user;
```

Replace ******* with the actual password (at least eight characters with uppercase and lowercase letters and numeric values).

Now let's look at how to import using the impdp data dump managed in SQL Developer Desktop. Let's get started.

1. Navigate to **View | DBA** and specify a new DBA connection by clicking on the green plus symbol (✛) in the **DBA** group.

2. Navigate to the cloud database connection and expand the cloud connection type. It consists of several administration and performance monitoring tools.

3. Navigate to the **Data Pump** section and right-click on **Data Pump**.

4. You should get two options: **Data Pump Import Wizard…** and **Data Pump Export Wizard….** Click on **Data Pump Import Wizard….**

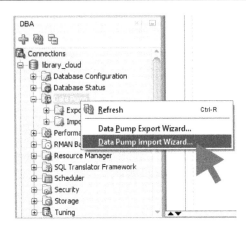

Figure 2.20 – Data Pump Import Wizard in the SQL Developer Desktop tool

By clicking on the **Data Pump Import Wizard…** option, a new process, and the wizard is launched. In the following screenshots, we will describe the process (defined by the wizard) step by step. Let's get started.

In **Step 1**, it is necessary to specify **Job Name** and source (**Data or DDL**) to be imported (data only, structure only, or both), **Type of Import** (**Full**, **Schemas**, **Tables**, **Tablespaces**), and source consisting of the credentials and the file URI created in the previous part:

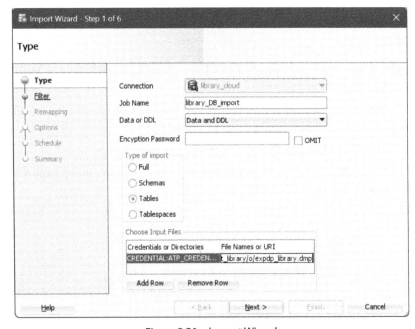

Figure 2.21 – Import Wizard

Step 2 is where we select the tables to be imported from a list of the available tables. **Step 3** allows remapping schemas or tablespaces (source and destination). In **Step 4**, options are defined – the number of threads, action, whether the table exists, and logging reference specifying the Oracle directory and the output data file:

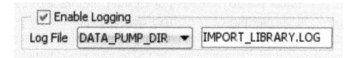

Figure 2.22 – Log reference definition

Step 5 deals with the job scheduling, and finally, **Step 6** provides the summary:

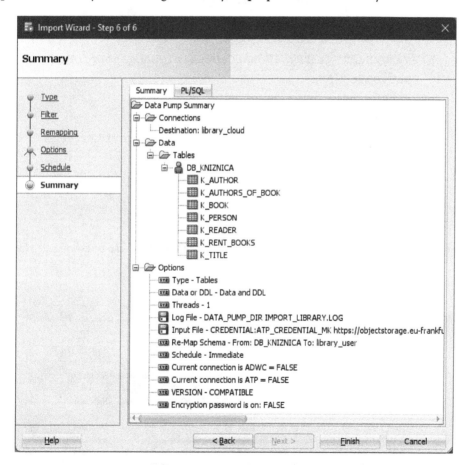

Figure 2.23 – Import Wizard Summary

After the successful execution of the import operation, the log file is automatically created. However, it is only accessible by the database (in my case, DIRECTORY: DATA_PUMP_DIR FILE: IMPORT_LIBRARY.LOG), and not through Object Storage.

So, how do we access the log file? Make it accessible using put_object of the dbms_cloud package specifying credential_name, the object_uri destination, the Oracle directory_name source, and file_name. Please note that object_uri can be composed of the pre-authenticated request at the bucket level extended by the name of the output file visible in Object Storage (**Write** privilege must be granted to do that). Please note that the file_name specification is case sensitive:

```
BEGIN
 DBMS_CLOUD.PUT_OBJECT(
  credential_name => 'ATP_CREDENTIAL_MK',
  object_uri => 'https://objectstorage.eu-frankfurt-
                 1.oraclecloud.com
     /p/PvVxyzBu2Ntwnm1Pyp4VUwMfCaD4zz1rIGBAUq1soUp1qU1-
        K9bBvD75bnxfYw8u
     /n/frrt85axbzme
     /b/bucket_library
     /o/log_library_import.log', -- destination file name
  directory_name => 'DATA_PUMP_DIR',
  file_name => 'IMPORT_LIBRARY.LOG'); -- source file name
END;
/
```

Then, the referenced log file is visible in the cloud interface of the bucket and can be downloaded locally:

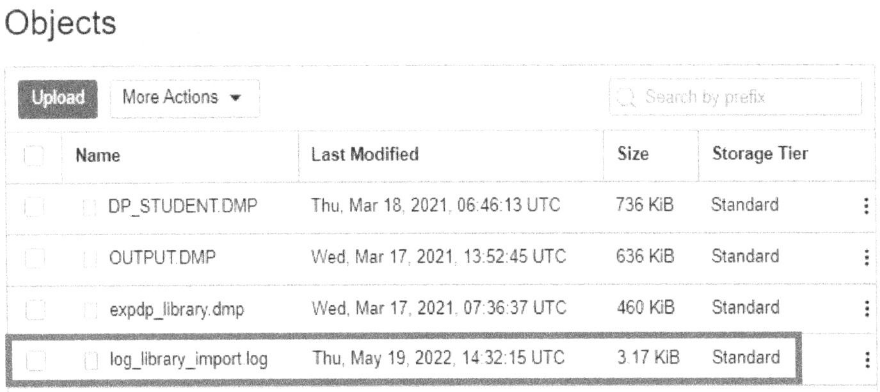

Objects

	Name	Last Modified	Size	Storage Tier	
	DP_STUDENT.DMP	Thu, Mar 18, 2021, 06:46:13 UTC	736 KiB	Standard	⋮
	OUTPUT.DMP	Wed, Mar 17, 2021, 13:52:45 UTC	636 KiB	Standard	⋮
	expdp_library.dmp	Wed, Mar 17, 2021, 07:36:37 UTC	460 KiB	Standard	⋮
	log_library_import.log	Thu, May 19, 2022, 14:32:15 UTC	3.17 KiB	Standard	⋮

Figure 2.24 – Log accessibility using the Oracle Cloud interface

In the next section, the opposite operation is described, highlighting the dump file export.

Export process using dump files

Data Pump export (expdp) is a new, more flexible, more reliable, and faster data export technique executed on the server side. It can be associated with various objects – the whole database, user schema, or specific tables.

The process of the export is analogous to the already-discussed impdp functionality. In the **DBA** section of the SQL Developer Desktop, expand **Connections**, navigate to **Data Pump**, and select **Data Pump Export Wizard**.

Data Pump export is a staged process consisting primarily of the data source. Commonly, the structure and data are exported together. However, the user can also choose only one of the options. First of all, the wizard window prompts the connection, data, or DDL to be exported, followed by the type (database, tablespaces, schemas, tables). Then, the filter can be applied to the objects or data themselves. Thus, it is not necessary to export all objects or all attributes. However, be aware of references when using the **Global Where** condition. Finally, individual options and output files can be specified.

For the options, the user can specify the number of threads and get the estimated number of blocks or statistics. Optionally, the read-consistent image can be obtained by referencing the **System Change Number** (**SCN**) or the date. SCN is a logical timestamp used by the Oracle database system to order the transactional events within the database. Every time the transaction is committed, the SCN value is incremented by the Oracle clock. SCN value is also used to build consistent data images in a parallel processing environment. When dealing with the dump file management process, it is strongly recommended to monitor the process by logging. Similar to the import, the Oracle directory and file name need to be specified.

The last step summarizes the action to be launched by listing the parameters, options, and data sources:

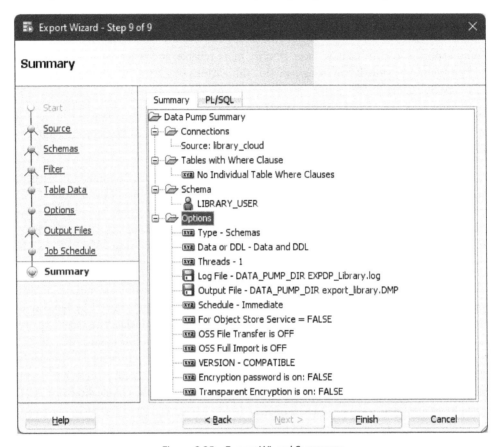

Figure 2.25 – Export Wizard Summary

On the **PL/SQL** tab, you can get the generated script, which will be used for the execution. It is recommended that you study it briefly to understand the available options and the way the definition (user selection) is transformed into code.

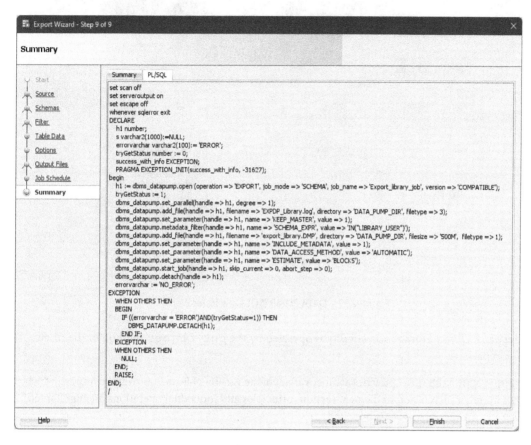

Figure 2.26 – PL/SQL code for export

The Data Pump import/export process can also be monitored by SQL Developer Web, pointing to **DATA PUMP** in the **Data Tools** subsection of the main menu:

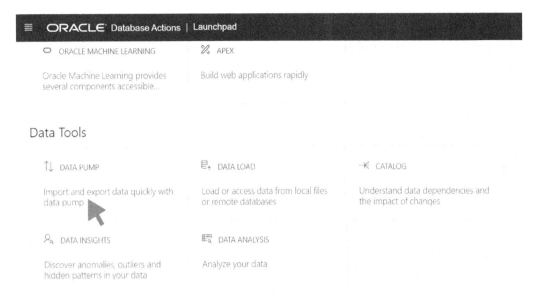

Figure 2.27 – DATA PUMP (SQL Developer Web)

Afterward, Object Storage accessibility is operated by the put_object method of the dbms_cloud package.

Transportable Tablespace provides another variant of the version of Data Pump import/export, which is commonly much faster. In the next section, principles and individual operations are highlighted, and a step-by-step approach is provided.

Understanding Transportable Tablespace Data Pump

Transportable Tablespace Data Pump is generally much faster than conventional Data Pump management. Data files forming the tablespace are simply copied to the new repository. Then, the Data Pump process transfers the metadata of the objects covered by the tablespace to the new database. The step-by-step sequence is as follows:

1. Prepare the database for the Transportable Tablespace Data Pump on the on-premise. Set the tablespaces to the READ ONLY mode:

    ```
    ALTER TABLESPACE <tablespace_name> READ ONLY;
    ```

2. Invoke Data Pump export (ExpDp) on the on-premises database.

3. Copy the tablespace data files to the destination securely.

4. Set the on-premise tablespaces to the READ WRITE mode.

```
ALTER TABLESPACE <tablespace_name> READ WRITE;
```

5. Prepare the destination database to import.

6. Invoke Data Pump import (ExpDp) on the destination database.

7. Set the destination database tablespaces to the READ WRITE mode.

8. Verify the data has been imported successfully. If so, you can delete dump files.

In this section, you got familiar with Transportable Tablespaces, by which it is possible to copy data from one repository to another. It is operated by copying data files, followed by metadata transfer. Next, let's take a look at Full Transportable Tablespaces.

Full Transportable Tablespace Data Pump

The Full Transportable Tablespace Data Pump operation extracts the metadata provided by the Data Pump and copying process of data and index. The performance improvement is based on limiting the need to interpret and extract individual rows and index entries. The Full Transportable Tablespace Data Pump is an extension of the Transportable Tablespace Data Pump. A callout mechanism and API are provided by Oracle components.

The Full Transportable Tablespace Data Pump is invoked by using the TRANSPORTABLE=ALWAYS and FULL=Y parameters. The process of using that technology requires multiple steps:

1. Make a physical directory on the source database host:

```
$ mkdir /u01/app/oracle/admin/orcl/dump_files
```

2. Connect to the database host using SQL*Plus. Log in as the SYSTEM user. Specify the password and, optionally, the connect identifier:

```
$ sqlplus system
```

3. Create an Oracle directory for mapping the physical directory (created in *step 1*) to the database on the source database host:

```
CREATE DIRECTORY data_pump_dir AS
        '/u01/app/oracle/admin/orcl/dump_files';
```

4. Set the READ ONLY mode for tablespaces to be migrated. In my case, I will manage the main_tblspc and index_tblscp tablespaces:

```
ALTER TABLESPACE main_tblspc READ ONLY;
ALTER TABLESPACE index_tblspc READ ONLY;
```

The list of available tablespaces and associated data files can be obtained by querying the dba_data_files data dictionary view:

```
SELECT tablespace_name, file_name FROM dba_data_files;
--> TABLESPACE_NAME        FILE_NAME
--> ----------------  --------------------------------
--> MAIN_TBLSPC
       /u01/app/oracle/oradata/orcl/main1.dbf
--> MAIN_TBLSPC
       /u01/app/oracle/oradata/orcl/main2.dbf
--> INDEX_TBLSPC
       /u01/app/oracle/oradata/orcl/index.dbf
```

5. Carry out the Data Pump export on the source database:

```
$ expdp system FULL=y TRANSPORTABLE=always
              DUMPFILE=expdat.dmp DIRECTORY=data_pump_dir
```

6. Create a new physical directory on the destination server node:

```
$ mkdir /u01/app/oracle/admin/orclpdb/dump_files
```

7. Use the **Secure Copy** (**SCP**) tool to copy the files. Ensure the SSH private key for access is savailable on the source host. The following command should be launched for each file associated with the tablespace to be migrated:

```
$ scp -i private_key_file
      /u01/app/oracle/oradata/orcl/main1.dbf
    oracle@dest_IP_address:
      /u01/app/oracle/admin/orclpdb/dump_files
```

8. Connect to the destination database using SQL*Plus. Log in as the SYSTEM user. Specify the password and, optionally, the connect identifier:

```
$ sqlplus system
```

9. Set the source tablespaces back to the READ WRITE mode:

```
ALTER TABLESPACE main_tblspc READ WRITE;
ALTER TABLESPACE index_tblspc READ WRITE;
```

10. Create an Oracle directory for mapping the physical directory (created in *step 6*) to the database on the destination node:

```
CREATE DIRECTORY data_imp_dir AS
        '/u01/app/oracle/admin/orclpdb/dump_files';
```

11. Start the Data Dump import using the FULL option. The TRANSPORT_DATAFILES clause refers to the source file to be copied:

```
$ impdp system@orclpdb FULL=y DIRECTORY=data_pump_dir
        TRANSPORT_DATAFILES=
        '/u01/app/oracle/oradata/orcl/main1.dbf',
        '/u01/app/oracle/oradata/orcl/main2.dbf',
        '/u01/app/oracle/oradata/orcl/index.dbf'
```

If all the steps were performed successfully, the created dump file can be deleted.

Data loading is just one piece of the whole strategy of moving data. Additionally, OCI offers you multiple techniques and tools to securely migrate data to the Oracle Cloud repository. There can be various data origins, such as transactional data, analytic data, structured data, or unstructured data; all mentioned database types can be treated as the source and robustly migrated. The upcoming section takes you through the migration technique summary.

Exploring migration techniques

The critical parts of information technology are data management, data access, and an internal repository. Without data, no relevant decisions can be made. When we shift systems to the cloud, it is inevitable that we need to transfer data as well. However, how do we do that if the data source is large? This section deals with the various techniques for data migration, so let's dive right in!

OCI Data Transfer Service

Databases can be really extensive, even in petabytes. Moving repositories from data centers to clouds can take weeks or even months. Moving data over the public internet can be expensive, maybe even impossible due to an unreliable network, low speed, or security aspects. OCI Data Transfer Service offers you the option to order the transfer, which is then managed directly by Oracle. It is operated by the Oracle Cloud console or the **Command-Line Interface** (**CLI**). The data to be imported is shipped to Oracle, commonly on encrypted disks. They extract the data and upload the files into the designated Object Storage bucket in your tenancy. Similarly, for the export, data is copied to form the export and sent to you. The data is encrypted using the AES-256 algorithm. The same algorithm is used for Object Storage data encryption as well. The whole process is fast. Data transfer is simple, cost-effective, flexible, and scalable – up to 150 TB per appliance, with multiple appliances per data transfer job. Data integrity is supervised by checksum at each migration process stage to ensure the reliability and correctness of the results. The whole process can be monitored and managed dynamically using the OCI console or CLI.

Oracle Zero Downtime Migration

Oracle **Zero Downtime Migration** (**ZDM**) allows easy and efficient migration of on-premises data to the Oracle Cloud using various technologies, such as Oracle Data Guard or Oracle GoldenGate to minimize downtime. The source databases can be from Oracle database 12c (CDB/PDB, Non-CDB) or Oracle database on AWS. The target databases use Exadata Cloud Service, Exadata Cloud at Customer, and Database Cloud Service using Oracle dataset 11g and newer.

OCI Database Migration

Database Migration Service (**DMS**) is formed of a set of concepts guiding the migration from on-premises to Oracle Autonomous Database on shared or dedicated infrastructure. It also covers third-party clouds as the source. The migration process can use online or offline methods, focusing on ZDM and an easy user interface. Data migration is done through the Object Storage or Oracle database link.

The process consists of the following steps:

1. **Autonomous Database provisioning** was discussed in *Chapter 1*. Navigate to the cloud console, then **Oracle Database** | **Autonomous Database**, and select the appropriate type for your workload.

2. Grant permissions to the database migration user. This is done by creating a group in OCI for a migration using the **Identity & Security** option and navigating to **Groups**.

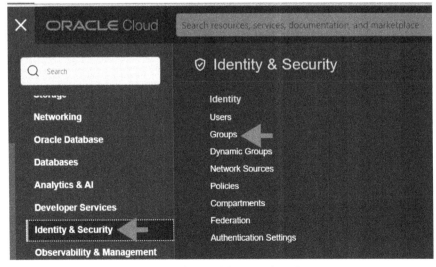

Figure 2.28 – Identity & Security then Groups

3. Configure `sudo` access to the database server by identifying the operating system user with SSH access to the source server and granting them `sudo` permissions.

4. Prepare the source database for the migration:

   *   Configure the streams pool memory structure by initializing its size (256 MB is the recommended lower limit):

   ```
   show parameter streams_pool_size;
   -->     NAME                          TYPE            VALUE
   -->     -------------------------- --------------- -----
   -->     streams_pool_size             big integer     0
   alter system set streams_pool_size=256M;
   ```

   *   Set the GLOBAL_NAMES parameter to `false` (if it is set to `true`):

   ```
   show parameter global_names;
   -->     NAME             TYPE      VALUE
   -->     ------------- ------- -----
   -->     global_names boolean FALSE
   alter system set global_names=false;
   ```

   *   Enable the ARCHIVELOG mode (note that it must be done in a mount stage of the database):

   ```
   shutdown immediate;
   startup mount;
   alter database archivelog;
   alter database open;

   select name,log_mode from v$database;
   -->     NAME      LOG_MODE
   -->     ----- ------------
   -->     ORCL      ARCHIVELOG
   ```

   *   Enable logging:

   ```
   select supplemental_log_data_min, force_logging
     from v$database;
   -->     SUPPLEMENTAL_LOG_DATA_MIN            FORCE_LOGIN
   -->     --------------------------------- -------------
   -->     NO                                   NO
   ```

```
alter database add supplemental log data;
alter database force logging;
```

- Choose the filesystem directory for holding the export files for the database migration process. This step is not applicable if the migration is done via a database link.

5. Prepare the target database for the migration by setting the `global_names` parameter to `false` (if it is set to `true`):

```
show parameter global_names;
-->     NAME            TYPE    VALUE
-->     ------------ ------- -----
-->     global_names boolean FALSE
alter system set global_names=false;
```

6. Register the source and target database in the Oracle Cloud console:

- Provide inputs in the **Migration | Registered Databases** sections

After the registration process, the databases are listed in the **Registered Databases** list inside the **Database Migration** submenu accessible from the Oracle Cloud console.

7. Create and validate the migration:

- Navigate to **Database Migration | Migrations** and click on the **Create Migration** button. Provide the required values for the source and target databases, migration types, encryption, database schema details, bucket name in Object Storage, and source database directory with the **Read** and **Write** privileges for the database owner. If the source database is visible from the cloud environment, then the **Direct connection to source database** option should be selected. Otherwise, **No direct connection to source database** needs to be used, followed by downloading and installing Agent (accessible via **Migration | Database Migration | Agents**).

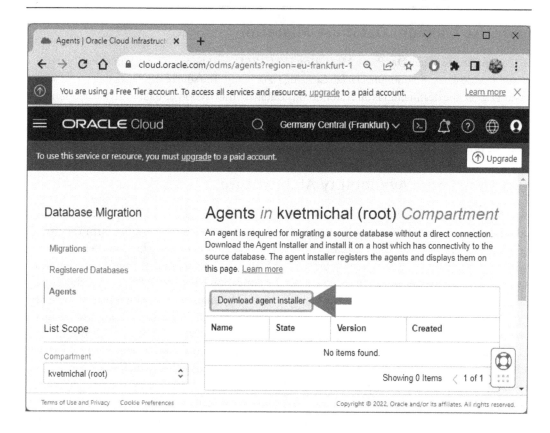

Figure 2.29 – Migration Agent download

- Check the compatibility options using **Cloud Premigration Advisor** (**CPAT**). Migration validation consists of these phases:

  - Source validation (SSH connectivity and database connection)

  - Target validation (connectivity over private endpoints and validation of the OCPU count for the parallelism)

  - Validation using CPAT (consisting of source database analysis for incompatibilities)

  - Validation of the Data Pump source settings (checking the Data Pump prerequisites and the directory object path validation)

  - Validation of the Data Pump target settings

8. Start migration.

Now let's move on to the next migration technique.

Enterprise Manager Database Migration

The **Enterprise Manager Database Migration** (EMDBM) workbench allows you to easily move database workloads to Autonomous Database using Data Pump. The source databases must be at least version 11.2.04, the destination is an Autonomous Database transaction or data warehouse-oriented using a shared or dedicated infrastructure. EMDBM is integrated with Autonomous Database Schema Advisor, which offers you tools for schema analysis to identify differences and limit incompatibility issues. The process of data migration can be monitored online in real time.

Oracle Maximum Availability Architecture

Oracle **Maximum Availability Architecture** (MAA) is based on Data Guard and provides various methods to ensure high availability. It is supervised by two approaches: Cross-Platform Database Migration and Oracle GoldenGate Cloud Service Migration. How does it work? How does it ensure maximum availability? What does it mean practically? Well, the whole solution is based on the features associated with the overall architecture. Oracle **Real Application Clusters (RAC)** is an architecture based on sharing Oracle databases across multiple servers by enabling horizontal scalability. Any outage can be immediately identified, and workloads can be rerouted to hide the impact of outages from the client's perspective. During the migration, a subset of servers can be associated with the process of moving data, while the rest serve the existing requirements. Oracle Active Data Guard is another feature that uses multiple databases with the same data perspective. Synchronization across the replicas is done using in-memory REDO logs. Thanks to this, the system can even survive disasters. By collapsing the primary data center, the secondary site can be used later on. Commonly, secondary standbys are located in a different region. A new repository location is created, followed by applying transaction REDO logs dynamically.

Oracle Application Continuity

By aggregating RAC and Oracle Active Data Guard, outages and failures can be automatically masked by remapping the processing to another node.

Other features related to the Oracle Maximum Availability Architecture are **Oracle Recovery Manager (RMAN)**, which can perform and automate recovery and backup strategies on Oracle databases, and Oracle Flashback technology (described in more detail in *Chapter 13*). Another option is *edition-based redefinition*, which enables online application upgrades with no accessibility loss. Current and upgraded applications can run simultaneously, allowing us to shift to a new version dynamically with no business interruption.

Move to Autonomous Database, Move to Oracle Cloud Database

Move to Autonomous Database (MV2ADB) moves data to the Autonomous Database, while **Move to Oracle Cloud Database (MV2OCI)** moves data to the Oracle Cloud database.

The principle of MV2ADB and MV2OCI is to create dump file exports (using ExpDp) in the on-premises server by prompting MV2ADB to interconnect it to the Object Storage by importing the dump (using ImpDP) to the Autonomous Database.

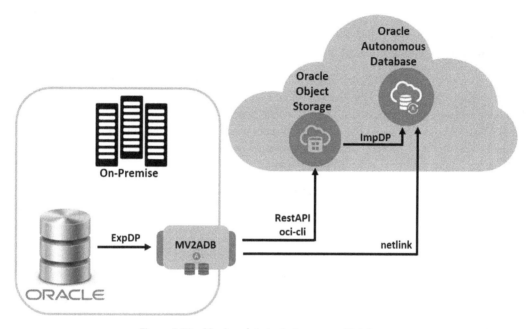

Figure 2.30 – Moving data to Autonomous Database

Thus, as can be seen in the preceding diagram, the whole operation consists of the following steps: expdp of the source, uploading the dump to Object Storage and applying the data to the Autonomous Database using ImpDp.

OCI GoldenGate

OCI GoldenGate is a complex platform for real-time data management. It is based on replicas to ensure robust data availability and real-time analysis. Users do not need to allocate compute environments. The data replication parameters, strategies, and overall solutions can be easily monitored, managed, and designed.

As you have seen, the Oracle database and Oracle Cloud environment offer you multiple techniques to transfer data from the data center to the cloud or generally between the repositories. Each of these techniques has its own characteristics, advantages as well as the environment for which it was developed. The goal is to ensure security, performance, and speed, but also continuous availability of systems, which is a critical business requirement.

Summary

In this chapter, the techniques of data migration were discussed. Third-party applications or non-Oracle databases can provide input files in various formats by getting the mapping instructions from the data layer to the database. Data loading can be operated by the SQL Loader consisting of the data (UNL) and control (CTL) files. The process is straightforward, and data rows that cannot be loaded can be easily identified. Raised exception details are present for each refused row.

The client side or the server side can manage data and shift it across Oracle platforms. In this chapter, we dealt with the principles and limitations of this approach by emphasizing the differences between client-side and server-side import and export operations. The main focus was on the Oracle Cloud as a destination by dealing with Object Storage and buckets. The server-side import and export types use the Oracle directory object to access the data storage layer. To make the files available through the cloud interface, the dbms_cloud package was used.

In the next chapter, you will learn about the roots of date and time standardization, also focusing on the concept of temporality.

Questions

1. Which import method can take CSV files as input?

    A. SQL Loader

    B. Imp

    C. ImpDP

    D. Object Storage reference

2. Select the correct statement related to the Exp and Imp methods:

    A. If the Conventional type option is specified, then the method is executed on the client side.

    B. If the Direct type option is specified, then the method is executed on the server side.

    C. The Exp and Imp methods operate on the client side.

    D. The Exp and Imp methods operate on the server side.

3. The put_object method of the dbms_cloud package is used:

    A. To make the file accessible via Object Storage

    B. To upload data to the database

    C. To import data to the database directly by mapping

    D. To create the database object specified in the file

4. What storage precision level is used for pre-authenticated requests?

    A. Bucket only

    B. Object only

    C. Object and bucket

    D. Whole cloud repository only

Further reading

- *Oracle Cloud Infrastructure Database Migration (DMS)*: This provides a step-by-step guide for offline logical migration, available here:

https://www.oracle.com/a/ocom/docs/oci-
database-migration-onprem-adb.pdf

- *Oracle Database Migration Methods: Learn the different ways to move an Oracle database* by Arun Kumar: This provides a summary of Oracle database migration methods using Oracle RMAN and Oracle Data Pump.

- *Expert Oracle Database Architecture: Techniques and Solutions for High Performance and Productivity* by Darl Kuhn and Thomas Kyte: Chapter 15 of the book deals with the data loading and unloading methods and provides an interesting discussion about identifying and dealing with errors, as well as best practices to optimize the data loading and unloading process.

Part 2:
Understanding the Roots
of Date and Time

This part aims to provide you with a detailed description of the roots of date and time modeling originating from the ISO 8601 standard. You will dive into relational databases and integrity related to temporal management, date-value definitions, as well as week management and ordinal date definition by focusing on specific aspects of Oracle Database. In this part, date and time values are initially treated separately by focusing on individual elements and value construction. By introducing periods of time, a discussion about validity interval modeling, unlimited validity, interval periodicity, as well as some enhancements is introduced.

Next, the concept of temporality is introduced. Are you familiar with the term **temporality**? And what about **summer** and **winter time**? Or **Daylight Saving Time**? Do you know the rules for detecting a **leap year**? Or oven **leap seconds**? Well, these are the key components that ensure proper date and time management. After reading this part, you will be familiar with all these terms and concepts, as well as the techniques for implementing them in real-world applications. Get inspired!

This part includes the following chapters:

- *Chapter 3, Date and Time Standardization Principles*
- *Chapter 4, Concepts of Temporality*

3

Date and Time
Standardization Principles

Currently, businesses focus on intelligent information systems supported by machine learning techniques, decision support systems, and prognosis management. Conventional principles are not suitable. The reference path over time needs to be covered. To cover the evolution over time, each state needs to be extended by the validity frames, represented by the date and time values, and even extended with multiple time zones and synchronization across regions. Unfortunately, there were no common standards and rules applicable in the first phases of development. Therefore, individual database system vendors decided to solve the problem in their own ways without creating a complex solution that would be generally applicable. Nevertheless, although individual systems do not apply the rules of **international standard (ISO)** standardization that were accepted later, we think it is very important to get a comprehensive overview of standardization, focusing on the roots and starting points. As you will see, later-accepted standardization concepts open up a wide range of possibilities but simultaneously point out the limitations of the given solution.

Oracle provides a date and time management system that reflects the user's time zone by differentiating between local and server (cloud) locations. This chapter covers the basic principles of standardizing date and time values. It separates the date and time values into two elements in the initial phase, focusing on the modeling and construction principles. An analogous solution is applied, for example, in MySQL. There is the `Date` data type, holding year, month, and day elements. The `Time` data type references hour, minute, and second elements. A specific data type, `DateTime`, puts the `Date` and `Time` values into one data attribute. A similar approach is used in SQL Server as well, allowing values ranging from `0001-01-01` to `9999-12-31`, expressing `year-month-day`. However, as we shall see, Oracle has its own method of implementation.

Although currently, some standards exist for particular goals, generally, database systems just use them as a reference (as ideas) but create their own solutions to meet their requirements. Consequently, each system has its own dialect, usability properties, and specific implementation requirements.

Soon after the release of the first database system, sophisticated techniques were required by users and developers. Thus, individual vendors had to specify functionality and management principles. Moreover, the whole process was not synchronized, resulting in various system representations.

In this chapter, we will cover the following main topics:

- Date and time references in relational databases

- Date and time standardization principles, focusing on ISO 8601

- Modeling date and time elements and composite values

- Modeling a period of time

Understanding relational database integrity

Relational databases were introduced in the 1960s, formed by the entities and relationships between them and influenced by relational algebra. The transaction support covered security, consistency, and integrity, ensuring atomicity, consistency, isolation, and durability.

In the field of databases, the concept of integrity is understood in terms of accuracy and correctness by ensuring data consistency. Moreover, it is often associated with data confidentiality. The integrity problem is, therefore, associated with ensuring any changes made to the database are correct. Errors and consequent data inconsistencies can be caused by data entry, operator errors, program errors, or even intentional database corruption attempts.

Integrity can be divided into the following categories:

- **Domain integrity**: This represents a set of rules shared by all column attributes assigned to a particular domain. In simpler terms, it refers to the data type, specifying the type, a set of allowable values, and sorting proportions.

- **Column integrity**: This extends the domain integrity principles by adding additional rules to reduce the values that can be contained within the dataset, focusing on duplications and undefined values (such as NULL or NOT NULL values).

- **Entity integrity**: This requires a *primary key* definition for the table, applying the rules of uniqueness and minimality.

- **Referential integrity**: This focuses on the foreign key, which can hold either a NULL value (if optional membership is used) or the referenced primary key (although generally, it can refer to the primary key candidate instead of the direct primary key).

- **User integrity**: This extends the previous integrity categories with specific application domain rules defined by the user.

Relational databases have always been strongly related to the conventional database approach of storing only currently valid data. Therefore, if the update operation is invoked, the original data is physically replaced with new versions. The database transactions supervised all activities. Thus, each object held just one version. Although the original state could be partially obtained by the transaction logs, generally, no complex history could be managed. With the globalization of trade and the need to optimize costs and create decision-making support systems, it was necessary to extend the principles. Namely, it became necessary to store historical states, as well as future plans, allowing us to track changes over time. Therefore, soon after the first releases of relational databases, it was evident that conventional databases did not have the required complexity to manage intelligent information systems and decision-making support.

With the creation of machine learning support, prognosis management, and overall optimization, it was clear that the individual states needed to be monitored over time. Namely, each object state needs to be extended by the time frame expressing its validity. Thanks to that, instead of using conventional principles, the temporality aspect is formed, storing not just currently valid states but also the whole evolution. This is where we get to the heart of the matter, which is effective time management. As it was necessary to define complex methods for date and time management, several attempts were made to standardize these methods. However, most of them were either rejected, left for later evaluation, or required an expansion of the solution with an emphasis on greater robustness, resistance, and complexity. As a result, there was no generic and generally accepted concept.

The concept of proper date and time processing is currently critical because data can originate from various sources and with varying amounts of precision. Data can also be produced in multiple formats. All the values need to be transformed, synchronized, analyzed, and evaluated. If the data is regionally specific, the problem can be even deeper. Thus, the standardization of its usage in a real-world context is important and this is where big data comes to the fore. Big data is collected from various sources with various speeds of generation and timeline references. It is characterized by **volume** (data amount), **variety** (types of data to be processed and stored), **velocity** (the speed at which data is generated), **veracity** (level of trust in the data), and **variability** (format). The most important part is the business value of the data obtained, processed, and stored, reflected by the **value** aspect, which is, however, strongly associated with the data format, precision, and date and time reference. Above all the data, there is an **analytical** interface, **data warehouses**, **marts**, and so on, which must consider the date and time reference, aggregating data into individual elements, such as days, weekly, or monthly reports.

The time between the first database systems and the acceptance of the ISO standard caused the creation of many streams for working with date and time. **ISO 8601** was accepted in 1988 and covers date and time element creation, definition, management, storage, and evaluation.

The ISO 8601 standard

ISO 8601 covers date and time-related data, including its exchange, transfer, and representation. It was first published in 1988, followed by multiple updates in 1991, 2000, 2004, and 2019. It aims to provide a system for the management of date and time data and the representation of individual elements to avoid numerical dates and times being misinterpreted, focusing on data transfer across countries and regions and emphasizing time zones and conventions that are specific to certain regions.

ISO 8601 divides the management into separate date and time models. The Date value is based on the *Gregorian calendar* (an international standard for civil use), while the *24-hour timekeeping system* expresses the Time. It uses the principle that the greatest temporal granularity element should be the left-most one. Each following lesser element comes to the right of the previous one. For example, the year must precede the month, the month definition must precede the day, and so on.

The next section will drive you through the Date value definition, describing the individual elements and their meanings. Let's dive in!

Date value definition

Based on the standards approved by the **American National Standards Institute (ANSI)**, values for the date are typically defined by the year, month, and day elements, ranging from 0001-01-01 through 9999-12-31, delimited by a specific symbol, such as a dot, comma, semicolon, or slash. Thus, we know that the value 2021-05-01 represents 1st May, not 5th January. The format for the Date value representation is YYYY-MM-DD, expressing the year (YYYY), month (MM), and day (DD) elements. To ensure correctness across the centuries, this standard requires a 4-digit year format to be applied precisely.

The available year definition ranges from 0000 to 9999, with 0000 being equal to 1 BCE (Before Common Era) and all subsequent values referring to CE (Common Era). ISO 8601 permits value management lower than 0000 if accepted by the sender and recipient, in which case, an expanded year definition with the + or − sign is present. Thus, +0000 represents *1 BCE* while −0001 references *2 BCE. In the past, Christian-centric terms were commonly used, AD referred to Anno Domini, Latin for the year of the Lord, while BC referred to the period before Christ.*

The MM denotation expresses the month value, ranging from 01 to 12. DD indicates the day value, ranging from 01 up to the last day of the month. Month and day elements are always expressed by two digits; thus, an initial 0 digit is applied where necessary.

The preceding format can be described as *extended*, delimiting individual elements by the predefined separator (dash in our case), such as 2022-05-31. The basic format removes the separator, providing an analogous value of 20220531. In that case, however, it is crucial to apply 2-digit values for the month and day. Otherwise, there may be a problem of misinterpretation, such as 2022111 representing either 2022-01-11 or 2022-11-01.

The standard allows us to specify a reduced precision by omitting the day element and just including the year and month elements. For example, 2021-04 represents the whole month of April 2021.

In 2000, the concept was significantly refined, including allowing users to specify a Date value that had an element missing. Thus, --06-09 represents 9th June, irrespective of the year, whereas a particular year element is not stated (expressed by the dash (-) symbol). Similarly, only one value specification for the date represents either the year (using 4 digits) or the whole decade (using 3 digits). To illustrate this, 2021 refers to the whole year 2021, whereas the value 201 refers to the inclusive decade from

2010 to 2019. Finally, the 2-digit value covers the whole century; for example, 20 refers to the entire 21st century. Other elements have been gradually generalized. The date can now be composed of the year, week, and day. This concept can be perceived differently across regions, whereas generally, different principles can be applied.

Instead of referring to individual month elements, weeks can be also referenced. The next section will walk you through the principles.

Week management

ISO 8601 uses the Gregorian calendar. YYYY expresses the year element. A week is modeled by the order from 1 to 53, prefixed by the letter W to convey the meaning. Then, the day element uses a weekday number from 1 to 7. 1 means *Monday*, while 7 means the last day of the week – *Sunday*. For example, New Year's Eve of 2017 can be expressed as 2017-12-31 or, in the form of week reflection, 2017-W52-7. Calendar week numbering is widely used in Europe to reference the time frame of an event.

Focusing on the standard, there are several modeling principles. How to define the first week, namely week 01, can be specified by the following:

- The week that contains the first Thursday of the year

- The week with January 4th in it

- The first week that covers at least 4 days in the processed year

- The first week that contains a Monday in the range of December 29th to January 4th

As a result, if January 1st is on a Monday, Tuesday, or Wednesday, it is part of the first week of the year and vice versa. If January 1st is on a Friday, Saturday, or Sunday, then it is covered by week number 52 or 53 of the previous year. So, there are two consequences:

- December 28th is always in the last week of the year

- Each week's number can be obtained by calculating the number of Thursdays in the year

ISO 8601 uses numerical representations for individual elements. This makes it more readable and language agnostic. For example, the words *January* and *February* would be understood by a native English speaker, but they might not understand *Janvier* and *Février*, which are used in French.

For the definition, it always uses the granularity flow from the grain to finer types. Larger units always precede smaller ones. Thus, the year definition precedes the month, the month precedes the day, and so on. The fixed order limits the opportunity for confusion and clearly declares the meaning of elements. Moreover, this approach is widely used across the globe. The 4-digit year representation limits the confusion related to the century and potential overflow problems.

Similar to the week representation, it is possible to refer to the serial number of the day within the whole year, as described in the following section.

Ordinal date values

The ISO 8601 standard can cover ordinal dates as well. This means a calendar date is represented by the easy extension of the year's day ranging between 1 and 365 (for a common year) or 366 (for a leap year). The first day uses the value 1, representing January 1st. The number is formatted as YYYY-DDD in the year (YYYY) and day (DDD) order. Individual representation models are then interchangeable and directly transformable. For example, 2021-12-24 can be represented by 2021-358.

An agreement was reached about trimming leading date components in the ISO 8601 version released in 2000. Instead of the whole year definition, just the 2 digits could be used and the century was added automatically. This decision, however, led to complaints and misunderstandings. Therefore, the consecutive version of 2004 removed this feature.

Now, you are familiar with the definition and modeling of date elements, so let's move on and look at the principles of time processing in ISO 8601 standardization.

Time element modeling and management

Generally, representations of the date cover the year, month, and day elements. Some of them can be omitted, or a transformation can be made to express the relevant value. This standard proposes a definition for time elements as well. The principles and model properties are similar. It uses a *24-hour clock* by pointing to the hour (HH), minute (MI), and second (SS) elements. The time value is prefixed with T. The basic format does not use padding; for example, T152217. Instead, each element is composed of 2 digits. The extended format is more user-friendly, splitting individual parts into categories – for example, T15:22:17. In both models, the character T expresses the time representation but can be omitted for extended representation by applying a colon delimiter.

The hour value ranges from 00 to 24. The minute value ranges from 00 to 59, and finally, for seconds, the available range is from 00 to 60 (due to the leap second discussed in *Chapter 6*). Note that if the hour value is 24, then the minute and second values must hold 00 values.

Like with dates, some elements can be omitted. For example, the value T15:22 does not represent seconds, and T12 focuses on the hour only. Higher-precision elements must always be present. Thus, it is impossible to specify a minute without an hour reference or a second with no minute or hour definition.

Version 2019 of the ISO 8601 standard pointedly fixed the beginning of a calendar day to value 00:00. Prior releases also accepted an analogous value of 24:00.

Here's an example of a specific solution using a decimal fraction: the minute value consists of decimal places, such as 10:30.5, which represents 10 hours and 30.5 minutes. Note that it is always possible to transform a particular value into the full specification, such as 10:30:30, in this case.

However, how do you apply hour notation that differs across regions? The following subsection explains how. Get inspired!

12-hour notation remarks

The 24-hour time model is commonly used across the world, but some English-speaking countries rely on the 12-hour format, extended by the *a.m.* and *p.m.* additions, expressing morning and afternoon. The 24-hour format is gradually replacing the 12-hour format, even in countries such as the US or in the fields of industry, military, or IT, where the 12-hour format was traditionally used. The 12- and 24-hour formats should be clearly mentioned during processing and evaluation to avoid any confusion.

The 12-hour time format requires additional space and storage for noting a.m. and p.m. references. The hour value cannot be compared to another hour value without taking the a.m. or p.m. definition into account. Namely, the value 3 in the 12-hour format can mean either 3 or 15 in the 24-hour format. Thus, without the a.m or p.m mark, a particular value cannot be compared and mapped directly. Moreover, one point in time can have multiple representations; for example, `00:00` or `24:00` represent *midnight*. Finally, the value is not easily comparable with the string reference.

In conclusion, I would like to note that many people do not understand the fundamentals of the transition between days and believe that a day ends at 12:59 a.m. and begins at 1:00 a.m. The correct moment of transition is from 11.59 p.m. to 12.00 a.m.

The Oracle database does not support pure `Date` (without a time reference) or `Time` data types. Instead, the `Date` data type in Oracle always models both components. This might cause problems when migrating data from and to an Oracle database. Compare the `Date` data type in an Oracle database with the `Date` data type in, for example, MySQL. The values are not directly compatible. Why? The time sphere is either ignored or must be added implicitly (mostly an issue when referring to midnight) if such elements are not originally present. Thus, the precision is changed during the transformation! The opposite problem also exists, where a `DateTime` value in MySQL cannot be directly transferred to Oracle, as this data type does not exist in Oracle. Instead, the `Date` data type should be used in Oracle. Last but not least, there can be a problem with the value precision of seconds. Oracle supports precision up to nanoseconds, while MySQL limits values up to microsecond precision.

Due to the manipulation of the date and time values, it is generally appropriate to model individual elements in a common structure. Imagine a situation where we need to add 5 hours to the value. If the date and time values were modeled separately as two attributes, in some cases, we would only need to load the time component, while in other cases, we need to load the date component as well. The extra hours may make the time value surpass midnight and thus cause the day value to change. When specifying duration through the start and end time references, we would always need both the date and time attributes. If we forgot one of them, the results would be incorrect. This was the motivation for creating a methodology for managing a composite `DateTime` value directly in the standard.

Composite date and time value

In the preceding sections, date and time values were treated separately. Some database systems, such as MySQL, use this approach by defining two data types—Date and Time data types. A DateTime data type is a combination of both values in a common block. The Date values are placed in the left-most part up to the second fractions. The predefined format is YYYY-MM-DDTHH:MI:SS[.FF]. Second fractions are optional. The T symbol is used to split the date and time values. Version 2019 allows you to omit the delimiter symbol, T. So, all of the following values would be the same: 2022-12-30T15:26:01, 2022-12-3015:26:01, 2022-12-30T15:26:01.00, and 2022-12-3015:26:01.00.

Each day starts and ends at midnight, covered by the values 00:00 and 24:00 to distinguish the timeline positions (start of the day or end of the day). However, the value 24:00 is the same as 00:00 for the consecutive day. Thus, both of the following values express the same point in time: 2022-05-20T24:00:00 and 2022-05-21T00:00:00.

In the previous sections, just one point in time was discussed and modeled with duration representation. In the next section, let us take a look at the **period of time** expressing the duration and time elapsed.

Periods of time

The last section described date or time point value representation principles. Periods of time can be modeled by two time points, expressing the beginning and end points of the validity, represented in different ways, delimiting whether the time point is exclusive or inclusive. The second principle is associated with just one attribute, commonly pointing to the beginning point, followed by the time interval duration (period). There are several designations to represent particular values. The general format for the period is P[n]Y[n]M[n]DT[n]H[n]M[n]S, where [n] is replaced by the numerical value of the physical element. P, Y, M, D, T, H, M, and S are the delimiters for individual element reflection, signifying the following:

- P expresses the period:

    - Y: Number of years

    - M: Number of months

    - D: Number of days

- T refers to the time components:

    - H: The hour value

    - M: The minute value

    - S: The second value

As stated, a common representation of periods is formed by the extended notation. For example, P3Y2M1DT3H12M55S represents 2 years, 2 months, and 1 day for the Date elements and 3 hours, 12 minutes, and 55 seconds for the Time elements.

Some designations can be missing. In that case, they include a 0 value for those elements (or their significance is not important, expressed by rounding or truncating them). For example, `P20DT10H` expresses 20 days and 10 hours; `P10M` signifies 10 months. The smallest value for the duration is 1 second; thus, the definition cannot be empty. Values such as `PT0S` and `P0Y` are not applicable. At least one element must have an associated positive value.

Be careful when dealing with time elements inside a period. M refers to both month precision as well as minute precision. So, what does the value `P6M` mean? Does it reflect 6 months or 6 minutes? Can it provide a false representation? Not at all. The `Date` and `Time` values are still separated by the T separator. Therefore, `P6M` refers to 6 months, while `PT6M` would be the correct statement for 6 minutes.

A period of time expressing a duration does not need to cover upper-value limitations for individual elements; the value can be spread to include other elements as well. To simplify it, the range for a minute is generally 0-59 seconds. If the value is greater, it would be impossible to state 70 minutes and must, therefore, be categorized by the hour element. Hence, you will have to express 1 hour and 10 minutes. Similarly, to express 36 hours, the correct definition is `P1D12H`, but the value `P36H` is also applicable. Keep in mind that those representations do not need to be the same due to Daylight Saving Time, which will be discussed in the next chapter.

Finally, decimal values for elements can be used as well, providing a universal solution. Half a year covers 6 months but can be expressed as `P0.5Y` or `P0,5Y`, depending on the regional usage of the decimal place marker.

Let's now move on and look at representing time intervals using two time points, expressing the **start point (SP)** and **endpoint (EP)** of validity. We will also discuss how to express that the endpoint is not yet known.

Validity interval modeling

A temporal timeline cannot be defined only with a period of time definition. This is because although the information about the temporal duration is present (that is, the event lasted for 2 days), the SP is not stated. Thus, additional contextual information is required to express when an event started. Validity intervals solve that problem.

The first solution delimits the SP and EP using a slash (`/`), for example, `2022-06-01T00:00:00/2022-06-30T23:59:59` expressing *June 2022*. A shortened representation can also be used by omitting elements from the EP that are the same as the SP. The preceding value can then be represented by the following shortened notation. Year and month elements are omitted in the EP by inheriting their values from the SP:

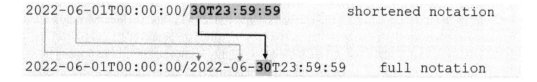

The shortened definition can be misinterpreted. What does the value 30 mean in the preceding definition? Does it really reference the day element precision? If it were treated as a month element, significantly different results would be obtained:

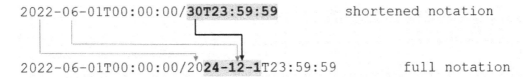

```
2022-06-01T00:00:00/30T23:59:59              shortened notation

2022-06-01T00:00:00/2024-12-1T23:59:59       full notation
```

Consequently, to ensure data transparency, shortened notations are used only for the time elements by omitting the `Date` part, such as `2022-06-01T10:00/11:00`.

The second solution deals with the SP point enhanced by the period of time (duration) differentiated by the slash. The first value expresses the SP. The second is the duration itself; for example, `2022-06-09T15:30:00/P1Y4M8DT16:00:00`. An analogous representation would be `2022-06-09T15:30:00/2023-10-17T16:00:00`.

The third solution, which is, however, not used so much, is delimited by the period of time, followed by the EP – for example, `P1Y4M8DT16:00:00/2023-10-17T16:00:00`.

Although individual systems and models are interchangeable, it is generally more useful to have a precise SP definition to which other database or application actions are often linked.

When dealing with the duration represented by any model, it is necessary to highlight the representation of the interval. The SP commonly references the first time point of validity. However, the EP can either reference the last time point part of the interval or the opposite, the first point, which is not covered. Proper references are crucial for processing reliability and accuracy. However, how can we find out from the values used which principle is being used? Generally, this is impossible to work out. It should just be specified in the data model description or covered by the application.

Moving on, another problem could arise related to unlimited validity. The next section explains unlimited validity management.

Unlimited validity

The preceding sections dealt with points in time and periods of time, precisely specified by SP and EP references. There are, however, many situations where the borders of the interval cannot be obtained. Let's look at the example of an employment contract. It is always evident when it was signed and when the contract entered into force, which is simply when the employee came to work for the first time. But what about its expiration? If the person is employed for a limited period, the solution is straightforward. For example, if a person is employed for a probationary period of 3 months, it is clear when the validity expires – that is, of course, if the SP of validity is known. On the other hand, how would you proceed in the case of indefinite validity? If a person is going to change jobs, we usually don't know when that will be, if it ever even happens. Most versions of ISO 8601 do not provide solutions for unlimited validity modeling.

Generally, although the EP value is unknown, it is clear that it will happen in the future. The workaround to undefined validity in models is based on its replacement with a large enough value positioned in the future or another user-defined and managed solution.

A natural question may arise here: how do we deal with a situation in which the SP of validity is unknown? Well, a similar approach to what was stated before is used. The implementation and management are left to the user to define. Similarly, it is clear that the SP was in the past.

> **Important note**
>
> Version 2019 proposed a solution for dealing with unbounded time interval modeling. The full description and reference are out of the scope of this book. To get a deeper insight, refer to the *Further reading* section, which provides a link to the full ISO standard.

Now that you know how to model duration, let's see how to model recurring events in the next section.

Interval periodicity

Individual events can occur periodically. After the key designation, R, the number of repetitions (optional value) and frequency are specified. The syntax is R[n]/[interval], where [n] expresses the number of repetitions and [interval] describes the period between consecutive events. The value 0 means no repetitions are present. If the number is unbounded, it is modeled by the value -1. The following example defines 10 repetitions, starting on February 1st, lasting 1 month: R5/2022-02-01T15:00:00/P1M.

So far, so good, so let's move on to a summary of date and time value representation. Then, we'll look at how to express date and time values in various object structures, such as **eXtensible Markup Language (XML)**, **JavaScript Object Notation (JSON)**, and **comma-separated values (CSV)**.

Representation enhancement remarks

As already stated, each date and time value consists of multiple elements, that is, the day, month, and year for the date value and the hour, minute, and second references for the time value. Except for the year, element values are represented by 2 digits, with the specific ranges forming relevant date and time values. Some elements, mostly related to the time sphere, can also hold 0 values. The extended version of the definition is robust and always correct. However, it requires more writing, more characters, and, last but not least, more storage capacity. Shortened versions are compressed and can omit some elements, mostly if they hold 0 values or their reference is clear. When using shortened versions, always be aware of the correct representation and reference to ensure reliability. Incorrect mapping can cause significant problems and system limitations.

The format definition using ISO 8601 standardization is clear. It can be used for date and time representation, as well as element mapping and duration modeling. However, what if a structure does not directly support date and time element modeling and management? How do we treat those values? The next section describes the CSV, XML, and JSON formats related to character string management.

Date and time value modeling in CSV, XML, and JSON format

The CSV format is a simple file format in form of plain text intended for storing tabular data. A CSV file consists of any number of records (rows), separated by a newline character. Each row contains columns separated by a specific character. Typically, a comma (,) is used, but a tab, semicolon, slash, or another character can also be used. The column value is a character string or numerical format. Thus, there is no fixed style for date and time management. It would lead to errors during the transformation, based on the element order and meaning.

XML was developed and standardized by the **World Wide Web Consortium** (**W3C**) as a generalization of its predecessor, the **SGML** language. It uses elements and tags and can manage various data sources. It is primarily intended for exchanging data between applications and systems. Although XML is a document and can be used generally, it can additionally be associated with a table attribute or a whole table can be XML-oriented. Then, the XML content is commonly validated against an XML schema, or a trigger can be created to ensure the format is retained.

JSON is the newest platform-independent format, used primarily for data transfer, but the JSON data type can also be used for the table attribute (column). In the past, it was operated as a **large object** (**LOB**); however, current versions of the Oracle database allow you to define the JSON data type directly. The input is any data structure, while the output is a textual format.

And that's just the point. The output of the described CSV, XML, and JSON formats is textually oriented. This means that the element order must be precisely specified for the consecutive mapping and evaluation. As JSON and CSV do not have built-in types for date and time values, the general solution is to store values in a string format compatible with the ISO standard. In XML format, such a rule referring to the ISO standard can be used as well, although the XML schema validation can ensure proper mapping and representation. Therefore, when dealing with these formats, always specify the style and structure.

As will be described later in *Chapter 6*, the transformation between date and time values and string formats is supervised by conversion function calls, or implicit conversions can be applied; however, the style is crucial and the same type should be used for all attribute values.

Summary

In this chapter, you learned the fundamental facts related to standardization. It is always good to understand these to start.

ISO 8601 was approved in 1988, and due to various consecutive enhancements, it provides a relevant summary of date and time management in information technology.

First, you learned about the `Date` value, formed by the day, month, and year elements, followed by the `Time` representation. After that, you learned how these values can be grouped into one specific `DateTime` reference, expressing one point in time. Afterward, solutions for modeling periods of time were proposed, focusing on durations and validity intervals. Until version 2019, there was no ability to model an unlimited validity interval. This was left to the user to manage and implement.

By accepting the roots defined by ISO 8601, the user gets an overview of the definition, management, and representation complexity, and is also aware of the limitations.

Although the Oracle database system does not support separate data types for `Date` and `Time` values, most approaches are applicable, providing reliable references to ISO standardization.

The next chapter continues to look at date and time concepts, focusing on time zone references, Daylight Saving Time, UTC, and leap years. It describes the principles and associated rules. These terms are really important to cover to understand the complexity of date and time management.

Questions

1. Which of the following is the correct ISO format for specifying the date?

   A. `YYYY-MM-DD`

   B. `MM-DD-YYYY`

   C. `DD-MM-YYYY`

   D. `YYYY-MM-DD HH:MI:SS`

2. Which character prefixes `Time` elements?

   A. `T`

   B. `D`

   C. `C`

   D. A space

3. Which calendar is primarily used for the ISO standardization reference?

    A. Roman calendar

    B. Julian calendar

    C. Gregorian calendar

    D. ISO calendar

4. Which of the following values references unlimited interval periodicity?

    A. `R5/2022-02-01T15:00:00/P1M`

    B. `R0/2022-02-01T15:00:00/P1M`

    C. `R/2022-02-01T15:00:00/P1M`

    D. `R-1/2022-02-01T15:00:00/P1M`

Further reading

A full description of the ISO 8601 standard is available at `https://www.iso.org/obp/ui/#iso:std:iso:8601:-1:ed-1:v1:en`.

4
Concepts of Temporality

In the last chapter, we dealt with the ISO 8601 standard, focusing on the data perspective, modeling principles, and individual element handling. This chapter goes deeper into the concepts related to **temporality** and the core background facts influencing temporality.

Everyone knows that when seasons shift, either to summer or winter time, the number of hours in a specific day also changes. The day can either change to 25 or 23 hours. In one case, we look forward to sleeping longer; in the other, we will sleep for fewer hours. This change factor is also related to time zones, which you will feel especially when traveling. Another real-world issue relates to Christmas. When do you celebrate Christmas, according to the Gregorian or Julian calendar? In fact, what are the key differences between calendars, and what do they actually affect? How does using a particular type of calendar affect leap years? These are practical questions to which we will find answers in this chapter, which is divided into several parts.

The first part of this chapter focuses on summer and winter times and describes the principles and impacts of **Daylight Saving Time (DST)**. The main idea behind this is to use sunlight to limit the costs related to energy consumption and the necessity for artificial light. In this chapter, the main pros and cons of DST are summarized, and we will look at the consequences. The next part of this chapter looks at the definition of **Universal Time Coordination (UTC)** and its principles, which are widely used across databases. Oracle databases also strongly recommend using UTC date and time format for the server, but the final decision is always up to you. We will walk you through its properties and limitations. The next part deals with the **Gregorian** and **Julian calendars**. The main difference is associated with the definition of a **leap year**, which does not occur exactly every four years as per the Julian calendar. Thus, currently, the Gregorian calendar is the right solution for practical usage. The last part introduces a relatively unknown topic, that is, **leap seconds**, introduced by the **International Earth Rotation and Reference Systems Service (IERS)** organization. Occasionally, a minute is composed of *61* or *59* seconds to account for the difference between UTC, the definition of a year according to the Gregorian calendar, and the Earth's rotational speed, which is not constant. The limitation of the leap second is *associated with its timeline unpredictability*. It is simply impossible to identify when the leap second will occur in the future. Consequently, individual database technologies do not cover it robustly.

In a nutshell, this chapter is going to cover the following main topics:

- What is temporality?

- DST principles

- UTC reference and database usage

- Time zone perspectives and shift calculations

- The Julian and Gregorian calendars – how they work and their principles, including the definition of a leap year

- The impact of leap seconds and overall consequences

The source code for this chapter can be found in the GitHub repository accessible via this link: `https://github.com/PacktPublishing/Developing-Robust-Date-and-Time-Oriented-Applications-in-Oracle-Cloud/tree/main/chapter%2004`. You may also access it via the following QR code:

> **Important note**
> Please note that UTC is often referred to as **Coordinated Universal Time**, but it means the same as Universal Time Coordination.

What is temporality?

What does the term temporality mean? Is it the logical concept or the physical model supported by the infrastructure? Or can it just be defined by a set of rules that must be applied? Well, it does not reflect a specific model or implementation. Instead, temporality covers any solution that deals with time. Namely, each object state (data tuple) can be positioned on a timeline. Thus, each object is delimited by multiple states valid during the specified period of time. Generally speaking, relational databases can be automatically considered to be partially temporal, whereas each data change is always encapsulated by the **transaction**. And each transaction generates **data change vectors** stored in the **transaction logs**. So, what does it actually mean? Such a transaction log structure clearly denotes the object state history, doesn´t it? Thus, the transactional databases can produce states valid in the past. However, it is not as simple as it may seem at first glance.

Obtaining the history from the transaction logs can be time- and resource-consuming. Moreover, if we don't have all the transaction logs, the system could produce incorrect values. Last but not least, transactional structures are primarily intended for a different purpose and cannot cover the evolution modeling of states valid in the future. Thus, an integral part of temporality is just the time itself, which can be represented in various ways. You could argue that the definition and processing of time are straightforward, typically represented by the hour, minute, and second, and optionally with the day, month, and year. But have you thought about the additional complexities? What about the winter and summertime references? What about the time shift and reference model coordinating time across time zones? Well, that's what this chapter is about. It will explain the basic principles and complexity of the problem.

Before digging deeper, it is useful to highlight the importance of storing the object states and evolution over time. Although there is a separate chapter dealing with temporal data modeling and principles (*Chapter 11*), some background aspects need to be highlighted.

The first relational databases were based on conventional principles. Only currently valid data was stored, with no time reflection. In the preliminary releases, date and time data types were not present, despite the fact that everyone knew that the date and time perspectives are important. Therefore, soon after the first releases, temporal concepts were discussed. However, what does it mean? It is not about storing multiple attributes delimited by the date or timestamp attributes (for Oracle databases) or other variants in different systems. Why? Because it still refers to currently valid states but is delimited by some time frame. Temporal databases are characterized by managing and storing object states over time. Thus, instead of replacing the original state during the update operation, a new state is created, enclosed by the valid time frame. Generally, temporality covers the entirety of date and time management in databases.

By using date and time attributes, sooner or later, it will be necessary to pay attention to time changes and time zones. The next section introduces DST concepts that deal with summer and winter time.

Introducing DST

DST (also referred to as **Daylight Time** in the US, Canada, and Australia and **Summer Time** in the United Kingdom and European Union) is a technique of time change during the summer months to adapt to natural light so that twilight occurs at a later time of the day (typically an hour later). As a result, one day, typically in October or November, has 25 hours (*fall back*), whereas one day in spring has only 23 hours (*spring forward*). This change brings economic benefits. But on days with time changes, issues often arise related to transport, industry, and non-stop working shifts. The **information technology** (IT) sphere commonly manages DST automatically without requiring any user intervention.

The correct definition of DST is Daylight Saving Time, but in older literature, the term **Daylight Savings Time** (with an *s* at the end of *Savings*) is used. It may also be seen in modern references, as many editors still prefer the original form. The other acceptable forms are Daylight Saving or Daylight Savings (without *Time*) or even Daylight Time. One way or another, it still has the same meaning.

DST was first introduced by Benjamin Franklin in a satirical letter in 1784. He suggested that an earlier hour of waking up would bring considerable savings and economic impacts. The concept was widely discussed, focusing on the pros and cons. However, it was not until 1908 that the idea was first implemented in Canada. Other countries followed suit over the decades. Specifically, some countries would not benefit from a 1-hour shift. Moreover, there are also countries, such as those near the equator (for example, Ecuador), where the benefits of DST cannot be significantly justified. Thus, there was generally a lack of consensus among countries regarding the time of sunrise. However, a significant milestone was the year 1970, when the energy crisis resulted in the spreading and implementation of the concept of DST in many parts of the world.

What are the different reasons for using DST? One of them has already been stated, which is its association with energy consumption decrease. Others are related to temperature and natural sunlight. Some drawbacks can be represented by the difficulty faced in activity planning (for example, a workplace may have a rule that one person can work no longer than 12 hours; however, during their shift, their workday would be 1 hour longer) or the health impacts.

The procedure of shifting the time commonly occurs on the weekend to minimize the impacts. Nowadays, it is a one-hour shift, but two hours were also applied in the past. Furthermore, in some regions, the time shift was minute-oriented (20 or 30 minutes). DST differs from the standard time (the time used throughout the rest of the year) and is used from spring to autumn. The exact time at which the change takes place differs across the world. For example, in the European Union, it is scheduled for 1:00 a.m. (UTC reference), which is 2:00 a.m. in the **Central European Time (CET)** time zone. This is equivalent to 3:00 a.m. in **Eastern European Time (EET)**. Thanks to this, the time shift between the European regions is still the same because the change occurs throughout Europe at the same time. A different principle is used in North America, where the shift is applied at 2:00 a.m. for each time zone separately. Thus, during that day, different time shifts (non-constant) can be perceived within multiple regions. Moreover, this can create issues with time zone management for countries in DST and non-DST regions that share borders.

That was a lot of technical detail. Now, it's time to apply the rules to the real world and examine the usage, overall consequences, and impacts of DST. To date, several studies have been done since the implementation of the DST. They have analyzed and balanced the costs, advantages, disadvantages, and consequences on all areas of life, including the human body. The next section provides a summary of the results of these studies.

The impact of DST

Several studies have been done on the impact of DST on the environment, industry, and electricity consumption. A meta-analysis was published in *The Energy Journal* by Tomáš Havránek et al. (`http://www.iaee.org/en/publications/ejarticle.aspx?id=3051`). It focuses on electricity consumption by collecting 162 estimates from 44 different studies, reaching the conclusion that DST brings a cost saving of 0.34%, suggesting it is not that great of a benefit.

Adam Cook published an analysis of the expanded implementation of DST in Indiana, US. It indicates that DST had reduced the mortality of elderly people (aged 65 years and older) uniformly across genders. The assumed reason for this is associated with greater exposure to sunlight. Moreover, applying DST led to a drop in robbery rates.

It can also have a natural impact on students and employees by optimizing time at work and increasing time for recreational opportunities before sunset. However, farmers argue that the behavior and routines of animals do not change like humans, which has a negative impact on their ability to carry out their work. In addition, employees who often travel, whether for meetings or conferences, must take into account individual regions and time zones. Another problem may be associated with IT, such as billing, transport, and medical systems, which must apply shifts and references correctly. From a health and security perspective, several studies have been carried out. The results do not observe any major change in traffic accidents, microsleep, and other similar factors, but the overall amount of criminal activity is slightly reduced (due to the reduction of natural light).

To conclude, several views have been presented with various pros and cons. It is not possible to identify an authoritative advantage that makes up for all the mentioned shortcomings and negatives of DST. Therefore, the European Union and other organizations are starting to consider the idea of canceling these time changes and setting other rules. And finally, a more light-hearted argument for why the implementation of DST should be ceased: it can be time-consuming to set many mechanical clocks.

Permanent DST usage

During the last decade, an idea of permanent DST usage has been proposed and discussed, resulting in its implementation in some jurisdictions, such as Argentina, Belarus, Iceland, Morocco, and Turkey. Interestingly, a country can even implement DST differently in individual regions. For example, Saskatchewan (a province of Canada) generally uses **Central Standard Time**, but the capital city, Regina, applies permanent DST. The United Kingdom and Ireland started to use permanent DST in 1968. The benefits were considered limited; thus, in 1971, they returned to the original principle.

The European Commission, as the executive body of the European Union, proposed to end the seasonal time shift in 2019. Individual members of the European Union got the opportunity to vote on implementing either DST or standard time with no shift present during the spring or autumn (the chosen rule would be applied permanently each year). Even though the European Parliament approved such a rule of canceling time shifts, the implementation period itself was postponed until 2021 as a general agreement could not be reached. Some countries prefer summer time, the rest winter time.

Finally, the European Union Council did not give the project the green light because they wanted to wait until further studies had been carried out, on both a global and individual state scale. As a result, the discussion has not ended yet, and it is assumed that it will take some more years before a decision is made or the concept of uniform time preservation is refused altogether. Namely, there is a discussion about the outcomes, impacts, and even implementation details. Therefore, the approval and practical implementation for the whole European Union is doubtful.

The term DST is generally used in the US, Canada, and Australia, while Summer Time is preferred in the United Kingdom or European Union states. In the next section, Summer Time will be described; we discuss its rules, concepts, and real-world usage.

Summer Time in Europe

Summer Time in Europe is a variant of the standard clock time applied in most European countries (except for Belarus, Iceland, Turkey, and Russia). It corresponds to DST with a one-hour shift during the period between spring and autumn. The day when the shift is applied follows these rules:

- Summer Time begins at 1:00 a.m. (UTC) on the last Sunday of March

- Summer Time ends at 1:00 a.m. (UTC) on the last Sunday of October

- The change is made across all the time zones at the same time

Summer Time lasts 30 or 31 weeks, depending on the calendar day of the last Sunday of March. That is, if it is after 28th March, then the duration is 30 weeks; otherwise, it is 31.

The concept of Summer Time is associated with the World Wars. It was introduced during the First World War and readopted during the Second World War, after which the concept was discontinued in the 1950s (as it was considered a reminder of the war).

Later on, regarding the principle of Summer Time definition, a rule was created by the European Union (Directive 2000/84/EC) requiring member countries to apply Summer Time, but it has not been accepted and used across individual regions generally.

The future of Summer Time and DST is uncertain. In Europe, individual institutions and the European Commission have not pointed to the conclusive benefits of individual solutions. This resulted in Finnish and Lithuanian parliaments prompting the European Union to reconsider the proposal. Later, it was also motioned by Poland and Sweden.

So, all these views and approaches must be taken into account when creating and maintaining information systems that are not just local to a particular region. Thus, to ensure the complexity, robustness, and reliability of information systems, it is important to familiarize yourself with time zone management, summer and winter time, and overall shift rules to reach the correct results from the client's perspective. Generally, business activities tend to spread across regions; therefore, synchronization across regions is important to ensure the reliability and usability of the provided data results. However, UTC provides a common reference. Let's take a look.

Introducing UTC

UTC is a time representation that applies zero offsets. It is then used as a reference for other time zones. UTC representation is denoted either by the symbol UTC placed after the time specification (for example, 08:30 UTC) or by using an empty time zone specification Z (for example, 8:30Z).

UTC is the primary standard for dealing with time management and representations. It regulates and synchronizes clocks and time. It references **Greenwich Mean Time (GMT)** and does not depend on DST. It was first used in 1960, followed by the standardization in 1963, introducing the official abbreviation – UTC. It was updated in 1970 to cover the leap second.

UTC is divided into individual day, hour, minute, and second elements. Days are identified by the Gregorian calendar reference, but the Julian day numbers can also be used (the transformation principles are described in the *Gregorian versus Julian calendar* section). Each day contains 24 hours. Each hour consists of 60 minutes. The number of seconds in a minute is commonly 60, but there can also be a leap second. Thus, the minute value can occasionally be 59 or 61.

UTC is stable and does not depend on the session parameters and other characteristic changes. It is not impacted by DST, but local or civil time can change. Each connection is characterized by the client on one side and the server on the other side. Between them, the session is established and used for communication and data exchange. Inside this environment, the client is characterized by the used region, time zone, formats, and other temporal parameters. UTC reference is independent of these characteristics. Therefore, to cover the complexity between the client and server, UTC is not enough to deal with the time zone shifts. Instead, DST should also be considered. For example, the local time in Brussels is two hours behind UTC during the winter and one hour behind during the summer season.

By extending the UTC reflection, the time offset can be appended to the time elements. Thanks to that, any region can be covered. Negative values define time zones west of the UTC reference, in which the common civil time is earlier than UTC. Conversely, positive values denote east of UTC and lead to a later time compared to UTC. *Table 4.1* shows the time zone abbreviations used in Europe with offset value representation:

Abbreviation	Time zone name	Offset
BST	British Summer Time	UTC +1
CEST	Central European Summer Time	UTC +2
CET	Central European Time	UTC +1
EEST	Eastern European Summer Time	UTC +3
EET	Eastern European Time	UTC +2
FET	Further-Eastern European Time	UTC +3
GET	Georgia Standard Time	UTC +4
GMT	Greenwich Mean Time	UTC +0
IST	Irish Standard Time	UTC +1
KUYT	Kuybyshev Time	UTC +4
MSD	Moscow Daylight Time	UTC +4

Abbreviation	Time zone name	Offset
MSK	Moscow Standard Time	UTC +3
SAMT	Samara Time	UTC +4
TRT	Turkey Time	UTC +3
WEST	Western European Summer Time	UTC +1
WET	Western European Time	UTC +0

Table 4.1 – Time zone abbreviations

As stated in *Chapter 3*, ISO 8601 proposes that date and time values be modeled and evaluated separately. It does, however, have some limitations if both values need to be used. That is, if the value shift over time needs to be applied, the resulting value can influence not only the time elements themselves but also the date component. For example, adding two hours to the time expressing one hour before midnight would cause the day to also shift, as illustrated here:

```
2022-06-09 23:00:00 + 2 hours --> 2022-06-10 01:00:00
```

Moreover, this shift can impact month or year elements too:

```
2022-12-31 23:00:00 + 2 hours --> 2023-01-01 01:00:00
```

Finally, leap years also need to be highlighted. The first example in the following snippet incorporates the leap year, while the second example does not. Thus, the month is changed from February to March:

```
2024-02-28 23:00:00 + 2 hours --> 2024-02-29 01:00:00
2023-02-28 23:00:00 + 2 hours --> 2023-03-01 01:00:00
```

Consequently, based on the ISO 8601 standard released in 2019, element values of the date and time can be grouped into a single unit. Date and time elements are separated by the letter T, for example, 2021-05-03T12:20Z. Z refers to the UTC format. However, generally, any time zone can be specified there. Explicit time zone management is covered by the right-most element, for example, 2021-05-03T12:20+04:00.

The next section dives deeper into time zone management, focusing on the database, client, and local time.

Time zone perspective

Oracle cloud database instances use UTC date and time references by default. Although it is strongly recommended to use UTC to simplify time zone management, shifts, and calculations and avoid data conversions, users can change the base time zone reference in the cloud console or by using APIs.

The standard usage of UTC arises from ISO 8601, which covers time zone management by using the values offset from UTC for local values.

When dealing with a complex system accessed across regions and multiple time zones, the *server* and *client* perspectives will inevitably need to be distinguished. The server, like the client, can be located anywhere in the world. However, the locations themselves are important for the correct determination of time zones and transformations between them. Thanks to that, it is possible to obtain and process time from the client's point of view, as well as the server's. Consequently, all values are linked to the time zone and shift.

The server time zone is also named the **database time zone** and is commonly modeled by UTC. The **client time** perspective can then be calculated very easily just by focusing on the UTC shift.

Local time does not have a *time zone* reflection. It simply does not need it. Why? Because it is used just by local company systems in a single region with no requirement to expand to other regions or time zones. Thus, it is the easiest and most secure solution, as no shift calculation is necessary (except possibly when changing to summer/winter time). Client and server time zones are always the same – no UTC relation is necessary. However, it is difficult to expand this system. Synchronization across multiple time zones is impossible. Note that it is not enough to deal only with the regions themselves and link them to geographic time zones. Emphasis should be made on DST and regional properties as well. Thus, to reach a robust solution, both client and server time zones must be taken into account even in local systems. This will allow you to expand the system to cover multiple regions if the need arises in the future.

Figure 4.1 shows the relationship between the database, client, and local time perspectives in a UTC database reference. The current database time is 11:47 UTC. The current client time is 13:47. Thus, it is evident that the time zone shift is 2 hours (+02:00). As a result, we know that the local time is 13:47, and the client time applies the time zone reference, getting the value `11:47 +02:00`:

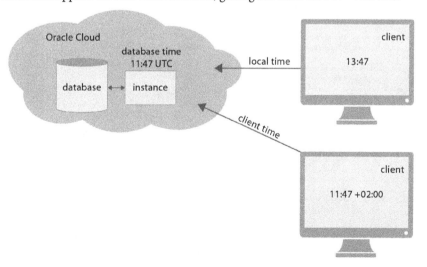

Figure 4.1 – Local and client time with UTC reference

More complicated situations are described in *Figure 4.2* and *Figure 4.3*. Let's suppose we have the same time, but the database time is expressed in CET.

Figure 4.2 shows the transformation between CET and **Australian Eastern Standard Time** (**AEST**). In this case, the local time is 21:47, and the client time applies the time zone reference, providing `11:47 +10:00`:

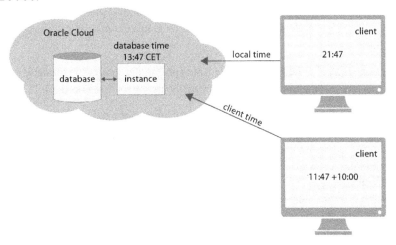

Figure 4.2 – Local and client time with CET reference – logical process

The transformation across multiple time zones is not, however, done directly between the time zones. It is always done through the UTC reference. *Figure 4.3* shows step-by-step processing and management by placing an additional module that provides the UTC reference:

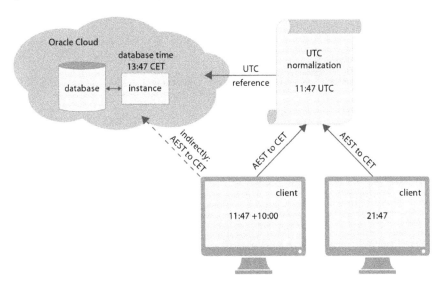

Figure 4.3 – Local and client time with CET reference – the physical process

As can be seen, the shift calculation and processing is, therefore, a staged process as follows:

1. The client provides the value by extracting the time zone.

2. The value is normalized to the UTC format.

3. The server time zone is extracted.

4. The client value is transformed to the destination (server) time zone.

Implementing DST in an Oracle database

Let's dig deeper into time zone management and references to the DST. It is strongly recommended to use a region name for the time zone definition instead of using an absolute time zone offset. The Oracle database never changes the time zone value parameter if DST occurs!

The region name is associated with the shift, and it applies the DST switch. The rules for the shift calculation are taken from the time zone specification and mapping files supplied with the database. These files are regularly updated to apply new rules, strategies, and changes in the DST definition. Thus, it is important to refer to the region name instead of the explicit numerical value representing the shift. To be more declarative, let's consider the year 2015, which is critical in terms of the DST management and shift in the US. During that year, the rules for identifying the DST end date changed. Prior to then, it referred to the last Sunday in October; later, it was represented by the first Sunday in November. But let's consider the consequences using the following value:

```
TIMESTAMP '2012-10-29 12:00:00 America/New_York'
```

The input data extended by the time zone definition is converted to UTC by applying the rules defined in the time zone specification file or by using an explicit time zone shift value. Thus, it can be expressed by the following value in the UTC reference:

```
'2012-10-29 17:00:00'
```

The preceding value can be extended by the identifier of the New York time zone reference. As evident, it carries out the UTC-05 value.

However, by applying new rules, consider the dates between 28th October and 4th November 2015. What about the results? Yes, it would reference UTC-04, correctly referring to the DST, which was valid at the time of the data collection.

Thus, when retrieving the data from the Oracle database, the output would be 13:00:00 for the time elements instead of the original 12:00:00. However, what does the 12:00:00 value actually mean? Does it refer to the local time or is it noon that is represented? So, what is the conclusion? Simply, it is always important to understand the meaning and point of the reference.

Besides, be aware that the dates when the DST applies do not need to be the same worldwide. For example, in the US, DST ends on 5th November 2023, while in Europe, it refers to 29th October 2023.

Now, we know how to refer to the date and time values, emphasizing the local, client, and database time zone perspectives, which can even make the day shift. Related to that, the leap year should be considered, arising from the Gregorian and Julian calendars. The next section describes these calendars, discussing the definition of a leap year in more detail.

Gregorian versus Julian calendar

The **Gregorian calendar** is currently commonly used, based on the already-specified principle of leap year management. The Gregorian calendar was introduced by Pope Gregory XIII in 1582. The aim was to apply date shift properly (one year does not take exactly 365.25 days, so the leap year should not be applied once in four years generally), respecting the Earth's speed. This way, proper day mapping over the centuries can be ensured.

The **Julian calendar** was established in 46 BC by Julius Caesar as an update to the **Roman calendar**, which used 29 days for most months, by introducing an extra intercalary month between February and March. Note that this intercalary month is not applied every year but only for those years that are marked as **intercalary years**). It was used up to 1582. Then, it was replaced by the Gregorian calendar. The main difference in these calendars is related to the year definition. The Julian calendar has two types of year – a standard year consisting of 365 days and a leap year that has 1 extra day. The main principle is associated with the regular rotation of two years and one leap year. Thus, according to the Julian calendar, each year has, on average, 365.25 days. However, the Earth's rotation does not take 365.25 days. Instead, one orbit around the Sun lasts precisely 365 days, 5 hours, 48 minutes, and 45 seconds.

Therefore, to apply real conditions, it is always important to be well-versed with the rules of standard and leap years. The next section outlines the history of the leap year, focusing on the Gregorian calendar. It details the rules by which it is easy to determine whether a given year is a leap year or not.

Leap years in the Gregorian calendar

The orbit of the Earth around the Sun takes a little more than 365 days (precisely 365.2422 days). As a result, some adjustments must be applied to keep the seasons correct. Namely, it is necessary to maintain correctness so that even in many years, decades, and centuries, it will be possible to apply the same rules for individual days and the position of the Sun from a calendar perspective.

Leap years are used periodically to decrease the difference between the calendar year and the Earth's real rotation and to ensure year-over-year mapping. How do we identify a leap year? Well, if the year value can be divided by four without any remainder, then that year can be considered a leap year. That is true just for the Julian calendar, but the Gregorian calendar applies additional rules that ensure that the number of leap years is lowered. Namely, the year number must be divisible by four, except for the first year of the century (divisible by 100 generally). Of these, only those years that can be divided by 400 without a remainder are still considered a leap year. So, the year 2000 was a leap year, but 1900 was not. The next leap years will be 2024, 2028, 2032, and so on. The first time a leap year was applied based on these rules was in 1752.

As is evident from the preceding rules, the Gregorian calendar applies a more complex rule to better match the value of the Earth's orbit around the Sun. Every 128 years, 1 extra day is gained by the Julian calendar. Thus, for the time frame 1901 to 2099, the difference between the Gregorian and Julian calendars is 13 days. For example, 24th December in the Gregorian calendar corresponds to 11th December in the Julian calendar.

In addition to the leap year, there is also a leap second, which exists due to the fact that the Earth's rotation is not constant (it is influenced by several external factors). The next section describes the term leap second, followed by the logical implementation and technical limitations.

Leap second

A leap second is characterized by a specific situation. Commonly, a minute lasts 60 seconds, but when a leap second occurs, 1 minute lasts either 61 seconds or 59 seconds. The leap second is applied occasionally without specific planning. This decision is made by the IERS to accommodate the difference between the high-precision atomic clock time and solar time that reflects the natural Earth's movement (commonly referred to by the abbreviation **UT1**). As stated, the common reference in civil environments is UTC; however, as the Earth's rotation speed varies according to climate and geological events, it is necessary to apply a leap second occasionally to ensure the difference between UTC and UT1 does not exceed 0.9 seconds. Typically, the leap second announcement is made approximately six months before its physical occurrence. It is commonly associated with the last day of the month. June or December is preferred if a leap second needs to be applied to a year. If it needs to be applied twice in a year, then the selection is in March or September. In the last few decades, a leap second was applied every 21 months on average (since 1972). Compared to the *leap day*, which always occurs just after *28th February*, the leap second occurs simultaneously worldwide. For example, it occurred on 31st December 2016 at 23:59:60 (UTC), which was 31st December 2016 at 16:59:60 in Vancouver (Canada), and 1st January 2017 at 9:59:60 in Sydney (Australia).

To get the elapsed time between two date points, leap seconds should also be reflected. It requires an update of the originally calculated difference, as common mathematical operations do not reflect leap seconds. Moreover, leap seconds are announced only a relatively short time in advance. The future occurrences of leap seconds cannot be computed/predicted, and even if we tried, we would get *inaccurate* results. However, the calculation regarding the leap seconds can be applied to identify the occurrences of leap seconds in the past years.

The principles and reasons for defining a leap second are clear. However, what about the technical details and implementation in the Oracle database system? The next section provides you with the answer.

The leap second in Oracle Databases

Oracle databases do not support leap seconds. The second element can hold values ranging from 0 to 59; otherwise, the ORA-01852: seconds must be between 0 and 59 exception is raised. The following code snippet illustrates the exception raised:

```
select to_date('31.12.2016, 23:59:60',
               'DD.MM.YYYY HH24:MI:SS')
  from dual;
```

Thus, getting the difference between two date (or timestamp) values covering a leap second provides incorrect outputs. The following example provides the value 1 (the number of elapsed seconds), but the correct value should be 2. The following code calculates the number of seconds between the timestamp values:

```
select EXTRACT( SECOND FROM
                (TIMESTAMP '2017-07-01 00:00:00.000'
                 - TIMESTAMP '2016-12-31 23:59:59.000')
               )
  from dual;
```

The recommended workaround of leap second management is to store the value as a character string (varchar) instead of storing it in a date and time format based on the precision (the Date or Timestamp data types). The following code shows you how. It creates a table where the occurrence attribute is associated with varchar instead of Date:

```
create table T(id integer, occurrence varchar(30));
insert into T values(500, '2016-12-31 23:59:60.000');
```

Although the preceding example is not the best solution and isn't very practical, it is now the only solution to cover leap seconds. The disadvantage, however, is that element management and overall arithmetic cannot be done directly. Moreover, methods taking the date or timestamp values as parameters cannot be used and their whole management is left to the user.

Scientists and government representatives from all over the world held a meeting in France in November 2022. They took the resolution of limiting leap seconds by 2035. The main reason was that the leap second is not commonly noticed by people. But mostly, whereas the presence of the leap seconds cannot be clearly identified for the future in advance, it brings significant problems in systems requiring uninterrupted time flow, such as satellite navigation, communication systems, software, or specific sensor networks.

Therefore, leap seconds do not have significant benefits from a personal perspective. They were introduced in 1972, and so far, 27 leap seconds have occurred, which does not make a big difference between the atomic and real-world time flow.

Summary

Until now, you may have felt that the world of time management was simple. A day has 24 hours, and a year consists of 365 days. In this chapter, we have shown that there is 1 day of the year that has only 23 hours and one that has 25 hours because of the transition to summer and winter time, described as DST. Similarly, although a standard year is made up of 365 days, there is also a leap year in which there is 1 extra day. The difference between the Julian and Gregorian calendars and the related leap year occurrence was also discussed.

Time zones are an integral part of the calendar and time. In the past, information systems were only local – server and client time were always the same. The client and server were in the same time zone within one small geographical area. With the globalization of systems, markets, and businesses, it is necessary to pay attention to time zones and transitions between them. We have described the concept of UTC, which is used for time normalization.

Finally, there was a reference to the leap second to cover the entire line of time specifications. Since the rotation of the Earth is not constant, it is necessary to apply additional rules from time to time. One of them is the case where a minute has either 59 seconds or 61 seconds.

Although this chapter was mostly theoretical, the concepts introduced are crucial for the development and understanding of the complexity related to the ISO standards.

The next chapter deals with the modeling and storage principles related to the constructor function, emphasizing data reliability and correctness. You will get an overview of the Oracle date and time management data types, constructor functions, arithmetic, and functions to get the actual date and time.

Questions

1. Which time zone perspective is commonly used for date and time value normalization with no offset?

   A. UTC

   B. CET

   C. CEST

   D. BST

2. If the database time is 11:47 UTC and the client time zone is +02:00, which of the following is the correct expression for the client time?

   A. 13:47

   B. 11:47 UTC

   C. 11:47 +02:00

   D. 13:47 +02:00

3. Which calendar's definition of a year (including leap years) most accurately represents the Earth's rotation around the Sun?

    A. Roman calendar

    B. Julian calendar

    C. Gregorian calendar

    D. Normalized calendar

4. Try to solve this problem before moving on to learning about the core elements of Oracle database implementation in the next chapter:

    My friend was 17 years old 2 days ago. Next year, he will be 20. What is the date today?

Further reading

*   Find out the current date and time for any town and time zone at `https://time.is/sk/UTC` or `https://www.timeanddate.com/worldclock/timezone/utc`.

*   *What Is Daylight Saving Time?* by Anne Buckle and Vigdis Hocken provides you with interesting historical details and principles of DST. By reading the text, you will gain a deeper understanding of DST specifications such as rules for identifying dates when winter time shifts to summer time or vice versa. It also provides an interesting background to the evolution of DST. For example, did you know the DST time change can be 30 or 45 minutes rather than 1 hour? Which country was the first to use DST? The answer can be found in the paper available at `https://www.timeanddate.com/time/dst/`. Get inspired.

- Technical details about the DST implementation in Java programming language can be found here: https://www.ibm.com/support/pages/java-daylight-saving-time-known-problems-and-workarounds, https://www.ibm.com/support/pages/java-sdk-daylight-saving-time or https://www.oracle.com/java/technologies/javase/timezones.html.

Part 3:
Modeling, Storing, and Managing Date and Time

As the title of this part denotes, we will be discussing how to model, store, and generally manage date and time values. First, the available date- and time-oriented data types in Oracle are introduced. The Oracle RDBMS provides a honed-in solution because the DATE data type applies second granularity. Besides that, the TIMESTAMP data type is available to serve finer precision, optionally extended by the **time zone** perspective.

Once you are familiar with the available data types, it is possible to dig deeper into date and time value definitions by focusing on the constructor functions and conversion and element extraction functions, specifically looking at reliability and integrity issues. Moreover, in this part, **National Language Support** (NLS) parameters, making your systems adaptable to regions, are discussed. Namely, what is the first day of the week? Is it Sunday or Monday? You will get insights into server as well as client NLS settings and their impacts on function results.

This part includes the following chapters:

- *Chapter 5, Modeling and Storage Principles*
- *Chapter 6, Conversion Functions and Element Extraction*
- *Chapter 7, Date and Time Management Functions*
- *Chapter 8, Delving into National Language Support Parameters*

5
Modeling and Storage Principles

In the past, conventional database principles were used, meaning only currently valid states were in operation and manipulated with no specification of their period of validity. Therefore, date and time values were significantly limited. Nowadays, the situation is different. One aspect of current information systems is state monitoring over time, and another is the length of time that the state is valid. For example, an employment contract specifies the time frame that it is valid, within which the employee is expected to be available and attend to their duties, as well as noting their activities, responsibilities, and so on. If the employee leaves the company, they should automatically lose access to the information systems or data storage repositories. However, data about the employee cannot be removed as many references to it exist. The history of the assignments and responsibilities must be stored. Another example is webshop registration. Even if a customer deletes their account, their orders must remain, although they will be anonymized. Thus, in systems dealing with validity, the physical *delete* operation is commonly replaced by a logical one by adding a validity time frame.

This chapter deals with the Oracle date and time data types. By using these values, each state can be limited by validity. Thanks to that, multiple states can be stored for the same object. If we return to the previous example related to employment, one person can have multiple employment contracts in the same company – the job position may evolve over time. However, what is always critical is the timeline reference of the states that expresses when the particular data was (or is) applicable.

Generally, date and time values can express and model one timeline point or can act as limits (borders) of the validity time frame. Value can be obtained from a numerical or character string representation by using constructor or transformation functions. In this chapter, a complex overview of applying the **ANSI constructor** and **conversion functions** in Oracle is provided, specifically the time zone definition modeled by the TIMESTAMP value extension, which can be normalized for the client or server perspective, creating an opportunity to share systems or applications worldwide. We will also look at the INTERVAL data type modeling duration. DATE arithmetic uses day granularity, while TIMESTAMP arithmetic refers to day-to-second precision using the INTERVAL data type.

To sum up, in this chapter, we're going to cover the following main topics:

- Data types for modeling date and time in Oracle RDBMS

- Constructor functions

- Limitations related to modeling date and time values using numerical representation

- Functions providing actual date and time values regarding the data type and client versus server perspective

- `DATE` arithmetic

- The `INTERVAL` data type for duration modeling

- `TIMESTAMP` arithmetic

The source code for this chapter can be found in the GitHub repository accessible via this link: `https://github.com/PacktPublishing/Developing-Robust-Date-and-Time-Oriented-Applications-in-Oracle-Cloud/tree/main/chapter%2005`. Alternately, you may scan the following QR code to access the repository:

Exploring data types

ISO 8601 proposes three characteristics related to date and time models:

- The `DATE` data type is used to deal with the granularities of the year, month, and day

- The `TIME` data type deals with the hour, minute, and second elements

- By combining those two types, the `DATETIME` data type is introduced

Typically, the precision is the *second* element itself, which can also be expanded to cover fractions of a second. This principle is used, for example, in the MySQL database system. Oracle Database has not adopted the preceding three types. It uses the `DATE` data type, which deals with the date and time elements up to second granularity. Thus, it is impossible to deal with the year, month, and day elements without the time elements. However, to avoid using the time elements, some time elements have the option to hold a 0 value. However, all those elements are always present, stored, and referenceable.

As a result, if the solution does not require time elements to be modeled and stored, developers try to reach their own solutions, typically representing dates through strings or numerical values. But, as we will show later, this decision entails several limitations, mostly related to the integrity and correctness of the values. It is always necessary to verify that the entered value is transformable to DATE. This ensures that the specified day exists. Oracle stores DATE in an internal numeric format. It requires 7 bytes for storage. The internal data type code is 12.

The second data type for handling date and time values is TIMESTAMP. It can also store second fractions up to nanosecond granularity. The precise definition of the data type is **TIMESTAMP[(n)]**, where the optional n parameter expresses the number of decimal places for the second. By default, microsecond precision is used. Thus, if the parameter value is not specified, it is automatically replaced by the value 6. The precision n ranges from 0 up to 9. The format symbol for the second fraction is FF. The meaning and representation slightly differ between the SQL and PL/SQL languages.

A TIMESTAMP data type can be extended by managing time zone reflections. **TIMESTAMP WITH TIME ZONE** is a variant of the TIMESTAMP format, also holding the time zone offset or time zone region. The time zone offset is expressed numerically as the difference between the local time and UTC or **Greenwich Mean Time (GMT)**, or the time zone region can be specified by its name.

TIMESTAMP WITH LOCAL TIME ZONE does not explicitly store the time zone; instead, the value is normalized by applying the time zone, thus the offset does not need to be stored. Oracle takes the value from the UTC and converts it to the database time zone using arithmetic. *Table 5.1* shows the available TIMESTAMP data type extensions with the defined precision, storage demands, and internal data type codes:

Data Type	Maximum Internal Length	Data Type Code
TIMESTAMP [WITH LOCAL TIME ZONE]	11 bytes	180
TIMESTAMP WITH TIME ZONE	13 bytes	181

Table 5.1 – TIMESTAMP versions

The following figure shows the value representation in Oracle Database:

Figure 5.1 – Data type reference

Oracle provides DATE and TIMESTAMP data types to reflect various precisions. It also allows specifying time zones. However, how is this data constructed? How do we create a new value from a string? The next section, dealing with the constructor functions, provides the answer.

Exploring constructor functions

How do you construct a DATE value? There are always two streams that can be used. The first type relates to the already described ANSI standard being adopted. The second stream uses explicit field mapping forming constructor functions. To ensure the processing is correct irrespective of the system and session format configuration, it is strongly recommended to always cover the mapping, that is, ensuring there is an unambiguous expression of the meaning of individual elements during the value construction. The DATE value cannot be treated as a string and individual values must be assigned to the elements, ensuring the correct meaning, representation, and management (mapping to the individual elements). For example, the value 12 can express the day, the month, or any time element (hour, minute, or second). Thus, without knowing the meaning of the values, two issues can be present. The first is that an exception will be raised if it is improperly specified. For example, let's take the value used in the following format: 12-31-2021. The element order is not specified. Thus, it can be assumed that 31 represents the day and 12 represents the month. However, if the system uses an implicit definition and assumes the day to be the left-most part, an exception is raised, as it would assume the middle part refers to the month and, of course, there is no month number 31. The second limitation is associated with improper mapping and processing. Let's have another simple value, 11-12-2021. It is not directly clear from the value which element defines the day and which defines the month.

Therefore, it is evident that the specification of the meaning must be present. Moreover, you can notice that the previously specified value (11-12-2021) does not meet ISO standardization. The representation, therefore, depends on the system instance settings or the session itself if an implicit constructor is used. If these settings are changed, the processing method will naturally also change. Thus, the value can be represented as *11th December*, but also *12th November*. To ensure the value is processed correctly, the format must be specified, or general rules must be applied.

Two approaches can be used to construct a DATE value from a string:

- Using the ANSI standard definition (DATE). It applies a strict format and element order.
- Using the Oracle date value constructor (TO_DATE) with the value as the first parameter and the mapping format as the second parameter of the function.

The next section deals with the analogous techniques applied to the TIMESTAMP data type. We will describe the principles of the ANSI constructor (TIMESTAMP) and conversion functions (TO_TIMESTAMP).

The ANSI DATE constructor

An **ANSI DATE constructor** must be strictly specified in the format YYYY-MM-DD representing the year (YYYY), month (MM), and day (DD). Values are defined by the numerical format, delimited by dashes (-). The element order is significant. The keyword DATE denotes the constructor function. The full syntax is the following:

```
DATE 'YYYY-MM-DD'
```

Practical usage is shown in the following code snippet. It is clear that the value 2021 represents the year, followed by the month (11), and the day by the value 12. Time elements cannot be stated, so they are replaced with 0 values for the hour, minute, and second elements. Hence, it is necessary to distinguish between the ISO DATE constructor function (managing only the year, month, and day elements) and its real mapping to the Oracle DATE data type, formed by the hour, minute, and seconds as well. In the following code snippet, the ANSI constructor is used. The value 2021 is mapped to the year value, the value 11 represents the month, and the value 12 expresses the day element:

```
select DATE '2021-11-12' from dual;
--> 12.11.2021 00:00:00
```

Note that the format of the output presented to the user can differ from how it is shown here. It depends on the session specification (the NLS parameter). A discussion about the principles and designations of the Date value processing influencing the output format will be presented later.

Evidently, it is not possible to omit any element. Thus, the following notation cannot be used because it raises the ORA-01861 exception:

```
select DATE '2021-12' from dual;
--> ORA-01861: literal does not match format string
```

The ANSI DATE constructor focuses on the pure elements – *year*, *month*, and *day*. The format of the value specified in the ANSI DATE constructor is strictly defined and the order of individual elements is fixed.

The TO_DATE function

Alternatively, the TO_DATE conversion function can be used for constructing the DATE value. It can have three parameters. For now, we will just deal with the first two types – input_value and format:

```
TO_DATE(<input_value>
          [, <format> [, <nls_date_language_format> ]]);
```

The first parameter deals with the input string to be transformed into the DATE representation. The second parameter is responsible for specifying mapping criteria to specific elements. Although this parameter is optional, it is strongly recommended to carry out the full specification of mapping to ensure correctness. The third parameter is also optional and specifies the date language for the particular select statement. The principles, importance, and usability will be described later.

The following code snippet shows the DATE construction from the character string. The second parameter defines the format and meaning of individual elements (mapping). It is clear that the month value is 11 and the day value is 10. By omitting the second parameter, the meaning of individual sub-values is not evident, leading to the possibility of processing the data incorrectly:

```
select TO_DATE('2021-11-10', 'YYYY-MM-DD') from dual;
```

If the mapping is not stated explicitly, the default option is used, delimited by the NLS_DATE_FORMAT session parameter. Thus, it works correctly only if there is a proper match between the specified value and element meaning (mapping). Otherwise, the exception is raised. Therefore, we do not recommend using an implicit transformation strategy. As you can see in the following snippet, the session parameter format does not fit the specified value, leading to the exception being raised:

```
select TO_DATE('2021-11-10') from dual;
--> ORA-01861: "literal does not match format string"
```

Note that many books on this topic that are published in the United States do not highlight these rules and assume that the format is always precisely specified and does not change. Based on real experience, this assumption is not correct. If the application is moved to another environment or server, the solution simply does not work. Many errors can be generated as a result of omitting format mapping. Due to the lack of explicit mapping, the system uses the default server values, which do not need to be as expected. Moreover, any NLS parameter change at the session or system level can cause incorrect data interpretation.

The main advantage of the explicit TO_DATE function call is the possibility of using any input data. The first element of the specification does not need to define the year at all (compared to the ANSI constructor). The positions of the DATE elements inside the first parameter value can be arbitrary. There is just one requirement defining the input and format correlation. Each element value is referenced by the format describing its meaning and mapping. Thus, all the following examples represent the same value:

```
select TO_DATE('2021-11-10', 'YYYY-MM-DD') from dual;
select TO_DATE('2021-10-11', 'YYYY-DD-MM') from dual;
select TO_DATE('11-10-2021', 'MM-DD-YYYY') from dual;
```

Moreover, individual element separation can be done by any appropriate symbol, such as a dash (-), slash (/), dot (.), or comma (,):

```
select TO_DATE('10.11.2021', 'DD.MM.YYYY') from dual;
select TO_DATE('11/10/2021', 'MM/DD/YYYY') from dual;
```

The following example combines the date and time elements in a common TO_DATE function constructor, resulting in providing the DATE data type value:

```
select TO_DATE('10-11-2021 10:12:24','DD.MM.YYYY HH:MI:SS')
    from dual;
--> 10.11.2021 10:12:24
```

Note that the delimiters used in each of the strings do not need to be the same. However, they must always match the format specified in the second parameter.

Using the TO_DATE function is a general solution and provides flexibility. The user can specify any order of elements. If some elements are missing, a 0 value is applied, if possible. Compared to the ANSI constructor, the TO_DATE function can also deal with the time elements.

Let's look at the case where an element is not part of the input string. Firstly, let's assume a time element, such as hour, is missing. In that case, either a particular value can be specified as 0 or an implicit strategy can be used. Both of the following function calls will provide the same values:

```
select TO_DATE('10.11.2021 00:12:24',
               'DD.MM.YYYY HH24:MI:SS')
    from dual;
```

```
select TO_DATE('10.11.2021 12:24',
               'DD.MM.YYYY MI:SS')
  from dual;
--> 10.11.2021 00:12:24
```

However, what about the situation where the day element is not present? Evidently, zero-value substitution cannot be used, as it would not provide the correct value. Instead, the smallest possible element value is used. Thus, if the day value is not present in the specification, the DATE value will display the first day of the month:

```
select TO_DATE('11.2021', 'MM.YYYY') from dual;
--> 01.11.2021 00:00:00
```

The preceding definition is just applied to the day element. If the month field is not specified, then the current value of the month is used. Thus, if the current month is December, then the missing value will correspond to the same value – December:

```
select TO_DATE('5.2021', 'DD.YYYY') from dual;
--> 05.12.2021 00:00:00
```

Finally, if the year element value is not present in the input value, then the current year is used instead.

In this section, you have learned, how to construct a DATE value using the TO_DATE constructor function. You became familiar with the format mapping and meaning of individual elements inside the input value. The next section digs into techniques for TIMESTAMP value constructing.

The ANSI TIMESTAMP constructor

Like DATE management, to specify the TIMESTAMP value from the character string, the ANSI constructor or the TO_TIMESTAMP constructor function can be used.

The ANSI constructor comprises the TIMESTAMP keyword followed by the individual element definition from the year to second fractions. Note that by default, the 24-hour format is used. The delimitation symbol for dates is a dash (-), and time fields are separated by a colon (:), with a dot (.) for fractions:

```
TIMESTAMP 'YYYY-MM-DD HH24:MI:SS:FF'
```

An example of this is as follows:

```
select TIMESTAMP '2021-05-01 16:20:15.1234' from dual;
--> 01.05.2021 16:20:15,123400000
```

If, for example, the day value is omitted, an exception (ORA-01861: literal does not match format string) is raised:

```
select TIMESTAMP '2021-05 16:20:15.1234' from dual;
```

Finally, note that if a smaller precision for the second fraction is used, zero values will be added to fit the definition. Conversely, if the precision is greater than the resulting repository, the value will be rounded (PUT_LINE of the DBMS_OUTPUT package provides the following output if run in a console environment):

```
declare
  value TIMESTAMP(3);
begin
  value:=TO_TIMESTAMP('2021-05-01 16:20:15.1235',
                      'YYYY-MM-DD HH24:MI:SS.FF');
  DBMS_OUTPUT.PUT_LINE(value);
end;
/
--> 01.05.2021 16:20:15,124
```

The main advantage of the ANSI constructor is its general applicability to any database system because most systems follow the ANSI norm. To ensure correct mapping to the elements, the ANSI definition requires a strict element order to be followed. Thus, the first value must be the year, followed by the month, day, and so on. The next section deals with the TO_TIMESTAMP function call, bringing you unlimited possibilities for defining the shape of the value. It is formed by the input value and the defined format.

The TO_TIMESTAMP function

A more general solution can be obtained by using the TO_TIMESTAMP constructor. The first parameter deals with the character string representation. The second parameter maps the elements by the defined format:

```
TO_TIMESTAMP(<input_value> [, <format> ])
```

The principles are the same as the TO_DATE function call. It is always recommended to specify the format for proper management. If omitted, the default representation specified by the system or overwritten by the session will be used. However, it is parameter value-dependent, and if changed, the application system will stop working or the evaluation will not be correct. The first parameter deals with the input string, while the second parameter defines the meaning of individual values specified by the first parameter – the mapping criteria. In the following example, 2021 is mapped to YYYY, expressing the year, 05 is mapped to MM, referring to the month, and so on:

```
select TO_TIMESTAMP('2021-05-01 16:20:15.1235',
                     'YYYY-MM-DD HH24:MI:SS.FF')
  from dual;
--> 01.05.2021 16:20:15,123500000
```

There are multiple format styles and representations, which can be part of the second parameter of the TO_TIMESTAMP or TO_DATE functions. These formats are self-explanatory: YYYY refers to the year, MM refers to the month, and DD refers to the day. From a time modeling perspective, HH24 defines the 24-hour format, whereas HH means the 12-hour format type is used. MI expresses minutes, and SS means seconds. Fractions can be used with the TIMESTAMP format FF. A list of the most commonly used formats can be found in *Chapter 6*.

So far, you have learned, how to construct the TIMESTAMP value using the ANSI constructor specified in the previous section, while this section was aimed at understanding explicit constructor functions. It used the TO_TIMESTAMP function and the focus was on proper element mapping. The TIMESTAMP extension can process values referencing time zones. The next section defines the rules of modeling and the representation of time zones.

Time zone enhancements

A TIMESTAMP data type can be extended by managing time zone reflections. TIMESTAMP WITH TIME ZONE is a variant of the TIMESTAMP format, also holding the time zone offset or time zone region. The time zone offset is expressed numerically as the difference between the local time and UTC or GMT. The time zone region can also be specified by its name. The syntax of TIMESTAMP WITH TIMEZONE is as follows:

```
TIMESTAMP[(<n>)] WITH TIME ZONE
```

The precision value expressed by n is optional and can range from 0 to 9. The ANSI constructor then has the following syntax:

```
TIMESTAMP 'YYYY-MM-DD HH24:MI:SS.FF <timezone_offset>'
TIMESTAMP 'YYYY-MM-DD HH24:MI:SS.FF <timezone_region>'
```

The time zone reference can be modeled by the numerical value expressing the hour and minute shift or the time zone name (region reference):

```
select TIMESTAMP '2021-05-09 14:25:12.1234 -6:00'
  from dual;
select TIMESTAMP '2021-05-09 14:25:12.1234 US/Pacific'
```

```
    from dual;
  select TIMESTAMP '2021-05-09 14:25:12.1234 Europe/Vienna'
    from dual;
```

The time zone region name is commonly specified by two parts – region location and capital city. The following query provides a list of time zones supported by the Oracle Database by providing the time zone name (`tzname`), such as `Europe/Berlin` or `Europe/Madrid`. The second attribute deals with the abbreviation (`tzabbrev`) and can hold, for example, the value GMT, EAT, or CET:

```
select tzname,tzabbrev from v$timezone_names;
```

`TIMESTAMP WITH LOCAL TIME ZONE` uses another principle. All `TIMESTAMP` values with this perspective are normalized to the server (database) time zone. Thus, there is no offset definition as part of the stored data. Instead, it is calculated dynamically during the data retrieval process. Oracle provides a local session time zone, which is calculated by working out the difference between the user time zone and UTC. The definition is provided by the `WITH LOCAL TIME ZONE` part of the following syntax:

```
TIMESTAMP[(<n>)] WITH LOCAL TIME ZONE
```

In terms of the definition, the same principles apply. The precision n refers to the precision specification.

More about time zone management and reflection can be found in *Chapter 16*, which provides a complex discussion on the topic.

To summarize the principles, `TIMESTAMP` data type values can be composed in the following ways:

- With the `TIMESTAMP` ANSI constructor
- With the `TO_TIMESTAMP`, `TO_TIMESTAMP_TZ` (explained later), and `TO_DATE` constructor functions
- With implicit conversion

The ANSI constructor and `TO_TIMESTAMP` conversion methods have already been discussed. Implicit conversion applies the system or session format, and it does not need to be robust and error-prone, just in case the format is changed. Therefore, it is not recommended to rely on implicit conversions.

In terms of the physical value, the `DATE` and `TIMESTAMP` values are really similar, their only difference being the accuracy to which they process second values. Thus, a common question that can arise, related to the date and time transformation, is: Is it possible to store a `TIMESTAMP` value in `DATE` format? Or vice versa? Let's find out in the next section.

DATE and TIMESTAMP transformation

`DATE` and `TIMESTAMP` values can be interchanged using implicit conversions automatically. Let's create a simple table consisting of two attributes in the `DATE` and `TIMESTAMP` formats extended by the identifier (`id`):

```
create table Tab(id integer,
                 date_val DATE,
                 time_val TIMESTAMP);
```

The common procedure of value specification is carried out by the `TO_DATE` function call for the `DATE` data type and `TO_TIMESTAMP` for the `TIMESTAMP` format, as used in the following code snippet:

```
insert into Tab
   values(1,
          TO_DATE('14/12/2021 15:24:12',
                  'DD/MM/YYYY HH24:MI:SS'),
          TO_TIMESTAMP('14/12/2021 15:24:12',
                       'DD/MM/YYYY HH24:MI:SS'));
```

The output of the `select` statement provides the default format for the data types. Namely, `DATE` uses second precision, while `TIMESTAMP` relates to the second fractions:

```
select * from Tab where id=1;
--> ID   DATE_VAL              TIME_VAL
--> 1    14.12.2021 15:24:12   14.12.2021 15:24:12,000000000
```

In this case, the input data is in the required format, so no transformation is done. However, values can be dynamically transformed on the fly. Let's insert one new row. The expected `DATE` data type will take the `TIMESTAMP` format and vice versa:

```
insert into Tab
   values(2,
          TO_TIMESTAMP('14/12/2021 15:24:12.9999',
                       'DD/MM/YYYY HH24:MI:SS.FF'),
          TO_DATE('14/12/2021 15:24:12',
                  'DD/MM/YYYY HH24:MI:SS'));
```

By converting DATE into a TIMESTAMP value, second fractions are added, as evident from the following code snippet:

```
select * from Tab where id=2;
--> ID  DATE_VAL            TIME_VAL
--> 2   14.12.2021 15:24:12  14.12.2021 15:24:12,000000000
```

As evident, in the case of lowering the precision, particular fraction elements are ignored and truncated from the results. Conversely, for the transformation from DATE to TIMESTAMP, missing values are replaced by zeros.

If the TIMESTAMP data type is extended by time zone management, the definition of the WITH TIME ZONE clause means that the time zone offset is provided in the result. Let's assume that there is a -7:00 session offset for the insert operation. Data retrieval is then processed in relation to UTC:

```
drop table Tab;
create table Tab(id integer,
                 date_val DATE,
                 time_val TIMESTAMP WITH TIME ZONE);
insert into Tab
  values(1,
          TO_DATE('14/12/2021 15:24:12',
                  'DD/MM/YYYY HH24:MI:SS'),
          TO_TIMESTAMP('14/12/2021 15:24:12.9999',
                       'DD/MM/YYYY HH24:MI:SS.FF'));
insert into Tab
  values(2,
          TO_TIMESTAMP('14/12/2021 15:24:12.9999',
                       'DD/MM/YYYY HH24:MI:SS.FF'),
          TO_DATE('14/12/2021 15:24:12',
                  'DD/MM/YYYY HH24:MI:SS'));
select * from Tab;
--> ID       DATE_VAL
-->          TIME_VAL
--> 1        14.12.2021 15:24:12
-->          14.12.21 15:24:12,999900000 -07:00
--> 2        14.12.2021 15:24:12
-->          14.12.21 15:24:12,000000000 -07:00
```

As is clear from the preceding code block, the time zone is added during the DATE value transformation as well. It is part of the value, thus it consists of the original value extended by the offset.

However, when using WITH LOCAL TIME ZONE in the definition, the offset between the server and the local session is applied directly to the data by transforming the timing:

```
drop table Tab;
create table Tab(id integer,
                 date_val DATE,
                 time_val TIMESTAMP WITH LOCAL TIME ZONE);
insert into Tab
   values(1,
          TO_DATE('14/12/2021 15:24:12',
                  'DD/MM/YYYY HH24:MI:SS'),
          TO_TIMESTAMP('14/12/2021 15:24:12.9999',
                       'DD/MM/YYYY HH24:MI:SS.FF'));
insert into Tab
   values(2,
          TO_TIMESTAMP('14/12/2021 15:24:12.9999',
                       'DD/MM/YYYY HH24:MI:SS.FF'),
          TO_DATE('14/12/2021 15:24:12',
                  'DD/MM/YYYY HH24:MI:SS'));
select * from Tab;
--> ID    DATE_VAL              TIME_VAL
--> 1     14.12.2021 15:24:12   14.12.21 22:24:12,999900000
--> 2     14.12.2021 15:24:12   14.12.21 22:24:12,000000000
```

While we have touched on the topic of time zones, it is also important to emphasize the issue of **Daylight Saving Time (DST)** related to the Oracle RDBMS, discussed in the next section.

Daylight saving time in Oracle

DST can be managed by the TIMESTAMP WITH [LOCAL] TIME ZONE data type in RDBMS Oracle and is handled automatically according to the specified time zone. Note that the TIMESTAMP value has no impact on DST.

DATE and TIMESTAMP values cannot trim time elements. Simply, Oracle cannot model DATE just by the year, month, and day elements. Instead, hour, minute, and second elements are always present and referenceable. Even if they were not explicitly specified, they are present, holding zero values. Therefore, researchers come up with their own solutions and implementations. In the next section, we will show you the limitations, risks, and consequences of this decision.

Storing the date as an INTEGER value

In some cases, I have seen date modeling implemented as an integer in business applications. It is mostly the result of formal date management in Oracle RDBMS, where time elements do not need to be considered. It is assumed that the storage requirements can be lowered by using appropriate numeric value sizing. Moreover, if function-based indexes are applied to numeric date modeling, then the performance of the element extraction can be ensured.

On the other hand, there can be significant issues related to the processing. Firstly, the representation strongly depends on the positional element models. Typically, when represented by the DATE value, separators (mappers) are present, dividing individual elements. However, let's look at a simple numerical value passing the format DDMMYYYY. The previously stated separator management principle cannot be used, resulting in a possible loss of reliability. Let's consider the example of 1st November 2021. Based on the previous principle, it will be covered by the value 01112021. But be aware, leading 0 values are trimmed from the numerical output, resulting in 1112021. That is just how it is. It is not evident whether this refers to *11th January* or *1st November*. Therefore, the format should be clearly presented by putting zero values for each internal element if the value is just a single digit. Thus, for 11th January, the value should be 11012021 instead of 1112021 by using 01 for the month. This approach can partially solve the issue. However, there are still complications related to element extraction, as the size can be different due to leading zeros. Direct division and separation cannot be applied directly. Moreover, this kind of value is often treated as a string by using SUBSTR to extract elements, which requires implicit conversion to a character string, bringing additional demands.

The solution to this is based on proper format definition. Typically, four-digit years are used, allowing year references from the 11th century and later. In that case, if the leading element is just the year, no problem occurs, and the representation is straightforward. As you may see, the order of the elements is very important. It is not possible to transform and store them in the database otherwise. Data integrity can be, in this case, compromised. It is not only the format itself that is important but also the values of the individual elements. Generally, whether the value to be processed is relevant must be checked in terms of date value transformation possibility. Specifically, the *month* must range from 1 to 12, which can be evaluated and ensured easily. However, what about the days? Some months have 30 days, while some have 31 days. These requirements must be confirmed and checked.

February can be even more demanding. Generally, it consists of 28 days. However, there are leap years, when February has 29 days. Clearly, ensuring that the date is correct is no trivial matter. To ensure this, several rules must be applied, typically forcing the system to convert a numerical value into a DATE value to check the correctness. This is done automatically during the conversion. To be honest, if you have obtained the DATE value, why not store it directly? That is because, if the DATE value is not stored directly, the user will lose the ability to use date-oriented functions.

Let's take, for example, New Year's Eve. There is no problem adding one day to the definition by using the DATE data type. Using a numerical representation, transitioning through the individual elements must be done to apply the addition of 1 day. Let's highlight the last day of the year (31st December 2021). Adding one day does not get 32nd December, but a transition of the month and year must be done, resulting in 1st January 2022. This must all be managed explicitly by the user. Again, the simplest solution in terms of implementation is to convert the value to DATE and rely on the transformation performed automatically by Oracle RDBMS.

To conclude, although date values can principally be modeled by the pure numerical format, far more additional efforts must be made to ensure that the provided data is correct and not treated as a sequence of random numbers. Naturally, another aspect of relational systems should be highlighted. Individual values should be atomic with no decomposition option. If the value is decomposable, then it should not be stored as a single common value.

Now, we know how to model date and time values and how to construct them from the character string. The next section walks you through the principles of getting an actual server or client value, emphasizing the time zone shifts.

Getting the actual date and time values

In the previous sections, we dealt with the date and time value composition from the character string format, providing examples of mappings to ensure the reliability and correctness of the processing. There are, however, two functions for each data type (DATE and TIMESTAMP) producing the actual date and time values.

We will intentionally not use the keyword current in this section, as current is a function name used in database processing. It reflects the client site by emphasizing region and time zone processing. The term actual is not used as a keyword by the client or server site and represents states or date and time values that are in effect at this moment.

The sysdate and current_date functions can be used when dealing with the DATE data type. sysdate is a SQL function that gets the actual date and time elements referencing the operating system of the database server. Thus particular values are normalized based on the server parameters in terms of time zone. Conversely, current_date produces the value referencing the session (client) time zone. When the server and client share the same region and time zone in an ideal environment, both values are the same. However, generally, they can differ, just focusing on the different perspectives.

The following code snippet shows an example of the calls and outputs. It is evident that there is an 8-hour shift between the server and client time zones. Generally, servers use UTC, but there can also be some user-specific implementations that violate that rule. Thus, based on the shift, it can be assumed that the client is located in *North America* (by taking the premise of UTC server value normalization):

```
select sysdate, current_date from dual;
--> SYSDATE                    CURRENT_DATE
--> 25.03.2022 12:55:13        25.03.2022 04:55:13
```

When dealing with the actual DATE values, mostly, the sysdate function was used in the past. The systems were rather local, referencing the same time zone and region, for both client and server sites.

Database systems were located in the company server room, and individual users were only locally devoted. Thus, to ensure the correctness of the processing, the server perspective (instead of the client perspective) was always referenced assuming that the server time zone is correctly specified, which was the responsibility of the **database administrator** (**DBA**). However, later, companies became widespread and a new layer synchronizing date and time values had to be created. As a result, the values did not need to reference server time. Just the client region perspective is enough. Therefore, we use the `current_date` function for referring client time instead of using `sysdate` that deals with the server reference.

Moreover, databases are not now managed locally, but the cloud environment is the best perspective to be focused on. Many DBA administration activities related to backups, patching, and updating are strongly limited, ensuring availability, reliability, and dynamic performance reflecting the workload. The DBA can then focus on other activities, and optimization rules can be autonomously supervised. When dealing with cloud architectures, several cloud services and locations are present, emphasizing the multiple regions. To ensure almost 100% accessibility, multiple architectures were proposed with the data located in numerous different availability domains. As a result, if there is a failure, processing can be shifted dynamically to another live standby database, which can be physically located anywhere across the world. Thus, regions and associated time zones should always be taken into account.

To get the actual `TIMESTAMP` value, either `SYSTIMESTAMP` or `LOCALTIMESTAMP` can be used. It uses similar principles to `DATE` management. Namely, `systimestamp` produces the server reference, while `localtimestamp` references the client region. The next query gets the system (server) and client (local) timestamps. The difference between them expresses the time zone shift:

```
select systimestamp, localtimestamp from dual;
```

If you consider the system to be spread across multiple regions, always use the `current_date` and `current_timestamp` function calls, even in reference to the cloud repository. This decision makes your applications and system independent of the database server's time zone, which can even be dynamically changed.

Figure 5.2 shows the local system perspective. Local and server date and time references are commonly the same; the available data types are `DATE` and `TIMESTAMP`.

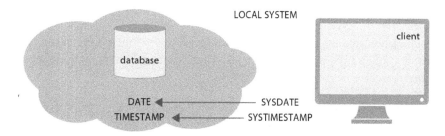

Figure 5.2 – Local system

Compared to *Figure 5.3*, which covers a global system applicable worldwide, there is no need for dealing with time zone management.

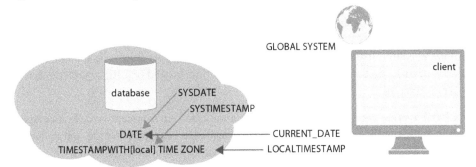

Figure 5.3 – Global system

Thus, in local systems, the client and server time zones are the same or do not change for individual clients. Global systems specify the time zone and point to the difference between the client and server. In global systems, client and server time zone references should be highlighted.

Now we know how to model date and time values in Oracle RDBMS. However, how do we add one day or one hour to the defined value? The solution is by applying arithmetic. The next section proposes the technique for the Oracle DATE data type.

Getting to know DATE arithmetic

Adding or subtracting a numerical value from the DATE value expresses the number of days (or a part of a day, for example, 0.5 expressing 12 hours):

```
select TO_DATE ('15.02.2022', 'DD.MM.YYYY') + 1 from dual;
--> 16.02.2022 00:00:00
```

Just to remind you, the TO_DATE conversion function in the preceding example deals only with the day, month, and year elements. In that case, the conversion causes undefined components (hours, minutes, and seconds) to be replaced with zero values.

Addition or subtraction automatically reflects the value consistency. Thus, if we're looking at the last day of the month, adding 1 day produces the first day of the consecutive month, as expressed in the following code:

```
select TO_DATE ('28.02.2022', 'DD.MM.YYYY') + 1 from dual;
--> 1.3.2022 00:00:00
select TO_DATE ('31.12.2022', 'DD.MM.YYYY') + 1 from dual;
--> 1.1.2023 00:00:00
```

The decimal part reflects the day fragment. Thus, the value of 1/24 represents adjusting the hour by 1:

```
select sysdate as now,
       sysdate + 1/24 as one_hour_later,
       sysdate -1/24 as one_hour_sooner
 from dual;
--> NOW                          ONE_HOUR_LATER
-->          ONE_HOUR_SOONER
--> 25.3.2022 23:55:12     26.3.2022 00:55:12
-->          25.3.2022 22:55:12
```

The result of subtracting two date values is the number of days elapsed between them (also with the decimal part). So, the resulting value will be 2:

```
select TO_DATE('15.02.2022', 'DD.MM.YYYY')
       - TO_DATE('13.02.2022', 'DD.MM.YYYY')
  from dual;
--> 2
```

Similarly, it can provide a day fraction:

```
select TO_DATE ('15.02.2022 12:00:00',
                'DD.MM.YYYY HH24:MI:SS')
       - TO_DATE ('15.02.2022 04:00:00',
                  'DD.MM.YYYY HH24:MI:SS')
 from dual;
--> 0.3333 expressing 8 hours (1/3 of the day)
```

Direct transformation of the result to a number of months or years would be incorrect because inexact values would be provided (some months have 30 days, some have 31 days, and February is a special case). Therefore, to deal with months or years, defined functionality does not provide exact results (for example, when booking flights or a visa, the exact age of the person must be provided).

To conclude, date arithmetic uses day granularity. To get the number of months elapsed, the available MONTHS_BETWEEN function should be used. Similarly, to add 1 month to the defined DATE value, the specification of 30 or 31 days is not relevant due to the variability of month lengths. Instead, ADD_MONTHS should be used. You will get a complex overview of these functions in *Chapter 7*.

The next section provides an overview of the INTERVAL data type expressing the duration. It does not have the timeline position delimiting the first or last point of the validity, only the elapsed time is expressed. Thus, you know how long the validity period lasts. However, without additional information, there is no evidence of when it started. The INTERVAL data type is split into two categories based on precision modeling.

Understanding the INTERVAL data type

INTERVAL data types are available to model the duration. There are two types based on granularity precision: dealing with the year and month precision (the first type) or digging deeper into time elements (the second type).

INTERVAL YEAR TO MONTH

INTERVAL YEAR TO MONTH consists of no more than two elements describing the year and month delimited by a dash (-). After the definition, the format is specified, with the optional specification of precision. *Figure 5.4* shows the syntax in graphical form. The first format element is mandatory, and the second is optional. Note that the default precision is 2, if not specified explicitly.

Figure 5.4 – INTERVAL YEAR TO MONTH

Table 5.2 shows examples of INTERVAL YEAR TO MONTH usage. The definition can be used independently or associated with the DATE or TIMESTAMP value, typically expressing the beginning point of the validity. The general definition consists of year and month element values:

Definition	Interpretation
INTERVAL '5-2' YEAR TO MONTH	An interval consisting of 5 years and 2 months. It is necessary to specify the leading precision if it is greater than the default two digits.
INTERVAL '567-2' YEAR(3) TO MONTH	An interval consisting of 567 years and 2 months. It is necessary to specify the leading precision if it is greater than the default two digits.
INTERVAL '5' YEAR	An interval of 5 years and 0 extra months.
INTERVAL '5' YEAR(2)	An interval of 5 years and 0 extra months.
INTERVAL '500' YEAR(3)	An interval of 500 years and 0 extra months. Note that when the precision should be higher than two digits, it must be explicitly specified.

Definition	Interpretation
INTERVAL '500' YEAR	Exception raised: ORA-01873: the leading precision of the interval is too small
INTERVAL '30' MONTH	An interval defining 2 years and 6 months. It is physically mapped to INTERVAL '2-6' YEAR TO MONTH, resulting in expressing the value +02-06.
INTERVAL '300' MONTH(3)	An interval of 25 years.
INTERVAL '300' MONTH(2)	An interval of 25 years. It does not raise an exception, as the representation can be split into years and months. The month definition does not exceed two digits.
INTERVAL '300' MONTH	An interval of 25 years. The default precision is 2.

Table 5.2 – INTERVAL YEAR TO MONTH usage

The second type of INTERVAL definition uses higher value precision and focuses on the day and time components.

INTERVAL DAY TO SECOND

Figure 5.5 shows the syntax of the INTERVAL DAY TO SECOND data type:

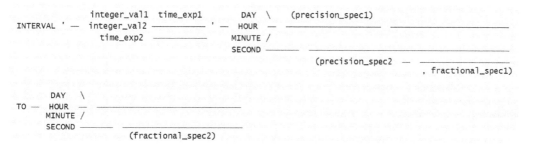

Figure 5.5 – INTERVAL DAY TO SECOND (source: docs.oracle.com)

Table 5.3 shows examples of `INTERVAL DAY TO SECOND`. The left column provides the definition, while the right provides the interpretation:

Form of Interval Literal	Interpretation
`INTERVAL` `'4 5:12:10.222'   DAY TO SECOND(3)`	4 days, 5 hours, 12 minutes, 10 seconds, and 222 thousandths of a second.
`INTERVAL` `'4 5:12'          DAY TO MINUTE`	4 days, 5 hours, and 12 minutes.
`INTERVAL` `'400 5'      DAY(3) TO HOUR`	400 days and 5 hours.
`INTERVAL '400' DAY(3)`	400 days.
`INTERVAL` `'11:12:10.2222222' HOUR TO SECOND(7)`	11 hours, 12 minutes, and 10.2222222 seconds.
`INTERVAL '11:20' HOUR TO MINUTE`	11 hours and 20 minutes.
`INTERVAL '10' HOUR`	10 hours.
`INTERVAL '10:22' MINUTE TO SECOND`	10 minutes 22 seconds.
`INTERVAL '10' MINUTE`	10 minutes.
`INTERVAL '4' DAY`	4 days.
`INTERVAL '25' HOUR`	25 hours.
`INTERVAL '40' MINUTE`	40 minutes.
`INTERVAL '120' HOUR(3)`	120 hours.
`INTERVAL '30.12345' SECOND(2,4)`	30.1235 seconds. The fractional second 12345 is rounded to 1235 because the precision is 4.

Table 5.3 – INTERVAL DAY TO SECOND

As stated, these intervals can be used directly as a separate structure, or these values can be associated with any time point (`DATE` or `TIMESTAMP` data values) modeling positional duration. To be more specific, let's execute the following queries. The first two statements provide separate interval management:

```
select INTERVAL '1 2:30:00' DAY TO SECOND from dual;
--> +01 02:30:00.000000
```

```
select INTERVAL '2-10' YEAR TO MONTH from dual;
--> +02-10
```

By applying the INTERVAL value to the DATE value, the particular shift is done. For the specified value, 1 day, 2 hours, and 30 minutes are added:

```
select sysdate,
       sysdate + INTERVAL '1 2:30:00' DAY TO SECOND
  from dual;
--> 25.03.2022 14:06:07        26.03.2022 16:36:07
```

Note that if the interval contains second fractions, they will be truncated because the DATE data type is used:

```
select sysdate,
       sysdate + INTERVAL '1 2:30:00.634' DAY TO SECOND
  from dual;
--> 25.03.2022 14:06:07        26.03.2022 16:36:07
```

The same principles would be applied for INTERVAL YEAR TO MONTH as well, even taking into account leap years:

```
select TO_DATE('1.2.2021', 'DD.MM.YYYY')
       + INTERVAL '28 5' DAY TO HOUR
  from dual;
--> 01.03.2021 05:00:00
select TO_DATE('1.2.2020', 'DD.MM.YYYY')
       + INTERVAL '28 5' DAY TO HOUR
  from dual;
--> 29.02.2020 05:00:00
```

Note that the INTERVAL data type does not need to cover all elements. If the minute and second values are not to be specified, they can be directly excluded from the format. However, this only applies if all smaller-granularity values are not specified. Thus, if you want to specify hour and second values for the interval, minute elements should be stated as well (such as zero, if not defined).

Multiple intervals *with the same type* of definition can be merged by applying mathematical operations, as follows:

```
select INTERVAL '2-10' YEAR TO MONTH
       + INTERVAL '3-3' YEAR TO MONTH
```

```
    from dual;
--> +06-01
```

The provided value can, generally, be negative as well, resulting in -00-05 (based on the following query result). It expresses 5 months being subtracted from the original value. It also merges two interval data types with the same representation, using a MINUS operation:

```
select INTERVAL '2-10' YEAR TO MONTH
          - INTERVAL '3-3' YEAR TO MONTH
   from dual;
```

Since the values to be subtracted have the same data type, this will also be used as the result. Thus, the preceding select statement produces the INTERVAL YEAR TO MONTH data type:

```
--> Name Null? Type
--> ---- ----- -------------------------
--> X            INTERVAL YEAR(9) TO MONTH
```

With this, we can conclude that it is impossible to merge different types of intervals. An exception occurs when attempting the mathematical processing of various models. Thus, merging both interval types cannot be done directly:

```
select INTERVAL '1 2:30:00' DAY TO SECOND
          + INTERVAL '2-10' YEAR TO MONTH
  from dual;
--> ORA-30081:
--> "invalid data type for datetime/interval arithmetic"
```

The only possible workaround is to apply both interval types to the DATE or TIMESTAMP element. Then, the resulting data type depends on the origin – DATE or TIMESTAMP – and the definition is not directly affected by the interval:

```
select sysdate, sysdate
                    + INTERVAL '1 2:30:00' DAY TO SECOND
                    + INTERVAL '2-10' YEAR TO MONTH
  from dual;
--> 25.03.2022 14:27:46        26.01.2025 16:57:46
```

Let's check the data type for the DATE value extension by INTERVAL DAY TO SECOND, as well as INTERVAL YEAR TO MONTH. Create a new table, as stated:

```
drop table tab_interval;
create table tab_interval
```

```
      as select sysdate + INTERVAL '1 2:30:00' DAY TO SECOND
                       + INTERVAL '2-10' YEAR TO MONTH x
          from dual;
```

By getting the table structure definition using the DESC command, it is evident that it produces the DATE data type:

```
desc tab_interval;
--> Name Null? Type
--> ---- ----- ----
--> X           DATE
```

If the interval is associated with the TIMESTAMP value, then the output data type of the processing is TIMESTAMP as well. Note that these data types take care of the second fractions:

```
select localtimestamp,
       localtimestamp
          + INTERVAL '4 5:12:10.222' DAY TO SECOND(3)
   from dual;
--> 25.03.2021 14:39:15,524000000
--> 29.03.2021 19:51:25,746000000
```

DATE arithmetic uses day granularity for the processing. In this section, we have been dealing with the INTERVAL data type, which is the output of **TIMESTAMP arithmetic**.

TIMESTAMP arithmetic

The TIMESTAMP data type and related arithmetic work differently from how it may seem at first glance. Let's take the TIMESTAMP value and add 1 day, represented by adding the value 1 to the particular value. However, what about the output? Will it even work? The following statement provides you with the answer:

```
select localtimestamp, localtimestamp + 1 from dual;
```

In addition, the following table shows the results:

LOCALTIMESTAMP	LOCALTIMESTAMP + 1
13.06.22 07:08:49,847000000	14.06.2022 07:08:49

Table 5.4 – Timestamp arithmetic

Yes, it took the `localtimestamp` value and added 1 day. The result is, however, a bit strange, isn't it? Namely, the second fractions are lost. What has happened? Well, particular addition and subtraction cannot be applied to the `TIMESTAMP` value. To prevent an exception from being raised, Oracle made an implicit conversion to the `DATE` data type, resulting in truncating the second fraction. Namely, the output of this arithmetic is not the `TIMESTAMP` value, but `DATE`. Please, be aware of that. In the following code snippet, a new table is created. Its structure is delimited by the associated `select` statement, so the data types are chosen by the system:

```
create table Tab as
   select localtimestamp as val1,
          localtimestamp + 1 val2
      from dual;
```

A pure mathematical operation done on the `TIMESTAMP` value forces the system to apply an implicit conversion to the `DATE` value, whereas it would be impossible to do that directly. The value 1 expresses *1 day*. The following code snippet gets the created table structure. Note that the `localtimestamp` function produces the `TIMESTAMP` value, but based on the mathematical operation, the particular value is transformed into the `DATE` value implicitly:

```
desc Tab;
--> Name Null? Type
--> ---- ----- -----------
--> VAL1        TIMESTAMP(6)
--> VAL2        DATE
```

To limit the implicit conversion (ensuring the output remains as the `TIMESTAMP` data type), mathematical operations cannot be used. Instead, another principle must be applied. Namely, `INTERVAL` has to be used to shift the value. To add 1 day, a solution is shown in the following snippet. The table consists of just one attribute holding a `TIMESTAMP` value:

```
create table Tab2(val timestamp);
insert into Tab2 values(systimestamp + INTERVAL '1' DAY);
```

Generally, any element can be manipulated:

```
insert into Tab2 values(systimestamp
                        + INTERVAL '10' MONTH);
insert into Tab2 values(systimestamp
                        + INTERVAL '10' YEAR);
insert into Tab2 values(systimestamp
                        - INTERVAL '10' MINUTE);
```

Moreover, multiple elements can be covered by one interval. The INTERVAL value can combine both types (INTERVAL DAY TO SECOND, as well as INTERVAL YEAR TO MONTH):

```
insert into Tab2 values(
  systimestamp - INTERVAL '4 5:12:10.222' DAY TO SECOND(3));
```

The calculation of the difference between the two TIMESTAMP values produces a result with the INTERVAL DAY(9) TO SECOND(6) data type, as seen in the following code snippet. It is a universal solution that can generally express any duration:

```
create table Tab3
  as select systimestamp - localtimestamp as result_val
       from Tab2;
desc Tab3;
--> Name            Null? Type
--> ----------- ----- ---------------------------
--> RESULT_VAL          INTERVAL DAY(9) TO SECOND(6)
```

However, what if the result of the difference is multiple years? For example, what about the result of the following code snippet? It should get 18 years, 5 months, and 7 days. How is it expressed?

```
select TO_TIMESTAMP('17.6.2018', 'DD.MM.YYYY')
       - TO_TIMESTAMP('10.1.2000', 'DD.MM.YYYY')
  from dual;
```

As stated, the result is INTERVAL DAY TO SECOND, thus the number of years and months is transformed and recalculated to the day granularity. The result is 6733 days expressed in this format: +6733 00:00:00.000000.

To conclude, DATE value management uses day granularity, while the TIMESTAMP arithmetic provides INTERVAL as an output of the difference between two TIMESTAMP values.

Summary

In this chapter, you learned about the date and time data types supported by RDBMS Oracle, focusing on the differences between the ANSI standard and practical usage. Unlike other database systems, such as MySQL, Oracle does not support time management in separate data types. Additionally, year, month, and day elements cannot be treated without the time elements (hour, minute and second). DATE and TIMESTAMP values can be created by the ANSI constructors focusing on the specific format. It takes the finer element granularities in the right part of the definition. Thus, the year element is preceded by the month, which is preceded by the day reference. Another solution is defined by the explicit constructor function from the string, referenced by the TO_DATE and TO_TIMESTAMP functions.

To ensure the reliability and correctness of the results, a discussion related to implicit conversions was presented. Storing date or time elements as a string or integer does not provide sufficient power. You must always ensure that a value can be transformed into the Oracle DATE data type. Moreover, you gained knowledge about the implicit conversions between DATE and TIMESTAMP values, with explanations of the arithmetic.

Finally, you were given a summary of the INTERVAL data type, expressing duration. This concept can deal with year-to-month or day-to-second precision by proposing a universal solution applicable to any granularity level.

The next chapter dives deeper into functions by emphasizing general conversion functions. The main emphasis is on element value extraction and reliability issues. Besides, you will also go through the differences between how TIMESTAMP is used in SQL and PL/SQL, concerning processed precision.

Questions

1. What is the default precision for TIMESTAMP second fractions?

    A. 0 decimal places

    B. 3 decimal places

    C. 6 decimal places

    D. 9 decimal places

2. What is the finest precision for the DATE value definition?

    A. Day

    B. Second

    C. Microsecond

    D. Nanosecond

3. Which data type can hold the value 2022-06-13 6:46:12.576000 GMT?

    A. DATE

    B. TIMESTAMP

    C. TIMESTAMP WITH TIME ZONE

    D. TIMESTAMP WITH LOCAL TIME ZONE

4.   What is the right syntax for the `ANSI TIMESTAMP` constructor?

   A.   `TIMESTAMP 'YYYY-MM-DD HH24:MI:SS.FF'`

   B.   `DATE 'YYYY-MM-DD HH24:MI:SS.FF'`

   C.   `TIMESTAMP 'YYYY-MM-DD'`

   D.   `TIMESTAMP 'HH24:MI:SS.FF YYYY-MM-DD'`

5.   Which function produces a local client `TIMESTAMP` value?

   A.   `sysdate`

   B.   `systimestamp`

   C.   `current_date`

   D.   `localtimestamp`

6.   What is the output of the subtraction of two `DATE` values?

   A.   Two `DATE` values cannot be subtracted

   B.   Number of seconds elapsed

   C.   Number of hours elapsed

   D.   Number of days elapsed

7.   What is the data type of the output of the subtraction of two `TIMESTAMP` values?

   A.   `INTERVAL DAY TO SECOND`

   B.   `INTERVAL YEAR TO MONTH`

   C.   `NUMBER`

   D.   Two `TIMESTAMP` values cannot be subtracted

Further reading

- *Managing Time in Relational Databases: How to Design, Update and Query Temporal Data* by Tom Johnston and Randall Weis. It provides a complex practical guide to the data modeling and query management of date and time-delimited tuples.

- *Time and Relational Theory: Temporal Databases in the Relational Model and SQL* by C. J. Date, Hugh Darwen, and Nikos A. Lorentzos. This book discusses the SQL:2011 standard focusing on temporal modeling and temporal management in the SQL language. It also focuses on temporal references and proper state coverage.

6
Conversion Functions and Element Extraction

So far, we have focused on the methods of DATE and TIMESTAMP value construction, delimited by the **American National Standards Institute (ANSI)** standardization and by using the TO_DATE and TO_TIMESTAMP functions. TO_DATE and TO_TIMESTAMP functions can also be used robustly for conversions and element extraction. This chapter summarizes the principles of these methods by focusing on the parameters and mapping formats. It also explains the extraction functions.

First, the TO_CHAR function is described. It produces a character string as a result, commonly defined by the specified format mapping. This function should be used, although the Oracle Database offers implicit conversions of the data types. Controlled transformation is always preferred, while implicit conversions can have a significant impact on performance. Second, the century reference is discussed. Namely, if the value of the year takes only two digits, the system must infer the missing values to provide the correct century reference. This chapter highlights the rules to do so. Third, the TIMESTAMP precision is discussed. The management and behavior of this function differ across the SQL and PL/SQL languages by the fraction precision of the *second* element. Fourth, the EXTRACT function is introduced. It is limited to returning just one element, which can produce values such as day, month, or year. It is impossible to refer to multiple elements with one EXTRACT function call.

Finally, reliability and integrity issues are discussed. They are based on the fact that visual representations of date and time values are in string format. As a result, some developers access and extract elements as strings. However, the format depends on the server or session parameters and can differ across platforms and servers. Therefore, the approach of treating date values as character strings is not relevant. Moreover, general implicit conversions can cause improper management, even if the CAST function is used for the transformation.

In a nutshell, this chapter will cover the following main topics:

- Understanding the TO_DATE, TO_TIMESTAMP, and TO_CHAR conversion functions
- Understanding the EXTRACT function

- Reliability and integrity issues related to Date and Time values being processed as character strings

- Converting values using the CAST function

The source code for this chapter can be found in the GitHub repository accessible via this link: `https://github.com/PacktPublishing/Developing-Robust-Date-and-Time-Oriented-Applications-in-Oracle-Cloud/tree/main/chapter%2006`. Alternately, you may scan the following QR code to access the repo:

Understanding the TO_CHAR and TO_DATE conversion functions

Each Date or Time value consists of individual elements to be handled. The format must be explicitly specified during the value construction, or the ANSI norm must be applied to ensure correct mapping and evaluation. Date and Time values can be composed of either character strings or numerals, depending on the context. Similarly, getting individual elements from a particular value can be done using a numerical or textual representation. The TO_CHAR and TO_DATE functions ensure the core functionality of obtaining elements and transforming values. As evident from the name of the function itself, the transformation from a string or number to a date is performed by the TO_DATE function. To obtain individual elements from the date, the TO_CHAR function is generally used. These functions behave similarly. Using them in a common structure, the value itself is defined in the first parameter. The second parameter determines the structure and format for mapping. In the previous chapter, we defined the principles of date composition from the individual numerical values of the elements, typically separated by a dot, dash, slash, or colon for the time spectrum. In this chapter, we will extend the principles to cover other elements, in both the numerical and string formats. However, several aspects affect the output, such as the system language, the format used, and the calendar.

Let's describe the available functions step by step, emphasizing the format that can be used for the representation. The following code shows the syntax:

```
TO_CHAR (<value> [, <format> [, <nls_parameter>]])
TO_DATE (<value> [, <format> [, <nls_parameter>]])
```

The first parameter (`value`) deals with the value to be transformed. The second parameter (`format`) deals with the format specifying the meaning and mapping of individual elements. A third parameter (`nls_parameter`) is also available, determining the language representation set for the particular function. Its scope is not the session but just the function itself influencing the output string. It will be described with a practical example in *Chapter 8*, in the *Embedding NLS parameters in the TO_CHAR function* section.

Tables 6.1, *6.2*, and *6.3* show lists of the available formats for date management. The first column of the table specifies the element and the second column provides an explanation. For numerical element processing, the year, month, week, and day, as well as the quarter, can be referenced. When manipulating the textual format, the year, month, and day can be expressed as a character string. *Table 6.1* explains the year and month elements:

Format	Explanation
Y, YY, YYY, YYYY	Year (the number of Y characters represents the number of digits); for example, 22 or 2022. Note that the year is commonly specified by two or four digits, but generally, Oracle allows you to use just one digit producing last numeral (Y), or the first digit of the year can be removed and triple Y used (YYY). If the year element value to be processed is 2022, then Y gives the value as 2, YY gives the value as 22, YYY gives 022, and YYYY gives 2022.
YEAR	Year in a textual format, for example, 2022.
MM	The number of the month of the year, from 01 to 12, for example, 01 for January, 02 for February, and so on.
RM	Number of the month of the year in Roman numeral format, from I to XII, for example, I for January, II for February, and so on.
MON	Abbreviation of the month (first three characters) in a textual format, for example, JAN for January, FEB for February, and so on.
MONTH	Month in a textual format, for example, JANUARY, FEBRUARY, and so on.

Table 6.1 – Format definition – year and month

The week element can be presented by relating it to the month or year reference, denoted by W or WW. The ISO standard reference is expressed by IW. The following table explains the week elements:

Format	Explanation
W	Week of the month. The first week starts on the first day of the month and ends on the seventh.
WW	Week of the year: 01-53. The first week starts on the first day of the year and ends on the seventh.
IW	Week of the year based on ISO standards: 1-52 and 1-53.

Table 6.2 – Format definition – week

The following table explains the day elements, with options for day, month, or year granularity:

Format	Explanation
D, DD, DDD	D refers to day of the week: 1-7. DD refers to day of the month: 1-31. DDD refers to day of the year: 1-366.
J	Julian day – the number of days since 1st January, 4712 BC
DY	Abbreviation of the day (first three characters) in textual format, for example, MON for Monday
DAY	Name of the day in a textual format, such as MONDAY

Table 6.3 – Format definition – day

The preceding tables deal just with the date elements; time elements are covered later, in *Table 6.4*.

Although the value is visually represented by the numerical value, physically, it provides the character string as well. As the name of the TO_CHAR function suggests, this function enables a transformation to a character string value:

```
drop table tab;
create table tab
        as (select TO_CHAR(sysdate, 'MM') x from dual);
desc tab;
--> Name Null? Type
--> ---- ----- -----------
--> X           VARCHAR2(2)
```

However, be aware that when dealing with complex data, a character string can be implicitly converted to a number on the fly. Take a look at the execution plan in *Figure 6.1* and *Figure 6.2* and focus on the implicit conversions. Namely, *Figure 6.1* denotes the implicit conversion from a character string to numerical format. *Figure 6.2* marks the year value at the right part of the equation as a string using single quote marks:

```
select * from employee
   where TO_CHAR(date_from, 'YYYY')= 2020;
```

In the preceding code snippet, the left part of the condition specifies a character string (provided by the TO_CHAR function), while the right part refers to a numerical value. Thus, a conversion must be done. The execution plan of the statement is shown in the following figure. It is evident that the character string was transformed into a numerical representation by using the TO_NUMBER function implicitly:

OPERATION	OBJECT_NAME	OPTIONS	CARDINALITY	COST
⊟ ● SELECT STATEMENT			1	3
⊟ ▦ TABLE ACCESS EMPLOYEE		FULL	1	3
⊟ ◐ Filter Predicates				
TO_NUMBER(TO_CHAR(INTERNAL_FUNCTION(DATE_FROM),'YYYY'))=2020				

Figure 6.1 – Implicit conversion – execution plan

If the particular value to be compared is denoted as a string – defined in single quote marks – then the evaluation can be done directly, and no implicit conversion is present. Note that implicit conversion can be used by the system dynamically. However, it can negatively impact the performance by refusing the defined index usage.

The following code snippet shows the use of the where clause. The left part of the condition produces a character string, and the right part is a character string as well because the numerical representation is enclosed by single quote marks:

```
select * from employee
   where to_char(date_from, 'YYYY')= '2020';
```

The following figure shows the execution plan of the preceding statement. The data types for both sides of the condition are the same (character strings), so no implicit conversion is necessary:

OPERATION	OBJECT_NAME	OPTIONS	CARDINALITY	COST
⊟ ● SELECT STATEMENT			1	3
⊟ ▦ TABLE ACCESS EMPLOYEE		FULL	1	3
⊟ ◐ Filter Predicates				
TO_CHAR(INTERNAL_FUNCTION(DATE_FROM),'YYYY')='2020'				

Figure 6.2 – Explicit data type management – execution plan

One statement can serve multiple date and time management function calls. The following statement gets the `sysdate` value and extracts the day of the week, the day number for the month, as well as the sequential number of the day inside the whole year. The D format represents the number of the day of the week, DD represents the numerical order referencing the whole month, and finally, DDD denotes the serial number of the day within the year:

```
select sysdate, TO_CHAR(sysdate, 'D'),
                TO_CHAR(sysdate, 'DD'),
                TO_CHAR(sysdate, 'DDD')
   from dual;
--> SYSDATE                     D    DD    DDD
--> 25.03.2022 07:54:31         6    25    084
```

So, breaking down the value in the previous code snippet, it is referring to the 25th day of the month, the 84th day of the year, and the 6th day of the week. Looking this day up on the 2022 calendar reveals it's a Friday. It is, however, important to emphasize the representation of the week. Numbering starts from a value of 1 by default. In Europe, the first day of the week is Monday, but in other regions, Sunday can be considered the week's starting day. The behavior and value representation depends on the parameters set on the client side or inherited from the system and must be covered properly to ensure correctness.

Multiple elements can be processed by one function call. In that case, the output elements should be separated by one of the approved delimiters – space, dot, comma, dash, colon, semicolon, or slash:

```
select TO_CHAR(sysdate, 'DD.MM.YYYY') from dual;
```

Because the delimiter is part of the output, naturally, it is processed as a character string.

The textual format can be obtained for each date element. The language used is inherited from the session or system parameter. The third (optional) parameter can change the behavior:

```
select sysdate, TO_CHAR(sysdate, 'DAY'),
                TO_CHAR(sysdate, 'MONTH'),
                TO_CHAR(sysdate, 'YEAR')
   from dual;
--> SYSDATE                     DAY       MONTH     YEAR
--> 25.03.2022 08:02:39         FRIDAY    MARCH     TWENTY TWENTY-TWO
```

The output format depends on the user-specified case (lowercase or uppercase):

```
select sysdate, TO_CHAR(sysdate, 'MONTH'),
                TO_CHAR(sysdate, 'month')
```

```
   from dual;
--> SYSDATE                     MONTH           month
--> 25.03.2022 08:03:57         MARCH           march
```

If a mixed case is used in the format definition, the first one or two letters are taken into consideration. This principle is visible in the following result set:

```
select sysdate, TO_CHAR(sysdate, 'mONTH'),
                TO_CHAR(sysdate, 'MoNTh'),
                TO_CHAR(sysdate, 'MONth')
   from dual;
--> SYSDATE                     mONTH      MoNTh      MONth
--> 25.03.2022 08:04:43         march      March      MARCH
```

The YYYY format denotes the entire year value, and YY denotes the century reference, providing just two digits:

```
select sysdate, TO_CHAR(sysdate, 'YYYY'),
                TO_CHAR(sysdate, 'YY')
   from dual;
--> SYSDATE                     YYYY        YY
--> 25.03.2022 08:07:15         2022        22
```

Instead of using YYYY and YY formats, the R symbol can be used, specifically RRRR or RR.

Note that a date-to-string conversion can add additional spaces to the result set. Even if you specified the spaces precisely, some extra ones may be added. This issue is shown in the following example. There is a specification of the day, followed by a comma and individual date elements (DD.MONTH.YYYY). However, by looking at the resulting value, after the day and month specification (in textual format), we can see that additional spaces are present. Thus, the mapping is not exactly as specified in the second parameter of the TO_CHAR function call:

```
select TO_CHAR(sysdate, 'DAY, DD.MONTH.YYYY') from dual;
--> FRIDAY    , 25.MARCH     .2022
```

Reflecting the provided output, additional spaces can be disturbing. Although the additional spaces can be solved by the replace operation or regular expressions to limit them, they often must be hardcoded. The FM keyword can be added to the format definition to strictly comply with the defined format, forcing the system to accept the prescribed format:

```
select TO_CHAR(sysdate, 'FM DAY, DD.MONTH.YYYY')
   from dual;
```

Spacing can then be directly defined in the format parameter.

Oracle DATE data types always consist of the time dimension, delimited by the hour, minute, and second elements. *Table 6.4* shows the format elements for time precision. An hour can be expressed in 12- or 24-hour format (HH and HH24, respectively). The MI element defines minutes. The second element can be defined with SS (for minute association) or SSSSS (for a whole-day reference – the number of seconds elapsed since midnight). For TIMESTAMP data values as well, seconds fractions can be referred to:

Format	Explanation
HH	Hour of the day in a 12-hour format: 1-12
HH12	Hour of the day in a 12-hour format: 1-12
HH24	Hour of the day in a 24-hour format: 0-24
MI	Minutes: 0-59
SS	Seconds: 0-59
SSSSS	Seconds since midnight: 0-86399
FF	Seconds fraction (this can only be applied to TIMESTAMP values)

Table 6.4 – Element format for the time precision

The following code snippet shows the difference between the 12- and 24-hour formats. HH represents the 12-hour format, while HH24 represents the 24-hour format:

```
select TO_CHAR(sysdate, 'HH:MI:SS'),
       TO_CHAR(sysdate, 'HH24:MI:SS')
  from dual;
--> 02:34:36    14:34:36
```

For the time and year components, the output format can be extended by the indicators listed in the following table:

Format	Explanation
AM	Meridian indicator
A.M.	
PM	
P.M.	

Format	Explanation
AD	AD indicator
A.D.	
BC	BC indicator
B.C.	

Table 6.5 – Element format indicators

The principles of the TO_DATE function call have already been covered in *Chapter 5*. Case sensitivity is not significant when using the textual format for individual elements. The database system automatically applies the rule:

```
select TO_DATE('15.October.2020', 'dd.MONTH.yyyy')
   from dual;
select TO_DATE('15.October.2020', 'dd.month.yyyy')
   from dual;
```

Note that an exception is raised if any input element value cannot be transformed to the correct DATE value. In the following case, the language is set to English. However, the provided month value is in French. Thus, the system does not recognize the value and raises an exception: not a valid month. As a result, the date value representation must be highlighted (individual parameters, their meanings, and impacts are discussed in *Chapter 8*):

```
select TO_DATE('15.FÉVRIER.2020', 'DD.MONTH.YYYY')
   from dual;
--> ORA-01843: not a valid month
```

Thus, it must be possible to evaluate the provided language of the input correctly – it must match the set session or system language. In the previous case, I used the French language convention. However, the server format uses an English set. Overriding such an option can be done at either the system, session, or statement level using the third parameter of the TO_DATE function.

The same exception is raised if the month value is out of the numeric value range of 1-12. Leading 0 values can be omitted:

```
select TO_DATE('15.13.2020', 'DD.MM.YYYY') from dual;
--> ORA-01843: not a valid month
```

The textual format is commonly applied to the month, as well as the year. Naturally, it does not make sense to provide a day in textual format, such as SATURDAY, as this value cannot be mapped to DATE value without additional information.

Although it is recommended to have the same delimiters in the input value and the format specification (which also brings clarity), Oracle is not so strict, and delimiters can differ, even in terms of spacing. Both of the following solutions will work well:

```
select TO_DATE('15.10.2020', 'DD/MM/YYYY')
  from dual;
select TO_DATE('15.    10   .   2020', 'DD/MM/YYYY')
  from dual;
```

So far, we have been dealing with the precise mapping and exceptions raised during conversion. We have focused on the available formats for the mapping, as well as conversion details. In the next section, dynamic exception handlers are discussed by introducing exact and flexible mapping formats.

Working with flexible format mapping

What will happen if the defined input string does not match the format (mapping)? What value will be produced? Or will an exception be raised? Let's consider the following example. The input value is defined by the day value followed by the textual representation of the month value and year. Even though the input value does not precisely fit the format mapping, conversion is done successfully:

```
select to_date('10.september 2022', 'DD.MM.YYYY')
  from dual;
--> 10.09.2022 00:00:00
```

Notice that even though the punctuation is not the same, the conversion can be done. However, take a look at the following example. In this case, the day value exceeds the total number of days in the month, resulting in raising an exception:

```
select to_date('30.february 2022', 'DD.MM.YYYY')
  from dual;
--> ORA-01858: a non-numeric character was found where a
numeric was expected
```

In a business environment, it is always necessary to handle and process exceptions. In the preceding example, it would be necessary to encapsulate the conversion function in a PL/SQL method and process the exception there.

The TO_DATE function can be extended by the `conversion error` clause, which applies a default value instead of raising an exception. Thus, if the conversion cannot be done, `default_value` is provided. The extended syntax of the TO_DATE function is shown in the following code block:

```
TO_CHAR (<value>
         [default <default_value> on conversion error]
         [, <format> [, <nls_parameter>]]
```

In the next example, the function attempts to process 30th February, which does not exist. It results in getting a NULL value:

```
select to_date('30.february 2022'
                default null on conversion error,
            'DD.MM.YYYY')
  from dual;
--> NULL
```

Generally, any valid string value can be used as a default option:

```
select to_date('30.february 2022'
                default '01.01.0001' on conversion error,
            'DD.MM.YYYY')
  from dual;
--> 01.01.0001
```

Besides, **flexible** or **strict format mapping** can be taken into account. Generally, the MM value can be replaced by the textual format (MON or MONTH) and conversion succeeds (under the assumption that the session language and values referred to correspond). The following code block illustrates the conversion opportunities:

```
select to_date('10.september 2022', 'DD.MM.YYYY')
  from dual;
--> 10.09.2022 00:00:00
select to_date('10.sep 2022', 'DD.MM.YYYY')
  from dual;
--> 10.09.2022 00:00:00
```

To conclude, the rule of available mapping refers to two principles:

- Any *punctuation* and any *separators* can be used
- The MM value can be replaced by the MON or MONTH value

Note, please, that the opposite conversion will not generally work. For example, if a textual format is expected and a numerical value is produced, a conversion cannot be done, resulting in raising the ORA-01843: not a valid month exception:

```
select to_date('12.05.2000', 'DD-MON-YYYY')
  from dual;
```

The FX value, part of the format mapping, forces the system to use exact conversion. In that case, therefore, the formats must be strictly the same. All of the following select statements will fail, raising the ORA-01861: literal does not match format string exception:

```
select to_date('10.september 2022', 'FX DD.MM.YYYY')
  from dual;
select to_date('31.dec 2022', 'FX DD.MM.YYYY')
  from dual;
select to_date('31.12.2022', 'FX DD.MON.YYYY')
  from dual;
```

The TO_DATE and TO_CHAR functions convert the input value from or to the string value. The mapping format is critical to ensure the possibility of conversion. The TO_DATE and TO_CHAR functions are used to manipulate DATE data type values. The next section deals with the finer precision by constructing TIMESTAMP values.

Constructing the TIMESTAMP value

The TIMESTAMP value is created by adding seconds fraction management, operated by TO_TIMESTAMP or TO_CHAR. The syntax of TO_TIMESTAMP is similar to TO_DATE. Additionally, seconds fractions and time zones can be dealt by this function. The syntax is shown in the following figure, in which input_char (the first parameter) is mandatory and can be extended by format (the second parameter) and the NLS parameter, NLS_TIMESTAMP_FORMAT_val (the third parameter):

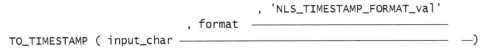

Figure 6.3 – The TO_TIMESTAMP conversion function syntax

The following code snippet shows the construction function getting the TIMESTAMP value, followed by the processing in the opposite direction – that is, constructing a character string format from the TIMESTAMP value. For processing and evaluation, it is always recommended to use the format mapping (the second parameter value) and not to rely on the implicit conversions covered by the set format:

```
select TO_TIMESTAMP('15.12.2021 15:24:14:535',
                    'DD.MM.YYYY HH24:MI:SS:FF')
   from dual;
--> 15.12.21 15:24:14,535000000
 select TO_CHAR(systimestamp, 'DD.MM.YYYY HH24:MI:SS:FF')
   from dual;
--> 23.02.2022 08:23:33:174000
```

Even in this case, the delimiters do not need to correlate with the format definition, but this makes it less readable from the user's perspective:

```
select TO_TIMESTAMP('15/12/2021 15.24.14.535',
                    'DD.MM.YYYY HH24:MI:SS:FF')
   from dual;
```

We have been dealing with the year definition, referenced by the RR, RRRR, YY, or YYYY format. As described in the next section, the shortened definition depends on the value itself, whereas the system must set the proper century reference.

Conversion functions – century reference

When dealing with the Date value by using the conversion functions, either the year can be fully represented by four digits or just two digits can be specified. Then, the system must apply defined rules to compose a full value referencing the century. The format for the numerical value of the year can be either YYYY or RRRR. These representations and meanings are the same. However, a significant difference can be identified when dealing with only two digits. The RR and YY values hold just two digits, and the century is omitted. Thus, in principle, the system must differentiate between the 20th and the current (21st) century. If the specified format is YY, then the value always references the 21st century, irrespective of the specified numerical value for the year.

However, when using the RR format, the representation depends on the provided value. Namely, if it fits the range *<0; 49>*, then the reference covers the current (21st) century. On the other hand, if the value is part of the range *<50; 99>*, then the 20th century is referenced. Thus, it influences the process of the value composition to be evaluated and stored. The data retrieval process in independent and RR or YY format models produce the same results.

```
create table Tab1(val DATE);
insert into Tab1 values(TO_DATE('1-1-99', 'DD-MM-RR'));
select TO_CHAR(val, 'DD.MM.YYYY') from Tab1;
select TO_CHAR(val, 'DD.MM.RRRR') from Tab1;
--> 01.01.1999
delete from Tab1;
insert into Tab1 values (TO_DATE('1-1-99', 'DD-MM-YY'));
select TO_CHAR(val, 'DD.MM.YYYY') from Tab1;
select TO_CHAR(val, 'DD.MM.RRRR') from Tab1;
--> 01.01.2099
```

Now, we know how to format, construct, and get the element values for the DATE and TIMESTAMP data types. The next section shows you how the approach for TIMESTAMP differs between the SQL and PL/SQL languages and how the default precision is maintained differently. Be aware of that to avoid raising an exception.

TIMESTAMP precision in SQL and PL/SQL

The management of the TIMESTAMP value differs in SQL and PL/SQL. As stated, the precision can range from one to nine digits expressing the *seconds fragments* up to *nanoseconds*. By default, the fractional part consists of six digits regardless of the source system. However, it is possible to reference higher precision than the system can really provide. The fractional part can be obtained by the FF format of the TO_CHAR function. Thus, the following select statement (characterizing the SQL statement) provides six digits (as a default format):

```
select TO_CHAR(systimestamp,'FF') from dual;
--> 801000
```

However, in PL/SQL, it works differently, and the highest precision is returned, consisting of nine digits:

```
begin
  DBMS_OUTPUT.PUT_LINE(TO_CHAR(systimestamp,'FF'));
end;
/
--> 984000000
```

That behavior in a PL/SQL environment can cause problems with the interpretation and assignment. The following code snippet shows the principles and consequences. A local variable (v_fractions) is created and delimited by the VARCHAR(6) data type. If the TIMESTAMP value is constructed by the PL/SQL function, it provides nine digits instead of the assumed six digits, resulting in raising an exception:

```
declare
  v_fractions varchar(6);
begin
  v_fractions := TO_CHAR(systimestamp,'FF');
end;
/
--> ORA-06502: PL/SQL: numeric or value error: character string
buffer too small
```

However, if you encapsulate the assignment command in a select statement, then it is treated as SQL inside PL/SQL and it works properly:

```
declare
  v_fractions varchar(6);
begin
  select TO_CHAR(systimestamp,'FF') into v_fractions
    from dual;
end;
/
```

Even if the value is assigned to a variable (v_timestamp) with restricted seconds fragment precision (six digits), the finest precision range (nine digits) is still processed:

```
declare
  v_timestamp TIMESTAMP(6):=systimestamp;
  v_fractions varchar(6);
begin
  select TO_CHAR(v_timestamp,'FF') into v_fractions
    from dual;
end;
/
--> ORA-06502: PL/SQL: numeric or value error: character string
buffer too small
```

For now, it is a known Oracle database issue. The workaround is to encapsulate the assignment in the `select` statement (direct processing) or to get only a specific amount of characters during the conversion extraction, as shown in the following example:

```
declare
 v_fractions varchar(6);
begin
   v_fractions := SUBSTR(TO_CHAR(systimestamp,'FF'),
                         LENGTH(v_fractions));
end;
/
```

Note that the described limitation is present in any TIMESTAMP enhancement (TIMESTAMP, TIMESTAMP WITH TIME ZONE, or TIMESTAMP WITH LOCAL TIME ZONE) and reference (systimestamp or localtimestamp). The solution is, however, the same:

```
declare
 v_timestamp TIMESTAMP(6) WITH TIME ZONE;
 v_fractions varchar(6);
begin
   select systimestamp into v_timestamp
     from dual;
   select SUBSTR(TO_CHAR(v_timestamp,'FF'),
               length(v_fractions))
           into v_fractions
     from dual;
   DBMS_OUTPUT.PUT_LINE(v_fractions);
end;
/
```

Now, you are familiar with the TO_CHAR conversion function and the mapping and available formats. Besides the TO_CHAR function, it is also possible to use the EXTRACT function. It is, however, limited to produce only one element from the source value.

Understanding the EXTRACT function

The EXTRACT function of the Oracle DBS is used to access only one element from the DATE or TIMESTAMP value. It can extract date elements (day, month, and year) that both the DATE and TIMESTAMP values can apply to. Time elements (hour, minute, and second) and time zone extensions (timezone_hour, timezone_minute, timezone_region, and timezone_abbreviation) can be applied only to TIMESTAMP values. Even in the Oracle Cloud 21c release version, time elements cannot be obtained by the EXTRACT function for DATE. So, in this specific situation, DATE values cannot be implicitly converted to TIMESTAMP. The syntax is as follows:

```
EXTRACT <field> FROM <source>
```

Here, the field value of EXTRACT can be day, month, year, hour, minute, second, timezone_hour, timezone_minute, timezone_region, or timezone_abbreviation. The following select statements express the principles by taking day and hour elements from the provided values:

```
select EXTRACT(day from sysdate) from dual;
--> 23
select EXTRACT(hour from systimestamp) from dual;
--> 8
```

The source for the EXTRACT function is commonly a DATE or TIMESTAMP value. However, it can also deal with the INTERVAL data type. In that case, the available element list depends on the granularity used. The following table shows the source data types and available fields:

Source data type	Available fields
DATE	YEAR, MONTH, and DAY
INTERVAL YEAR TO MONTH	YEAR and MONTH
INTERVAL DAY TO SECOND	DAY, HOUR, MINUTE, and SECOND
TIMESTAMP	YEAR, MONTH, DAY, HOUR, MINUTE, and SECOND

Table 6.6 – The TO_TIMESTAMP conversion function syntax (source: docs.oracle.com)

A time zone reference is available for TIMESTAMP if the time zone specification is stated. Therefore, if such a component cannot be extracted, an exception is raised:

```
select EXTRACT(timezone_hour from sysdate) from dual;
--> ORA-30076: invalid extract field for extract source
select EXTRACT(timezone_hour from systimestamp) from dual;
--> OK. Provided value can be, for example 1.
select EXTRACT(timezone_hour from localtimestamp)
  from dual;
--> ORA-30076: invalid extract field for extract source
```

There is a single rule for the usage of the EXTRACT function – the EXTRACT function will need to be called separately for each element. Thus, from a code point of view, the TO_CHAR method is more appropriate and less code-demanding.

The TO_CHAR and EXTRACT functions provide relevant solutions for getting particular element values. The next section deals with the technique in which the DATE value is treated as a character string and thus the SUBSTR function is used to obtain element values. However, as we will show, such a solution is not safe, robust, or universal. By changing the date format parameters at the server or session level, the values obtained would be incorrect.

Reliability and integrity issues

It is important to comply with this rule: never rely on implicit conversions, and never treat Date and Time values as character strings by applying the SUBSTR function. By changing the session format, the provided input value does not need to fit the mapping. Imagine a system migration to another server or the cloud. The original formats and assumed parameter values would not be retained. It can result in obtaining an improper value (such as changing the value of month and day, if applicable) or even raising an exception. Thus, it would be necessary to revise and rebuild the whole application! Avoid this situation before it occurs! It is enormously time- and money-consuming and requires the involvement of many developers. In addition, after the process, the whole system must be retested.

Thus, extraction using a substring is insufficient. If the session date format is changed, the extracted value does not need to be relevant. The developer cannot ensure that the server system format will always remain the same. Although the definition can be used on the session level, if it is changed by another application process, the system would collapse.

Similarly, later code enhancements in terms of the session parameter changing would stop the system from working. The following example expresses the problem of referencing the same code in two instances. As evident, the same code produces different outputs. The first system uses the DD.MM. YYYY HH24:MI:SS format, while the second system uses the YYYY/MM/DD HH24:MI:SS format:

```
--> SYSTEM 1
  select sysdate, SUBSTR(sysdate, 4,2) from dual;
  --> SYSDATE                SUBSTR(SYSDATE,4,2)
  --> 25.03.2022 08:40:26    03
--> SYSTEM 2
  select sysdate, SUBSTR(sysdate, 4,2) from dual;
  --> SYSDATE                SUBSTR(SYSDATE,4,2)
  --> 2022/03/25 08:40:55    2/
```

Conversion between the individual data types can also be done by using the CAST function. This can cause reliability issues too, where the format of the mapping is not stated. Thus, it can be said that an implicit conversion is done, but managed explicitly.

Investigating the CAST function

The CAST function gets the input value and converts it into the specified data type in the definition. The input value can be provided by the attribute, function result, or expression, or a NULL value can be used. A CAST function can be used where expressions are permitted, using the following syntax:

```
CAST ( {<expression> | NULL } as <output_datatype> )
```

In this section, we will explain the DATE and TIMESTAMP data types, which originated from the character string and numerical values. We will focus on the transformations across all data types. Finally, there will be a note related to the undefined value, NULL.

Casting a character string to a DATE value

An example of the usage of the CAST function is reflected in the following code snippet. In the CAST function, a character string is taken by providing the DATE value as the output. The first element is assumed to represent the day followed by the month and year specification. Thus, the following operation can be done:

```
select CAST('15/1/2000' as DATE) from dual;
--> 15.01.2000
```

As evident, there is no specific format or mapping definition. Thus, the management and element association depend on the session definition. By changing the format, implicit conversion cannot be done:

```
select CAST('2000/1/15' as DATE) from dual;
```

The CAST function uses implicit conversion and can cause reliability or even usability issues. Namely, if the session format specification is changed, or the value is not properly formatted, an exception is raised and the transformation fails:

```
--> ORA-01861: "literal does not match format string"
```

Simply, the specified format must always match the session definition.

Casting a numerical value to a DATE value

Similar to the preceding example, a numerical value can also be considered the source of the transformation. As evident, it is treated as a string (where single quote marks enclosing the value are mandatory). The order of the elements is critical to ensure proper mapping; otherwise, an exception is raised (you will learn more about conversion formats in *Chapter 8*):

```
select CAST('02032022' as date) from dual;
--> 02.03.2022 00:00:00
```

If the session parameter is changed or the transformation is done incorrectly, the processing fails:

```
select CAST('20220302' as date) from dual;
--> ORA-01861: "literal does not match format string"
```

As a result, using an input format that respects the session parameter is crucial! Be aware of that!

Casting DATE to TIMESTAMP

Another example can be covered by converting DATE to TIMESTAMP, or vice versa:

```
select sysdate, CAST(sysdate as TIMESTAMP) from dual;
--> DATE value             TIMESTAMP value provided by CAST
--> 02.03.2022 07:51:30    02.03.2022 07:51:30,000000000
```

The time zone reflection can be processed as well:

```
select CAST(sysdate AS TIMESTAMP WITH LOCAL TIME ZONE)
  from dual;
--> 02.03.22 08:35:24,000000000
```

Casting TIMESTAMP to DATE

DATE and TIMESTAMP values can be directly transformed. Here, the focus is on casting TIMESTAMP to the DATE format. Look at the following statement pointing out the seconds fraction. It is truncated during the transformation:

```
select systimestamp, CAST(systimestamp as DATE) from dual;
--> TIMESTAMP value: 02.03.2022 07:38:58,722000000 +01:00
--> DATE value provided by CAST: 02.03.2022 07:38:58
```

Casting a DATE value to character string format

In the case of converting a DATE value to the character string format, the output data type must be of the correct length. If not, the output value cannot fit the range and an exception will be raised:

```
select CAST(sysdate as char(20)) from dual;
--> 25.03.2022 07:58:52
select CAST(sysdate as varchar(30)) from dual;
--> 25.03.2022 07:58:52
select CAST(sysdate as char(1)) from dual;
--> ORA-25137: "Data value out of range"
select CAST(sysdate as varchar(3)) from dual;
--> ORA-01801:
--> "date format is too long for internal buffer"
```

Casting a TIMESTAMP value to character string format

The principles of TIMESTAMP casting to a character string are the same as DATE value management. However, for completeness, we will also show this process:

```
select CAST(systimestamp as char(100)) from dual;
--> 25.03.2022 08:17:13,874000 +01:00
select CAST(systimestamp as varchar(100)) from dual;
--> 25.03.2022 08:17:57,046000 +01:00
select CAST(systimestamp as char(1)) from dual;
--> ORA-25137: "Data value out of range"
select CAST(systimestamp as varchar(3)) from dual;
--> ORA-01801:
--> "date format is too long for internal buffer"
```

Casting NULL

As stated earlier, the NULL value can be converted to a DATE or TIMESTAMP data type value, as well. The output value is still NULL, but the data type is shifted:

```
select CAST(null as DATE) from dual;
select CAST(null as TIMESTAMP) from dual;
```

The CAST function is one of the options for transforming the input value to the required data type format. It takes the value followed by the required data type to be the output. Whereas there is no explicit rule for the transformation specified, processing relies on the session or database (server) format. Throughout this chapter, we have introduced the principles of character string and numerical value transformation; however, there is also a relevant discussion related to the DATE and TIMESTAMP value transformations emphasizing the precision frame. To cover the complexity, this section covered the processing of the NULL value as a source of the transformation

Several techniques for converting the value from the input to output data types have been introduced, either as constructor functions or conversion types. It is necessary to ensure the input value can be mapped to the resulting format; otherwise, an exception will be raised. The next section provides a method for validating conversions.

Validating conversions

The data to be stored in the database can originate from various sources. For example, by loading the values from CSV files, a textual representation is present, requiring the system to convert the data to the proper format and data type. Thus, before processing and storing values, it is always necessary to ensure the value can be mapped to the desired data type. Moreover, for the date and time values, the format mask (mapping) must also be applicable. To test the conversion opportunities, instead of hardcoding using PL/SQL and exception handling, the validate_conversion function, introduced by the Oracle database, can be used. It takes an expression and the output data type (output_data_type), optionally extended by the format mask (format_mapping) and NLS parameters (nls_parameters) for the date and time values. The syntax of the validate_ conversion function is shown in the following code block:

```
validate_conversion
  (
    <expression> as <output_data_type>,
    [ <format_mapping> ],
    [ <nls_parameters> ]
  )
```

It produces the value 1 if the conversion can be done; otherwise, it produces the value 0. The validating can be optionally enhanced by the format mapping or NLS parameters for the DATE values.

The first, second, and third statements will succeed, as flexible mapping is used. The fourth and fifth statements require strict mapping, thus they will fail, resulting in getting the value 0. Finally, the last statement fails because September has only 30 days. This can be seen in the following code block:

```
select validate_conversion('12-05-2000' as date,
                           'DD-MM-YYYY')
  from dual; --> 1
select validate_conversion('12.05.2000' as date,
                           'DD-MON-YYYY')
  from dual; --> 1
select validate_conversion('12-SEPTEMBER-2000' as date,
                           'DD-MON-YYYY')
  from dual; --> 1
select validate_conversion('12-SEPTEMBER-2000' as date,
                           'FXDD-MON-YYYY')
  from dual; --> 0
select validate_conversion('12-SEPTEMBER-2000' as date,
                           'FXDD-MM-YYYY')
  from dual; --> 0
select validate_conversion('31-SEPTEMBER-2000' as date,
                           'DD-MM-YYYY')
  from dual; --> 0
```

Generally, the validate_conversion function can be applied to any data type. It checks whether the conversion can be done or not.

To conclude, this section aimed at validating conversions, which are commonly used if the data originates from various third-party systems and needs to be transformed into a specific data type by checking an assumption that the value format is valid and transformable. As shown, the validate_ conversion function can provide sufficient power by providing a Boolean value expressing the ability to convert the input value to the desired data type.

Summary

In this chapter, you learned about the TO_DATE, TO_TIMESTAMP, and TO_CHAR conversion functions, with the available formats being highlighted. They can reference any elements and formats, with precision ranging from the day to fractions of a second. Century reference rules were described, including when the year is specified by just two digits and the century assignment can be calculated by the system.

You saw how the opposite of the TO_CHAR function is covered by the supplementary EXTRACT function, which can take only one element.

Finally, you learned about the reliability issues if date and time values are treated as a character string. From the visual perspective, the date and time values seem to be represented textually, right? However, that does not mean we should treat these values as strings, because they strongly depend on the format, which can be changed on the client and server sides anytime. Moreover, by migrating the solution to another server, different rules can be applied to influence the output format. Thus, you learned that proper format mapping is crucial for providing general solutions that are applicable worldwide.

The next chapter deals with the available functions for date and time management and manipulating the day, month, and year shifts. It also covers the rounding and truncating principles of the Date and Time values.

Questions

1. Which format displays the number of the month in Roman numeral format?

    A. MM

    B. RM

    C. MR

    D. MON

2. Which parameter mapping value provides you with the name of the day in a textual format?

    A. DAY

    B. DY

    C. D

    D. DDD

3. Which parameter expresses the output of the hour extraction in 24-hour format?

    A. HH

    B. HH12

    C. HH24

    D. HHEurope

4. Which data type can be specified as the source of the EXTRACT function?

    A. DATE only

    B. TIMESTAMP only

    C. INTERVAL only

    D. DATE, TIMESTAMP, and INTERVAL

Further reading

-   *Datatype conversions – strange internal function* by Connor McDonald. This is a practical guide dealing with implicit conversions and related performance aspects. It can be found at `https://connor-mcdonald.com/2021/01/27/datatype-conversions-strange-internal-function/`:

-   *When Implicit Date Conversions Attack* by the Oracle-Base team.

-   This paper deals with the consequences of implicit date conversions and cascading. It can be found at `https://oracle-base.com/blog/2020/07/08/when-implicit-date-conversions-attack/`:

7
Date and Time Management Functions

Date and time arithmetic is not suitable for ensuring the complexity of element management and arithmetic shift. Specifically, although it is possible to add or remove a defined number of days (or hours) for a specific date, processing at a monthly level would be quite complicated. If we want to reference an event that takes place in a month, it is not enough to add 30 days, because some months have 31 days, and February only 28 or 29. Also, changing the value of the month element does not provide a relevant solution. For example, if it is January 31, adding one month to the month element would result in February 31, which does not exist. Moreover, transitioning through the years should be emphasized.

This chapter deals with the available functions related to date and time management. The first function is ADD_MONTHS, which accurately solves the described problems. A similar problem is then related to the number of months that have elapsed between two date values. The mathematical difference relates to the number of days that have elapsed; the month reference can be obtained by using the MONTHS_BETWEEN function.

First, you will learn how to get the next day based on the specified input value and weekday. You will find out that the existing functionality is not sufficient in terms of **NLS session parameters** defining the rules for date and time value specification and management. In response to that, we bring our own implementation. This chapter also deals with date value **rounding** and **truncating**. Additionally, you will get an overview of the specific PERSONAL_ID value used in Central Europe for unique person identification. It consists of a hash value composed of someone's date of birth and gender.

Finally, we will discuss random DATE value composition. As is evident, separate element management is not suitable and many rules must be applied to ensure correctness and integrity.

In a nutshell, we're going to cover the following main topics in this chapter:

- The `ADD_MONTHS` function, dealing with the month granularity
- Obtaining the last day of the defined month
- Evaluating the elapsed time using month granularity
- Getting the next day based on the input `DATE` value and referenced weekday
- Truncating and rounding the `DATE` value
- Identifying people using the `PERSONAL_ID` value
- Getting a random `DATE` value

The source code for this chapter can be found in this book's GitHub repository, which can be found at `https://github.com/PacktPublishing/Developing-Robust-Date-and-Time-Oriented-Applications-in-Oracle-Cloud/tree/main/chapter%2007`. You can also access the repository via the following QR code:

> **Important note**
> Throughout this chapter, two basic concepts will be used, which might not be obvious to you at first glance. The expression *arithmetic shift* refers to the addition of a specific time to the defined `DATE` value, such as adding 1 day, 3 months, or a quarter. In this chapter, we will introduce many available functions to serve you the shift properly. In addition, we will use the term *element management*, which is based on the fact that each date and time value is made up of elements. In this sense, the element can refer to the day, month, year, hour, minute, or second.

Getting to know the ADD_MONTHS function

Mathematical arithmetical operations that are done on the `DATE` value express the day granularity. Thus, adding a value of `1` to the `DATE` value results in tomorrow being reflected while the original time values remain the same.

The ADD_MONTHS function has two parameters – the date value (date_val) and the number of months (number_months) to be added or subtracted. The result is the input date with processed months. Such functionality can also be done explicitly by parsing the DATE attribute. However, the transition between years and months must be handled explicitly.

The syntax of the ADD_MONTHS function is stated in the following line:

```
ADD_MONTHS(<date_val>, <number_months>)
```

The following code snippet highlights its usage. The shift of the month, denoted by the second parameter of the ADD_MONTH function, can be positive as well as negative, expressing adding or subtracting months. The first statement adds 7 months to the defined date, while the second statement provides the appropriate date 7 months ago:

```
select ADD_MONTHS(TO_DATE('15.02.2022', 'DD.MM.YYYY'), 7)
  from dual;
--> 15.09.2022 00:00:00
select ADD_MONTHS(TO_DATE('15.02.2022', 'DD.MM.YYYY'), -7)
  from dual;
--> 15.07.2021 00:00:00
```

Notice that the function automatically correctly manages transitions through years, also recognizing leap years. In the following example, 17 months are to be added:

```
select ADD_MONTHS(TO_DATE ('15.02.2021', 'DD.MM.YYYY'), 17)
  from dual;
--> 15.07.2022 00:00:00
```

The following example shows the principle of adding a month to the DATE value regarding the last day of the month. If you want to add one month to *January 31, 2022*, what will be the result? One month is added, so February is reflected, namely the last day of that month:

```
select ADD_MONTHS(TO_DATE ('31.01.2022', 'DD.MM.YYYY'), 1)
  from dual;
--> 28.02.2022 00:00:00
```

If the year is a leap year, then the provided result value would reference February 29:

```
select ADD_MONTHS(TO_DATE ('31.01.2024', 'DD.MM.YYYY'), 1)
  from dual;
--> 29.02.2024 00:00:00
```

Simply, the whole month is always added. However, be aware of subtraction:

```
select ADD_MONTHS(TO_DATE ('31.03.2022','DD.MM.YYYY'), -1)
   from dual;
select ADD_MONTHS(TO_DATE ('30.03.2022','DD.MM.YYYY'), -1)
   from dual;
select ADD_MONTHS(TO_DATE ('29.03.2022','DD.MM.YYYY'), -1)
   from dual;
select ADD_MONTHS(TO_DATE ('28.03.2022','DD.MM.YYYY'), -1)
   from dual;
```

All the preceding statements result in getting a value of 28.02.2022 00:00:00. Vice versa, taking 1 month away from *March 27* retains the day element – just the month is changed:

```
select ADD_MONTHS(TO_DATE('27.03.2022', 'DD.MM.YYYY'), -1)
   from dual;
--> 27.02.2022 00:00:00
```

Note there is no function called ADD_YEARS or another variant. However, adding 1 year means adding 12 months. So, to provide such functionality, ADD_MONTHS can be multiplied by 12:

```
create or replace function ADD_YEARS(dat_val DATE,
                                      number_years number)
 return DATE
is
 begin
  return ADD_MONTHS(dat_val, 12*number_years);
 end;
/
```

Another approach to getting the same results is by adding or subtracting the INTERVAL data value from the date value, like so:

```
select sysdate + INTERVAL '1' MONTH from dual;
```

Coding such functionality explicitly requires many techniques to ensure correct results. The proposed solution (discussed in the following paragraphs) consists of two parameters. The first one reflects the source date. The second parameter delimits the number of months to be processed and applied to the source value. For simplicity, it will be expressed as an integer value. If the second parameter's value is zero, no change is made, and the original value can be directly returned. To be able to better explain the principles in the next few paragraphs, we will process individual elements separately. The following command results in the extracted values being stored in the v_day_orig, v_month_orig, and v_year_orig variables:

```
select TO_CHAR(p_date, 'DD'),
       TO_CHAR(p_date, 'MM'),
       TO_CHAR(p_date, 'YYYY')
  into v_day_orig, v_month_orig, v_year_orig
from dual;
```

Based on the provided input values, the output year element (enhanced by the shift across the years) is obtained in the first phase. The original (input) month value (v_month_orig) is extended using the shift (the second parameter, p_month_shift) by extracting whole years. It is done by dividing the sum of the original month element and month shift by a value of 12. Naturally, the resulting value can also be 0. In that case, it should reflect a 1-year shift in a negative manner (-1):

```
select v_year_orig
       + decode(TRUNC((v_month_orig + p_month_shift) / 12),
                0,
                -1,
                TRUNC((v_month_orig + p_month_shift) / 12)
               )
       into v_year_shifted
  from dual;
```

Similarly, the output value for the month is obtained as the remainder after dividing the sum of the original value of the month and the shift value by 12. If the provided value is less than or equal to zero, the resulting value is transformed into the positive spectrum by adding a value of 12, meaning a whole-year shift:

```
v_month_shifted:=mod(v_month_orig + p_month_shift, 12);
  if v_month_shifted<=0
     then v_month_shifted:=v_month_shifted + 12;
  end if;
```

Finally, the day element should be handled. This is the most complicated part of the evaluation. It is necessary to emphasize the last day of the month to ensure correctness. Thus, if the original day is the last day of the month, then the output value will also represent the last day. If the value is less, it is necessary to check whether that particular day exists in the resulting month. If not, the last day is used. In ordinary conditions, the original value of the day can be copied. For evaluation purposes, original and output values are managed separately for each element:

```
v_last_day_orig:=TO_CHAR(last_day(p_date), 'DD');
 v_last_day_shifted:=
     to_char(last_day(TO_DATE('01.'
                                 ||v_month_shifted||'.'
                                 ||v_year_shifted,
                      'DD.MM.YYYY')),
             'DD');
 if v_day_orig=v_last_day_orig
     -- last day of the month
   then v_day_shifted:=v_last_day_shifted;
 elsif v_day_orig>v_last_day_shifted
   then v_day_shifted:=v_last_day_shifted;
 else v_day_shifted:=v_day_orig;
 end if;
```

The complete function can be accessed by going to this book's GitHub repository. Note that the time elements are truncated; however, it is no problem to copy the original hour, minute, and second elements to the produced output.

For the complete code for the ADD_MONTHS_OWN function, you can refer to ADD_MONTHS_OWN. sql in the GitHub repository for this chapter.

As is evident, the own function definition can be demanding to ensure the same results. Note that the provided code only works correctly for the integer representation of the month shift:

```
select sysdate,
       ADD_MONTHS_OWN(sysdate, 1),
       ADD_MONTHS(sysdate, 1)
 from dual;
--> 31.01.2022 11:10:52
--> 28.02.2021 11:10:52
--> 28.02.2022 11:10:52
select ADD_MONTHS_OWN(TO_DATE('28.2.2022', 'DD.MM.YYYY'), -1)
```

```
    from dual;
 --> 31.01.2021 00:00:00
 select ADD_MONTHS_OWN(TO_DATE('28.2.2022', 'DD.MM.YYYY'),  -63)
    from dual;
 --> 30.11.2017 00:00:00
```

In the implementation of the own function for handling the month shift, the LAST_DAY function has been used. The next section introduces the implementation and usage details of the LAST_DAY function.

Identifying the number of days in a month using LAST_DAY

The LAST_DAY function returns the last day of the month based on the input date value (date_val). This is easy, right? January has 31 days and December has 31 days. However, what about February? Refer to the leap year. Thus, as is evident, this particular function is really useful and it must be available.

The syntax of this function can be seen in the following code block:

```
LAST_DAY(<date_val>)
```

The following statement provides you with the last day of October:

```
select LAST_DAY(TO_DATE('10.1.2022', 'DD.MM.YYYY'))
   from dual;
--> 31.01.2022
```

Naturally, it also manages leap years. The year *2023* is not a leap year, but the year *2024* is a leap year:

```
select LAST_DAY(TO_DATE('15.2.2023', 'DD.MM.YYYY'))
   from dual;
--> 28.2.2023
select LAST_DAY(TO_DATE('15.2.2024', 'DD.MM.YYYY'))
   from dual;
--> 29.2.2024
```

When using the TIMESTAMP value, second fractions are ignored (truncated):

```
select systimestamp,
       LAST_DAY(systimestamp),
```

```
        LAST_DAY(sysdate)
 from dual;
--> 23.02.2022 13:29:04,760000000 +01:00
--> 23.02.2022 13:29:04
--> 23.02.2022 13:29:04
```

This section aimed to identify the last day of the month by emphasizing leap years. It can be obtained very easily using the LAST_DAY function. In the next section, we will look at the time that has elapsed. So far, we have shown that if you subtract two dates, the number of days is obtained. But how do you get the number of months? By calling the MONTHS_BETWEEN function.

Understanding the usage of the MONTHS_BETWEEN function

As the function's name indicates, the MONTHS_BETWEEN function returns the number of months between two defined dates. The syntax of the MONTHS_BETWEEN function is stated in the following block:

```
MONTHS_BETWEEN(date_val1, date_val2)
```

Let's look at the results. To get a positive value, the first parameter value (date_val1) must be greater than the second parameter (date_val2). If not, a negative result will be provided:

```
select MONTHS_BETWEEN(TO_DATE ('15.12.2022', 'DD.MM.YYYY'),
                      TO_DATE ('15.2.2022', 'DD.MM.YYYY'))
  from dual;
--> 10
select MONTHS_BETWEEN(TO_DATE ('15.2.2022', 'DD.MM.YYYY'),
                      TO_DATE ('15.12.2022', 'DD.MM.YYYY'))
  from dual;
--> -10
```

The result of this function can also be the real number expressing the fraction of the month that has elapsed:

```
select
  MONTHS_BETWEEN(TO_DATE('15.12.2022 6:22:12',
                         'DD.MM.YYYY HH:MI:SS'),
               TO_DATE('13.2.2022 5:13:12',
                       'DD.MM.YYYY HH:MI:SS'))
```

```
    from dual;
--> 10,066
```

Dividing the difference between two DATE values by 30 does not provide relevant data, due to leap years and the total number of days in an individual month. However, the MONTHS_BETWEEN function is robust and works correctly, so to get the actual period elapsed (such as the age of the person from their birth date to the current date), always rely on the MONTH_BETWEEN function call's result.

Note that there is no function called *YEARS_BETWEEN* or another variant. However, the MONTHS_BETWEEN result can be divided by 12 to provide an analogous result.

The following section introduces two functions for getting the next day of the week according to the specification. The difference between those methods is based on the possibility of general applicability. Namely, the first function (NEXT_DAY) is generally available but is limited by the NLS parameters of the session, while NEXT_DATE is user-defined and overrides the session parameters. This makes the NEXT_DATE function more general and applicable.

Exploring the NEXT_DAY and NEXT_DATE functions

It often happens that you need to find the next free day in your calendar, the date of the next Wednesday, or the date of the second Sunday of the month when system updates are to be performed. In this section, we will take a closer look at two functions. The first of them is directly available within the database system. It refers to the NEXT_DAY function. However, as we will show, its evaluation and processing depend on several factors, specifically **NLS parameters**. Therefore, in the second part, we will propose an own function implementation that is resistant to changes in terms of the format and local parameters.

Exploring the principles of the NEXT_DAY function

The output of the **NEXT_DAY** function is the first weekday greater than the defined input date (date_val):

```
NEXT_DAY(<date_val>, <weekday>)
```

The following table shows the mapping principles.

Weekday	Description
SUNDAY	First Sunday greater than the input date (`date_val`)
MONDAY	First Monday greater than the input date (`date_val`)
TUESDAY	First Tuesday greater than the input date (`date_val`)
WEDNESDAY	First Wednesday greater than the input date (`date_val`)
THURSDAY	First Thursday greater than the input date (`date_val`)
FRIDAY	First Friday greater than the input date (`date_val`)
SATURDAY	First Saturday greater than the input date (`date_val`)

Table 7.1 – Weekday representation

For illustration purposes, we will also get the day of the processed date. The day of the first example is WEDNESDAY; the second example deals with SUNDAY. The output value should always reference SUNDAY:

```
select actual_date,
       TO_CHAR(actual_date, 'DAY'),
       NEXT_DAY(actual_date, 'SUNDAY'),
       TO_CHAR(NEXT_DAY(actual_date, 'SUNDAY'), 'DAY')
  from (select TO_DATE ('23.2.2022', 'DD.MM.YYYY')
                                      as actual_date
          from dual);
--> ACTUAL DATE: 23.02.2022 00:00:00  WEDNESDAY
--> NEXT_DAY:    27.02.2022 00:00:00     SUNDAY
select actual_date,
       TO_CHAR(actual_date, 'DAY'),
       NEXT_DAY(actual_date, 'SUNDAY' ),
       TO_CHAR(NEXT_DAY(actual_date, 'SUNDAY' ), 'DAY')
  from (select TO_DATE ('27.2.2022', 'DD.MM.YYYY')
                                      as actual_date
          from dual);
--> ACTUAL DATE: 27.02.2022 00:00:00 SUNDAY
--> NEXT_DAY:    06.03.2022 00:00:00 SUNDAY
```

The NEXT_DAY function takes two parameters – input date and weekday specified by the character string. It returns the first weekday later than the defined input parameter value. It always returns DATE, even if the provided parameter value is TIMESTAMP. The second parameter defining the day of the week must be in the date language of the session, delimited either by the full name or the abbreviation. The minimum number of letters applies to the abbreviated version. Any characters immediately following the valid abbreviation are ignored. The day, month, and year elements are shifted based on the specification; time (hour, minute, and second) elements remain the same.

The story behind the NEXT_DAY function relies on textual representation and particular language references. As will be shown in the next section, if the processed language is changed, mapping and evaluation can fail and an exception will be raised.

The impact of language definition on the NEXT_DAY function

In the following example, the English language is referenced. However, you probably already feel its strengths and limitations. If you're using a different language, naturally, the reference value will differ. To describe this function's usage, let's assume that the system uses the English language (denoted by the NLS_DATE_LANGUAGE parameter set to English):

```
alter session set nls_date_language='English';
```

New Year's day of 2022 was a Saturday, so the following statement returns January 3:

```
select NEXT_DAY(TO_DATE('1.1.2022', 'DD.MM.YYYY'), 'MON')
  from dual;
--> 03.01.2022
```

If the same day is defined as the corresponding DATE value in the parameter, 1 week is added:

```
select NEXT_DAY(TO_DATE('1.1.2022', 'DD.MM.YYYY'), 'SAT')
  from dual;
--> 08.01.2022
```

As stated, time elements remain the same:

```
select NEXT_DAY(TO_DATE('1.1.2022 17:24:12',
                        'DD.MM.YYYY HH24:MI:SS'),
            'SAT')
  from dual;
--> 08.01.2022 17:24:12
```

Second fragments for the `TIMESTAMP` format are ignored:

```
select systimestamp, NEXT_DAY(systimestamp, 'MON')
  from dual;
--> SYSTIMESTAMP: 05.02.22 14:26:45,884000000 +01:00
--> NEXT MONDAY:  07.02.2022 14:26:45
```

For individual weekdays in English, three characters are used, so the rest are ignored:

```
select NEXT_DAY(TO_DATE('1.1.2022', 'DD.MM.YYYY'),
                'MONXXX') from dual;
--> 03.01.2022
```

Therefore, it can be said that the full name is logically treated as an abbreviation by ignoring the rest of the string.

In terms of dealing with the `NEXT_DAY` function call, there is a relevant point related to reliability. Proper management and evaluation by forming the result set depend on the character string format. It relates to the session date language specification. If it is changed, it does not provide sufficient power and results in an exception being raised. Namely, by changing the language to *German*, the *Tuesday* representation no longer works. Instead, the term *Dienstag* should be used:

```
alter session set NLS_DATE_LANGUAGE='German';
```

Evaluation based on the textual representation always depends on the `NLS_DATE_LANGUAGE` value. Thus, by using the *German* language, the reference should be `DIE` for Tuesday, instead of the English variant, `TUE`:

```
select NEXT_DAY(TO_DATE('1.1.2022', 'DD.MM.YYYY'), 'TUE')
  from dual;
--> ORA-01846: "not a valid day of the week"
select NEXT_DAY(TO_DATE('1.1.2022', 'DD.MM.YYYY'), 'DIE')
  from dual;
--> 04.01.2022
```

Naturally, for some days, it can apply the English rules, mostly if the abbreviations are the same in the leading part, such as `Monday` in English and `Montag` in German:

```
select NEXT_DAY(TO_DATE('1.1.2022', 'DD.MM.YYYY'), 'MON')
  from dual;
--> 03.1.2022
```

Note that the weekday abbreviation's length can differ across languages. For example, the English variant takes three characters, but the German language requires only two characters.

As is evident, the NEXT_DAY function does not ensure robustness. If the session language is changed, a particular function call does not need to provide a relevant value, generally raising the already-stated exception. Except for the string values specified in the second parameter, it is also possible to use a numerical representation of the days of the week:

```
select NEXT_DAY(TO_DATE('1.1.2022', 'DD.MM.YYYY'), 1)
  from dual;
```

The preceding query returns January 3. The second parameter value can range from 1 to 7 (inclusive). If the value is out of range, the *ORA-01846* exception is raised.

Deliberately, in the preceding example, I emphasized the result provided to me. However, the value provided to you may differ. It is really strange, isn't it? What does a value of 1 as the second parameter represent? It should correspond to the weekday number. However, what about the ordering? In some regions, the first day of the week is Sunday. In the rest of the world, it references Monday. So, it depends on another parameter, NLS_TERRITORY. Thus, the preceding code will generally provide a value of January 3, but January 2 can also be stated if a value of 1 denotes *Sunday* instead of *Monday*.

In conclusion, regarding the NEXT_DAY function's representation, it only provides correct results if NLS_DATE_LANGUAGE and/or NLS_TERRITORY are not changed, and the user is strongly aware of their values. Embedding that function call inside the application can be risky if the definition can be enhanced. Therefore, it is pertinent to discuss the own representation and definition, which does not depend on the session parameters at all. The NEXT_DAY function is owned by the user SYS, enclosed by the STANDARD package. For each such function, a *global synonym* is available so that the user can reference the function directly without stating the package name and owner:

```
select SYS.STANDARD.NEXT_DAY(sysdate, 'MON') from dual;
select NEXT_DAY(sysdate, 'MON') from dual;
```

However, the *synonym* specification can be overridden by the local function defined by the user. As a result, the developed function can have the same name and parameters but must be owned by the local user.

In the next section, we will implement our own function to ensure that the results are not influenced by the NLS parameters and that the proper value is always obtained.

Implementing the NEXT_DATE function

For explanatory and descriptive purposes, this particular developed function is named NEXT_DATE instead of the original NEXT_DAY. Similar to the original function, it takes two parameters.

p_date_val specifies the original date to be processed, while p_spec sets the day to be shifted to. To calculate the shift, the weekday of the provided value as well as the result value should be obtained.

The numerical *weekday* representation of the input date is extracted and stored in a local variable, v_input_day_nr. It is treated in two phases so that processing is independent of the NLS session parameter values. Firstly, the name of the day is obtained by using the TO_CHAR function call, extended by the NLS_DATE_LANGUAGE definition for its usage. Secondly, a numerical representation is provided. In the following code snippet, the American version of the day is obtained and stored in the v_weekday variable, which is then transformed into a numerical reference using case conversion. If this conversion cannot be done, an exception is raised:

```
select TO_CHAR(p_date_val, 'DAY',
                          'NLS_DATE_LANGUAGE=American')
          into v_weekday
  from dual;
case trim(v_weekday)
    when 'MONDAY'    then v_input_day_nr:=1;
    when 'TUESDAY'   then v_input_day_nr:=2;
    when 'WEDNESDAY' then v_input_day_nr:=3;
    when 'THURSDAY'  then v_input_day_nr:=4;
    when 'FRIDAY'    then v_input_day_nr:=5;
    when 'SATURDAY'  then v_input_day_nr:=6;
    when 'SUNDAY'    then v_input_day_nr:=7;
    else RAISE_APPLICATION_ERROR(-20000,
              'Input format not correctly specified');
  end case;
```

To ensure the correctness of the results compared to the existing solution, NEXT_DAY, either a character string or numerical day representation can be used. One way or another, in our case, the *English* abbreviation representation is always used. Moreover, in our case, *Monday* is always considered the first day of the week for mapping. As a result, a numerical representation is stored in a v_output_day_nr local variable.

Using the v_input_day_nr and v_output_day_nr variables, the day shift can be calculated as follows:

```
shift = (7 - i + j) mod 7
```

The i expression denotes v_input_day_nr, while j denotes v_output_day_nr. If both values are the same, then 1 week is to be added, so a value of 7 is returned. These shift principles are shown in the following figure:

	1	2	3	4	5	6	7
1	7	1	2	3	4	5	6
2	6	7	1	2	3	4	5
3	5	6	7	1	2	3	4
4	4	5	6	7	1	2	3
5	3	4	5	6	7	1	2
6	2	3	4	5	6	7	1
7	1	2	3	4	5	6	7

Figure 7.1 – GET_SHIFT – transformation table

For the shift calculation, the local function must be specified:

```
function GET_SHIFT(I integer, j integer)
   return
     integer
 is
   begin
     if i=j then return 7;
            else return mod(7-i+j, 7);
     end if;
   end;
```

For the complete code for the NEXT_DATE function (with two parameters), please refer to NEXT_DATE - 2 params.sql in the GitHub repository for this chapter.

The solutions we've discussed are complex and ensure proper mapping in a textual format. However, what about the numerical representation for the days of the week? The next section aims to provide an answer.

Numerical day-of-week representation related to the NEXT_DATE function

Now, let's emphasize a numerical representation of the days of the week. In the preceding implementation, we assumed that the first day of the week is *Monday*, represented by a value of 1. However, that is not generally true. Some regions start the week on *Sunday*. Therefore, the developed function can be extended by the third optional parameter, representing the assigned region. The p_sunday_first parameter defaults to *0*, meaning that the week starts on *Monday*. If set to *1*, then a value of 1 for the day number represents *Sunday* instead of Monday.

The transformation to the original solution can be done by subtracting a value of 1 if the specified second parameter value is numeric unless it is a Sunday (1). In that case, the value is replaced with a value of 7. The header of the function can be extended using the third parameter:

```
create or replace function NEXT_DATE
   (p_date_val DATE, p_spec varchar,
    p_sunday_first integer default 0)
```

Then, the local function implementation is stated. In the first phase, parameter value correctness is checked, followed by the transformation. If the specified day reference is 1, it should be transformed into a value of 7, representing Sunday. In other cases, 1 day is subtracted, shifting the processing of the solution, where Sunday is represented as the first day of the week:

```
function IS_NUMERIC(p_spec varchar)
  return boolean
 is
  v_integer integer;
  begin
   v_integer:=to_number(p_spec);
   return true;
   exception when others then return false;
  end;
begin
 if p_sunday_first=1 and is_numeric(p_spec) then
     return NEXT_DATE(p_date_val,case
                               when p_spec=1 then 7
                               else p_spec-1
                          end, 0);
 end if;
```

For the complete code for the NEXT_DATE function (with three parameters), please refer to NEXT_DATE - 3 params.sql in the GitHub repository for this chapter.

Getting the second Sunday of the month

The NEXT_DAY function, or its generalized variant, NEXT_DATE, provides the nearest day based on the weekday specification. However, it is often necessary to process more complex conditions related to the first day of the month. For example, backups are obtained on the *second Saturday* of the month, and updates are then applied on the *second Sunday*. The already discussed functions can be directly combined to ensure a more complex evaluation. Specifically, any date can be obtained when the first day of the month is extracted in the first phase. Thus, to get the second Sunday of the month, the following evaluation conditions can be used:

1. Get the first day of the month.

2. Evaluate the weekday of the first day of the month:

   A. If it reflects *Sunday*, 1 week is added (*7 days*), a particular value is returned, and the processing ends.

   B. If it is not *Sunday*, then the nearest next Sunday is obtained (using the NEXT_DAY or NEXT_DATE function), and 1 week is added. In this phase, the processing ends.

In the following code snippet, the previously described steps have been implemented, forming the GETSECONDSUNDAY function. Individual steps have been marked for reference:

```
create or replace function GETSECONDSUNDAY(p_date_val DATE)
 return DATE
is
 v_date DATE;
 v_weekday varchar(20);
begin
-- step 1
 v_date:=TRUNC(p_date_val, 'MM');
-- step 2
 select trim(TO_CHAR(v_date, 'DAY',
                            'nls_date_language=American'))
        into v_weekday
   from dual;
-- step 3
 if v_weekday = 'SUNDAY'
    then return v_date+7;
```

```
  else return NEXT_DATE (v_date,'SUN')+7;
 end if;
end;
/
```

The source code for the GETSECONDSUNDAY function can be found in the GitHub repository for this chapter.

For March 7, 2022, the first day of the month is Tuesday, so the first Sunday is March 6, and the second Sunday is then March 13:

```
select GETSECONDSUNDAY(TO_DATE('7.3.2022', 'DD.MM.YYYY'))
  from dual;
--> 13.03.2022 00:00:00
```

This can be seen in the following figure:

Figure 7.2 – Month calendar – getting the second Sunday of the month

Individual elements can be removed for less precision. For example, for employment contract validity, time element values are typically unimportant. The TRUNC function, which limits individual elements, will be discussed in the next section.

Investigating the TRUNC function

The TRUNC function, for dealing with DATE type values, removes all parts with smaller granularity than defined. It consists of one obligatory and one optional parameter. The first one is the DATE attribute value (date_val), while the second reflects the granularity (format). If the second parameter is not defined, the default value for the day is used (the time element value will be 00 in that case). Thus, smaller granularity values (*day*, *hours*, *minutes*, and *seconds*) are removed if the second parameter defines a *month*. The second parameter value can be DD, MM, YY, HH, or MI. The format unit (the second parameter) can also be Q (quarter), W (week of the month), or WW (week in the year). The syntax of the function is stated in the following code block:

```
TRUNC(<date_val>, [<format>])
```

Let's assume this is the actual time to be set by this value:

```
select sysdate from dual;

--> 2022.02.20   12:04:36
```

The following code shows examples of the TRUNC function's usage and provided outputs:

```
select TRUNC(sysdate) from dual;

select TRUNC(sysdate, 'DD') from dual;

--> 2022.02.20   12:04:36   YYYY.MM.DD HH:MI:SS

                              RESULT: 2022.02.20 00:00:00


select TRUNC(sysdate, 'MM') from dual;

--> 2022.02.20   12:04:36   YYYY.MM.DD HH:MI:SS

                              RESULT: 2022.02.01 00:00:00

select TRUNC(sysdate, 'YY') from dual;

--> 2022.02.20   12:04:36   YYYY.MM.DD HH:MI:SS

                              RESULT: 2022.01.01 00:00:00
```

What about the difference between the result of the LAST_DAY function and getting the last day based on the TRUNC and ADD_MONTHS functions? Does it provide the same results? Let's consider the following examples.

The first piece of code is based on invoking the LAST_DAY function. Compared to the second solution, the time spectrum is not trimmed away:

```
select LAST_DAY(sysdate) from dual;
--> 28.02.2022 13:10:19
select TRUNC(ADD_MONTHS (sysdate,1), 'MM')-1 from dual;
--> 28.02.2022 00:00:00
```

The second parameter of the TRUNC function (the format unit) can hold other specific values (CC, Q, W, IW, or WW). Most of them are self-explanatory. Therefore, we're just showing a simple example. Let's consider sysdate to be *February 23, 2022*. The next two code blocks deal with CC and Q parameter value usage.

If using the CC value for the second parameter, the output value is trimmed to the beginning point of the particular century:

```
select TRUNC(sysdate, 'CC') from dual;
--> 01.01.2001 00:00:00
```

The Q parameter is used for trimming the quarter, returning the first date of the particular quarter:

```
select TRUNC(sysdate, 'Q')
  from dual;
--> 01.01.2022 00:00:00
select TRUNC(to_date ('15.6.2020', 'DD.MM.YYYY'), 'Q')
  from dual;
--> 01.04.2020 00:00:00
```

The next section looks at the weekly management of the DATE value using the TRUNC function. It will introduce week of month and year references, as well as *ISO week* reference importance.

The TRUNC function and week management

Week management, as shown by the second parameter of the TRUNC function, requires a more in-depth description. The following figure shows the calendar representation for January and February 2022:

January 2022

Mo	Tu	We	Th	Fr	Sa	Su
27	28	29	30	31	1	2
3	4	5	6	7	8	9
10	11	12	13	14	15	16
17	18	19	20	21	22	23
24	25	26	27	28	29	30
31	1	2	3	4	5	6

February 2022

Mo	Tu	We	Th	Fr	Sa	Su
31	1	2	3	4	5	6
7	8	9	10	11	12	13
14	15	16	17	18	19	20
21	22	23	24	25	26	27
28	1	2	3	4	5	6
7	8	9	10	11	12	13

Figure 7.3 – Month calendar – January and February 2022

A value of W trims the week, so now it starts with the same weekday as the first day of the particular month:

```
select TRUNC(TO_DATE('10.1.2022', 'DD.MM.YYYY'), 'W')
  from dual;
--> 08.01.2022 00:00:00
```

Refer to the situation expressed in the following figure:

Figure 7.4 – Month calendar – week management

Thus, in the preceding example, the week does not start with Monday, but it is delimited by the reference to the weekday of the first day of the month. To normalize week management (starting with Monday), the *ISO week standard* can be used, represented by the `IW` parameter value. The ISO week is independent of the `NLS_territory` parameter setting:

```
select TRUNC(TO_DATE('12.1.2022', 'DD.MM.YYYY'), 'IW')
  from dual;
--> 10.01.2022 00:00:00
```

This can be seen in the following figure:

Figure 7.5 – Month calendar – ISO week

The `W` parameter option delimits the week of the month, while `WW` defines the representation of the week for the year. Thus, the first day of the week is associated with the first day of the year. So, for the year *2022*, the first day is *Saturday* and the first full week of *February* starts on *February 5* and ends on *February 11*, covering the requested value of *February 10*:

```
select TRUNC(TO_DATE('10.2.2022', 'DD.MM.YYYY'), 'WW')
    from dual;
--> 05.02.2022 00:00:00
```

This can be seen here:

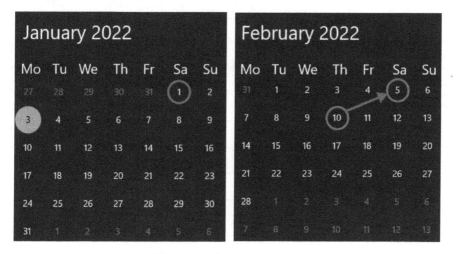

Figure 7.6 – Month calendar – week of the year (1)

Similarly, *February 18* is mapped to *February 12*:

```
select TRUNC(TO_DATE('18.2.2022', 'DD.MM.YYYY'), 'WW')
  from dual;
--> 12.02.2022 00:00:00
```

Here is the result:

Figure 7.7 – Month calendar – week of the year (2)

The list of all available parameters, along with their representations, can be seen in the following table:

Format	Representation
CC, SCC	Century
SYYYY, YYYY, YEAR, SYEAR, YYY, YY, Y	Year
IYYY, IY, I	ISO year
Q	Quarter
MONTH, MON, MM, RM	Month
WW	Week of the year (the same day of the week as the first day of the year)
IW	Week of the year (the same day of the week as the first day of the ISO year)
W	Week of the month (the same day of the week as the first day of the month)
DDD, DD, J	Day
DAY, DY, D	Starting day of the week
HH, HH12, HH24	Hour
MI	Minute

Table 7.2 – TRUNC format models

In addition to trimming values, rounding can be used. This principle is explained in the following section.

Understanding the usage of the ROUND function

The ROUND function returns a DATE value rounded to a specific granularity. It takes two parameters. The first parameter (input_date) is mandatory and specifies the DATE value to be rounded. The rounded unit is defined by the second parameter (format). The second parameter is optional. The default value has day precision. Thus, if the value for the second parameter is not stated, the input date is rounded to the nearest day.

The syntax of the ROUND function is as follows:

```
ROUND (<input_date> [, <format>])
```

The following code snippets show examples of the ROUND function's usage related to the DATE value's management. The value is rounded down by the day precision specified. As it reflects the morning hours, the returned value is 26.01.2022 00:00:00:

```
select ROUND(TO_DATE('26.1.2022 10:15:22',
                     'DD.MM.YYYY HH24:MI:SS'), 'DD')
   from dual;
--> 26.01.2022 00:00:00
```

If the time is in the afternoon, then the value is rounded to the beginning of the next day:

```
select ROUND(TO_DATE('26.1.2022 16:15:22',
                     'DD.MM.YYYY HH24:MI:SS'), 'DD')
   from dual;
--> 27.01.2022 00:00:00
```

Noon is also rounded up:

```
select ROUND(TO_DATE('26.1.2022 12:00:00',
                     'DD.MM.YYYY HH24:MI:SS'), 'DD')
   from dual;
--> 27.01.2022 00:00:00
```

Similarly, the rounding process can be done for each granularity level:

```
select ROUND(TO_DATE('17.6.2022', 'DD.MM.YYYY'), 'YYYY')
   from dual;
--> 01.01.2022 00:00:00
```

OK, but what about month precision? How do you round the value? Let's show this principle using examples.

February 2022 consisted of 28 days. So, how would you round February 15? Would it be February 1 or March 1?

```
select ROUND(TO_DATE('15.2.2022 00:00:00',
                     'DD.MM.YYYY HH24:MI:SS'), 'MM')
   from dual;
```

The solution is based on the time that elapsed between the first day of the month and the specified date, compared to the number of days to the last day of the month.

The time that elapsed between the specified date and the last day of the month is 13 days, whereas the time that elapsed between the first day of the month and the specified date is 14 days:

```
select LAST_DAY(TO_DATE('15.2.2022 00:00:00',
                         'DD.MM.YYYY HH24:MI:SS'))
          -(TO_DATE('15.2.2022 00:00:00',
                    'DD.MM.YYYY HH24:MI:SS'))
  from dual;
--> 13
select (TO_DATE('15.2.2022 00:00:00',
                'DD.MM.YYYY HH24:MI:SS'))
          -TRUNC(TO_DATE('15.2.2022 00:00:00',
                         'DD.MM.YYYY HH24:MI:SS'), 'MM')
  from dual;
--> 14
```

As a result, it is rounded down:

```
select ROUND(TO_DATE('15.2.2022 00:00:00',
                      'DD.MM.YYYY HH24:MI:SS'), 'MM')
  from dual;
--> 01.02.2021 00:00:00
```

Even though the value is extended by the time elements, the same value would be still provided as a result, as seen in the following snippet:

```
select ROUND(TO_DATE('15.2.2022 23:59:59',
                      'DD.MM.YYYY HH24:MI:SS'), 'MM')
  from dual;
--> 01.02.2021 00:00:00
```

Specification of the format model holds the same principles as defined for the TRUNC function.

In this section, we showed the basic functions for working with date and time. Specific date and time values can also be used to identify a person. In the following section, we will explain the concept of a birth number, which uniquely identifies a person.

Understanding the PERSONAL_ID concept and birthday management

PERSONAL_ID refers to a specific value assigned to people in Slovakia and the Czech Republic. It is a unique personal identification number formed of the person's date of birth, gender, and a terminal number, which is a differentiating number for people born on the same day. In total, it consists of six digits followed by the slash symbol and another four digits.

The first two digits of PERSONAL_ID are the last two digits of the person's year of birth. The second two digits express the numeric designation of the person's month of birth (the value is increased by *50* for women). The third two-digit set represents the numerical designation of the person's day of birth. This is an example of a man and woman:

```
--> 90 06 23 / 1234             90 56 23 / 1239
--> MAN                         WOMAN
--> Year = 90                   Year = 90
--> Month = 6                   Month = 6
--> Day = 23                    Day = 23
```

The whole personal identifier can be divided by 11 without a remainder, which means its reliability and correctness can be checked. However, the division rule is not so strict. Mostly for very old records, occasionally, it can happen that the division rule does not work:

```
--> 9056231239 / 11 = 823293749
```

Typically, PERSONAL_ID is formed of the four digits after the slash symbol. However, for people born before *January 1, 1954*, the numerical designation after the slash symbol consists of only three digits.

As the year of birth is given by its last two numbers, in the current format, the birth number would cease to be a unique identifier in 2054. However, it is assumed that before then, it will be replaced by another identifier, or a suitable modification will be introduced. One way or another, nowadays, it forms a relevant person identification. It is unique and not very demanding storage-wise. Moreover, the PERSONAL_ID value is a composite – date of birth, as well as gender, can be directly obtained. In most state information systems and communication with the state, PERSONAL_ID is referred to as a unique person identifier.

The following code snippet shows the function definition for transforming the PERSONAL_ID value into a date of birth. It takes the PERSONAL_ID value as a parameter and extracts individual day, month, and year elements. This is done by getting the substring SUBSTR function call. The day element can be taken directly. For the month, a particular value must be checked to subtract 50 for women (using the MOD function). Dealing with the year, the century reference depends on the length of the whole value. The whole functionality may look like this. There is also an exception handler to ensure that no exception is raised outside the function. Instead, a fictive date of birth is returned:

```
create or replace function PIDTOBIRTH(p_pid varchar2)
 return DATE
is
 val varchar2(11);
begin
 val:=substr(p_pid,5,2)||'-'              -- DAY
     ||mod(substr(p_pid,3,2),50)          -- MONTH
     ||case length(trim(p_pid))
         when 10 then '19'
         else '20'
       end
     ||'-'
     || substr(p_pid,1,2);                -- YEAR
 return TO_DATE(val, 'dd-mm-yyyy');
   EXCEPTION WHEN OTHERS
     THEN return TO_DATE('01-01-0001', 'dd-mm-yyyy');
end;
/
```

The source code for the PIDTOBIRTH function can be obtained from the GitHub repository for this chapter.

Just to remind you of the principles of the evaluation, to get a list of people who have a birthday today, do not forget to evaluate just the day and month comparison:

```
select name, surname
  from personal_data
    where to_char(PIDTOBIRTH(PERSONAL_ID), 'DD.MM')
                    =TO_CHAR(sysdate, 'DD.MM');
```

Comparing the whole date format would produce a list of people who were born on that specific date, but also, the time elements (meaning the hour, minute, and second) must be the same. Be aware of that! The following code snippet provides the list of people who were precisely born right now – using second precision:

```
select name, surname
  from personal_data
    where PIDTOBIRTH(PERSONAL_ID)=sysdate;
```

Even removing the time elements is not suitable. It gets the people born today, not a list of people who have a birthday today! Although it seems to be clear, based on my expertise, programmers sometimes forget to process the elements correctly. The following code snippet removes the impact of the time element processing. It results in getting the list of people born today (irrespective of their time of birth):

```
select name, surname
  from personal_data
    where to_char(PIDTOBIRTH(PERSONAL_ID), 'DD.MM.YYYY')
                    =TO_CHAR(sysdate, 'DD.MM.YYYY');
```

PERSONAL_ID is an interesting concept of person identification defined by the date of birth and gender. For testing purposes, it can be necessary to generate valid PERSONAL_ID values. For females, the value of the birth month is increased by 50. To ensure the value is applicable, there are additional rules. Namely, the correctness of the value itself is protected by its divisibility by a value of 11. However, how can you obtain a random date? How can you ensure uniform distribution? How can you ensure that the value is consistent and such a DATE value exists? The number of days in particular months varies; leap years must also be taken into account. The next section provides answers and points to various techniques and their limitations, mostly related to managing individual elements explicitly.

Generating random dates

When dealing with date management and testing, a crucial task is associated with generating dates. Based on my experience in business, it is inevitable to focus on relevance and correctness. Each date consists of day, month, and year elements. There is also time denotation, but it can be ignored in this phase.

I often encounter my students modeling individual elements separately, and then the resulting value is achieved by putting the elements together. However, how do you generate and process it? No problem – the year can be defined by the range to be generated. The same approach can be applied to month reflection. However, to ensure overall consistency, the day elements must be emphasized. Namely, what about the right border for day processing? Some months have 31 days. The rest have only 30 days. Moreover, there is a specific month (February) with a different number of days. It consists of either 28 or 29 days (if it's a leap year). The processing can be complicated and demanding to apply all the rules, resulting in managing just 28 days for each month generally. As a result, consistent date composition is ensured, but such decisions can negatively impact individual dates that are not uniformly distributed.

The solution for generating birth dates by applying all rules can look like the code stated in the following part. First, the year element is generated from the defined range (bordered by the two function parameters). Then, the month value is generated from a range of < 1 ; 12 >. Finally, the day itself is provided by getting the last day of the month as a right border, stored in the v_day_limit local variable. The particular value of the day is then generated from the range of 1 to v_day_limit. To create the data values, the DBMS_RANDOM package is used. It is commonly part of the database system method set. Generally, the VALUE function of the package produces numbers in a range of < 0 ; 1) with 38 digits of precision. Another version returns a random NUMBER value, x, that is greater than or equal to the specified low value and less than the specified high value. In our case, just the integer value is used. The implicit conversion uses a rounding operation so that the right border can be used, as well:

```
DBMS_RANDOM.VALUE
   RETURN NUMBER;
DBMS_RANDOM.VALUE(
   <low>  IN  NUMBER,
   <high> IN  NUMBER)
RETURN NUMBER;
```

The complex solution for generating DATE is as follows:

```
create or replace function GENERATEBIRTH(L_year integer,
                                         R_year integer)
  return DATE
   is
  v_date DATE;
  v_year integer;
  v_month integer;
  v_day_limit integer;
  v_day integer;
begin
    -- getting year element
  v_year:=DBMS_RANDOM.VALUE(l_year,r_year);
    -- getting month element
  v_month:=DBMS_RANDOM.VALUE(1,12);
    -- getting right border for the day of month
  v_day_limit:=TO_CHAR(
```

```
              LAST_DAY(TO_DATE(v_month||'.'||v_year,
                              'MM.YYYY')),
                    'DD');
    -- getting day element
  v_day:=DBMS_RANDOM.VALUE(1, v_day_limit);
    -- providing output
  return TO_DATE(v_day||'.'||v_month||'.'||v_year,
              'DD.MM.YYYY');
end;
/
```

As usual, the source code for the GENERATEBIRTH function can be found in the GitHub repository for this chapter.

It uses the LAST_DAY function to provide the date of the last day of the particular month. If that function was not present, a definition would be even more complicated.

Digging deeper into the DBMS_RANDOM package, there are several methods for getting random values that differ from provided values in terms of data types, limits, and so on. There is no explicit function to generate random dates. However, to apply the same functionality in comparison with the proposed solution, a random value can be added or subtracted from the defined DATE value. The following code can be used to generate random dates covered by the year *2022*. That year is not a leap year, so the total number of days is *365*. The first part gets the first day of the year; then, the random amount of days is added. In this case, time management can be covered, whereas the value is numerical:

```
select TRUNC(TO_DATE(2022, 'YYYY'), 'YYYY')
        + DBMS_RANDOM.VALUE(0,364)
  from dual;
--> 10.04.2022 13:09:13
--> 22.08.2022 00:01:12
--> 27.06.2022 07:37:38
```

Thus, there is just a necessity to limit the range of days. If the time does not need to be processed, the generated value can simply be rounded.

Looking at this solution more generally, the date of birth can be generated using the year borders as well. The addition or subtraction of the numerical value from the date is robust and recognizes leap years, transitions between years, and other limiting factors. In the following case, date limits are obtained by transforming the input parameters. The start_limit local variable is obtained directly by converting the numerical input value representing the year into the DATE format. Only the year element is specified, so it produces the first day of the current month of the particular year.

A similar approach is used for `end_limit`. Firstly, it gets the first day of the next year, with the original input value incremented by one. From the DATE perspective, 1 day is subtracted, providing the last day of the specified year:

```
create or replace function GENERATEBIRTH(L_year integer,
                                         R_year integer)
   return DATE
is
   start_limit DATE;
   end_limit DATE;
   difference integer;
begin
   start_limit:=TRUNC(TO_DATE(L_year, 'YYYY'), 'YYYY');
   end_limit:=TRUNC(TO_DATE(R_year+1, 'YYYY'), 'YYYY')-1;
   difference:=end_limit-start_limit;
   return start_limit
           + ROUND(DBMS_RANDOM.VALUE(0,difference));
end;
/
```

As is evident, the DBMS_RANDOM package does not primarily focus on date and time management. It does not simply provide functionalities to generate such values. Instead, its primary intention is to manage numerical values, as well as character strings. However, as explained, a random numerical value can be applied to the DATE value to express the number of days to be added or subtracted. If the value is decimal, processing can result in even higher precision – hours, minutes, and seconds. Thus, by combining random numerical values and a precisely specified reference DATE point, random values from a date and time perspective can be obtained.

Summary

In this chapter, you got familiar with the existing functionalities in DBS Oracle, mostly related to day and month granularity. We emphasized the ADD_MONTHS and MONTH_BETWEEN functions for dealing with month precision.

Mathematical arithmetic applies a day precision shift. Therefore, this chapter referenced the LAST_DAY function for providing the last day of the month and the NEXT_DAY function, which provides the closest larger DATE value according to the specified weekday. The weekday value is set by the string abbreviation; alternatively, the serial number of the day within the week is used. However, as we have shown, these values depend on the region, language, or territory. Therefore, our own implementation for getting the next day was proposed and discussed.

In addition, we learned the principles of DATE value rounding and truncating. Then, the PERSONAL_ID value was introduced, which can be used as the primary key for a person table, providing composite information about the gender and date of birth of a person. The PERSONAL_ID value itself is divisible by 11, making the solution resistant to errors and typos. Finally, we summarized the possibilities of generating a random date, both by explicitly managing individual elements and by applying a numerical shift.

While describing the various methods and parameters, we pointed out the meaning of the parameter value, which can be influenced by the parameters of the session or server. In the next chapter, we will guide you through the complexity of the NLS parameters while pointing to the individual values, impacts, and data dictionary to get the current set (or inherited) value.

Questions

Answer the following questions to test your knowledge of this chapter:

1. What is the output of the following statement?

```
select ADD_MONTHS(TO_DATE('31.1.2023',
                            'DD.MM.YYYY'),
           1)
   from dual;
```

   A. February 28, 2023

   B. February 29, 2023

   C. February 31, 2023

   D. An exception is raised

2. Select the right statement:

   A. The MONTHS_BETWEEN function can provide a negative value if the second parameter value is greater than the first

   B. The MONTHS_BETWEEN function provides only an integer value, not the numerical representation, generally

   C. If the output value of the MONTHS_BETWEEN function is negative, an exception is raised

   D. The internal implementation of the MONTHS_BETWEEN function takes both values and swaps them, if necessary, so that the result is always a positive value

3. Which NLS parameter impacts the NEXT_DAY function?

    A. NLS_DATE_FORMAT only

    B. NLS_DATE_LANGUAGE only

    C. NLS_REGION only

    D. NLS_DATE_LANGUAGE and NLS_TERRITORY

4. You would like to trim the value based on the quarter. Which format mask should be used to do so?

    A. CC

    B. Q

    C. QQ

    D. IW

5. Which values can be separated from the PERSONAL_ID value?

    A. Name and surname

    B. Date of birth

    C. The region in which the person was born

    D. Date of birth and gender

Further reading

To learn more about the topics that were covered in this chapter, take a look at the following resources:

- Personal ID definition and usage in Slovak Republic: `https://www.oecd.org/tax/automatic-exchange/crs-implementation-and-assistance/tax-identification-numbers/Slovak-Republic-TIN.pdf`

- Date and time functions used in MySQL are listed in the documentation. It provides an interesting point of view on date and time separation, as well as methods for managing and extracting individual elements: `https://dev.mysql.com/doc/refman/8.0/en/date-and-time-functions.html`

8

Delving into National Language Support Parameters

It is obvious that individual clients working with applications can be located anywhere in the world. It is therefore necessary to prepare the platform so that all clients can use their own customary rules and regional customs. **National Language Support** (**NLS**) parameters determine the locale-specific runtime behavior on clients and servers. They can be set either for the server instance, specified in the initialization file applied during the instance startup (INIT.ORA), or for the client via environment variables overriding the default (server) values. It is done by executing the alter session command. Additionally, the definition can be enhanced by the statement level.

NLS parameters enhance the processing by focusing on local definitions, regional specifics, and rules. This chapter mainly focuses on the parameters related to date and time management, but also summarizes the main principles in a general manner, including local language adoption for error messages, sort order, and calendar conventions. By applying NLS parameters, multilingual database support can be provided.

In this chapter, we're going to cover the following main topics:

- NLS_DATE_FORMAT parameter, usage, and impacts
- NLS_DATE_LANGUAGE parameter, usage, and impacts
- NLS_CALENDAR parameter, usage, and impacts
- NLS_TERRITORY parameter, usage, and impacts
- Embedding NLS parameter definition into the TO_CHAR conversion functions

The source code for this chapter can be found in the GitHub repository at `https://github.com/PacktPublishing/Developing-Robust-Date-and-Time-Oriented-Applications-in-Oracle-Cloud/tree/main/chapter%2008`. Alternatively, you can access the repository by scanning the following QR code:

NLS parameter overview

Various principles are used around the world for date and time management. Regional rules are applied, such as considering either Monday or Sunday the first day of the week, using a specific format for presenting date and time values (such as element delimiters), using specific time zones, and so on. To make a date and time system generalizable and applicable to any region by implementing local principles and rules, it is necessary to provide an interface that's appropriate for the database, as well as the client side. And this is exactly where NLS parameters come onto the stage. These parameters can be set for the whole database, or each client can configure the parameters for themselves. Throughout the following sections, you will learn the proper definitions of individual parameters, values used, formats, and the impacts these have on the processing and output.

The following table shows a list of the individual parameters with their descriptions and default values:

Parameter	Description	Default value
NLS_CALENDAR	Calendar system	Gregorian
NLS_COMP	SQL operator comparison	Binary
NLS_CREDIT	Credit accounting symbol	NLS_TERRITORY
NLS_CURRENCY	Local currency symbol	NLS_TERRITORY
NLS_DATE_FORMAT	Date format	NLS_TERRITORY
NLS_DATE_LANGUAGE	Language for day and month names	NLS_LANGUAGE
NLS_DEBIT	Debit accounting symbol	NLS_TERRITORY
NLS_ISO_CURRENCY	ISO international currency symbol	NLS_TERRITORY

Parameter	Description	Default value
NLS_LANG	Language, territory, and character set	American_America .US7ASCII
NLS_LANGUAGE	Language	NLS_LANG
NLS_LIST_SEPARATOR	Character separating items in a list	NLS_TERRITORY
NLS_MONETARY _CHARACTERS	Monetary symbol for dollar and cents (or their equivalents)	NLS_TERRITORY
NLS_NCHAR	National character set	NLS_LANG
NLS_NUMERIC _CHARACTERS	Decimal character and group separator	NLS_TERRITORY
NLS_SORT	Character sort sequence	NLS_LANGUAGE
NLS_TERRITORY	Territory	NLS_LANG
NLS_DUAL_CURRENCY	Dual currency symbol	NLS_TERRITORY

Table 8.1 – NLS parameters

The following table shows the scope-level applicability – the value can be set on the server covered by the initialization file (INIT.ORA), the client level defined by the environment variable, and the alter session command:

Parameter	Server (INIT.ORA)	Client (environment variable)	Alter session command
NLS_CALENDAR	✓	✓	✓
NLS_COMP	✓	✓	✓
NLS_CREDIT		✓	
NLS_CURRENCY	✓	✓	✓
NLS_DATE_FORMAT	✓	✓	✓
NLS_DATE_LANGUAGE	✓	✓	✓
NLS_DEBIT		✓	
NLS_ISO_CURRENCY	✓	✓	✓

Parameter	Server (INIT.ORA)	Client (environment variable)	Alter session command
NLS_LANG		✓	
NLS_LANGUAGE	✓		✓
NLS_LIST_SEPARATOR		✓	
NLS_MONETARY_CHARACTERS		✓	
NLS_NCHAR		✓	
NLS_NUMERIC_CHARACTERS	✓	✓	✓
NLS_SORT	✓	✓	✓
NLS_TERRITORY	✓		✓
NLS_DUAL_CURRENCY	✓	✓	✓

Table 8.2 – Scope of the NLS parameters

We have had a brief introduction to the NLS parameters by looking at the available options, default values, and applicability levels. Now, it's time to go deeper and explain the individual parameters and their impacts.

Exploring NLS parameters and their impact

In order to be able to prepare a universal solution that can be deployed anywhere in the world, it is necessary to know the regional characteristics and formats and adapt the solution to your specific environment and customs. The aim of this section is to introduce you to individual NLS parameters and show you how to create a robust solution for various specific elements of date and time processing and representation.

We will look closely at parameters directly related to date and time processing. Toward the end of the chapter, we will see a summary of a few other parameters that are outside of the scope of this book.

We will start with NLS_DATE_FORMAT, followed by NLS_DATE_LANGUAGE, NLS_CALENDAR, and NLS_TERRITORY. By learning about them, you will be able to implement sophisticated solutions directly tailored to specific clients' requirements in any region and on a global scale.

NLS_DATE_FORMAT parameter

For date and time management, the most relevant parameter is NLS_DATE_FORMAT, which is applied by the implicit conversions, but mostly as a string representation in the output. If you query any date value, physically, in the result set, it must be formatted in textual form. This parameter denotes the visual format of the date and time:

```
alter session set NLS_DATE_FORMAT='DD.MM.YYYY HH24:MI:SS';
select sysdate from dual;
-->   25.03.2022 08:51:12
alter session set NLS_DATE_FORMAT='DD/MM/YYYY';
select sysdate from dual;
--> 25/03/2022
```

Please, do not be confused if you encounter a value with just the date elements (lacking the precise time). Often, the default setting is something like DD.MM.YYYY. It simply depends on the set NLS_DATE_FORMAT value. However, it does not mean that the time fields are not stored or specified! They are just not visualized in the output format.

As stated, NLS_DATE_FORMAT can be applied on all levels. The lower levels override the definition. Namely, the definition of the server format can be overridden by the session. Similarly, the session definition can be replaced for the particular function call level, if necessary (for example, by TO_CHAR). Vice versa, if the values are not explicitly specified, they are inherited across all levels, so the server specification is always present.

On the server instance level, the definition is performed by the alter system statement, as follows:

```
alter system
   set NLS_DATE_FORMAT='DD.MM.YYYY HH24:MI:SS' SCOPE=SPFILE;
```

Note that the parameter is static and can be applied after the instance restart. For now, it is impossible to change the instance behavior dynamically. The selection is then stored in SPFILE for the particular instance. The example of the SPFILE content is shown in *Figure 8.1*, highlighting the set NLS_DATE_FORMAT.

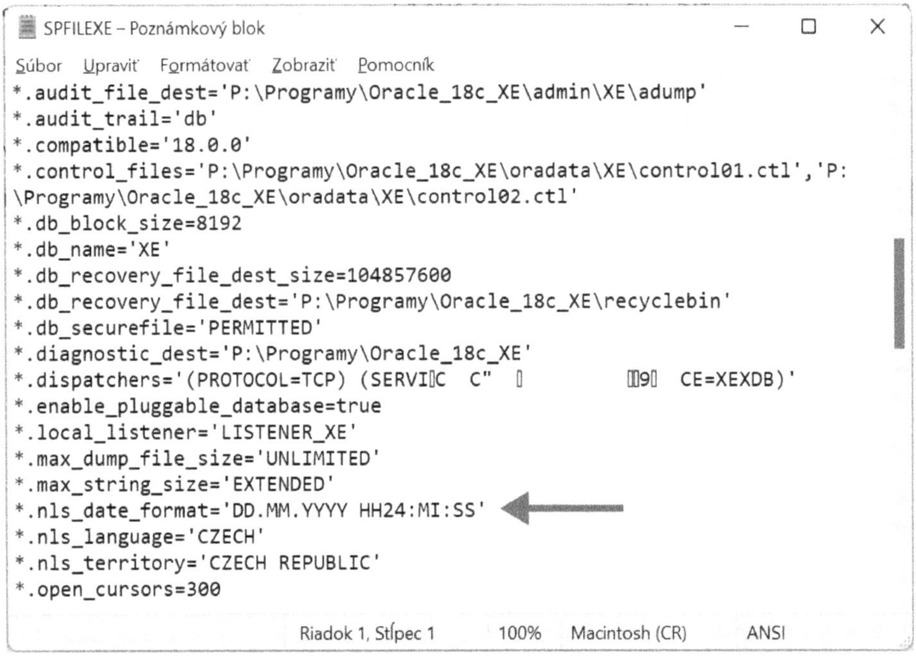

Figure 8.1 – SPFILE content

The `alter session` command works on a lower level compared to the system (database) definition and therefore, it overrides the current definition of the latter for the particular session. The example of the command in the following code snippet defines the default `DATE` format used in the session:

```
alter session set NLS_DATE_FORMAT='DD.MM.YYYY HH24:MI:SS';
```

Besides this, NLS parameters can be set using environment variables. In the next section, we will walk through the definitions for the Windows operating system; however, the principles we will cover can be applied to any environment.

Setting properties using environment variables

Client properties can be enhanced by environment variables. Their names must be the same as the given parameter name. For Windows operating systems, navigate to **Control Panel** | **System** | **Advanced system settings** (in the left panel):

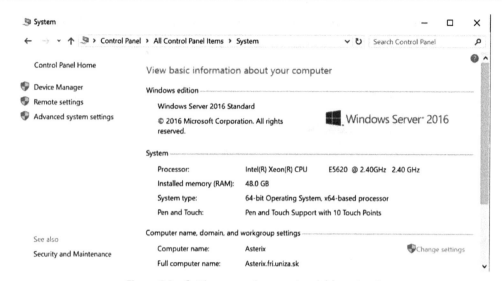

Figure 8.2 – Setting an environment variable – step 1

Click on the **Environment Variables…** button:

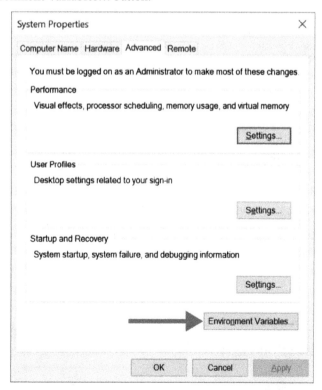

Figure 8.3 – Setting an environment variable – step 2

A new window will be launched consisting of a list of system variables. Click on the **New…** button to add a new environment variable:

Figure 8.4 – Setting an environment variable – step 3

The variable's name is the same as the particular NLS parameter value – the *name* is strict, so be aware when specifying it.

The *value* is the representation to be set (see *Figure 8.5* for reference). In my case, for **Variable name** I enter NLS_DATE_FORMAT and for **Variable value** I enter FM DAY - DD.MM.YYYY HH24:MI:SS.

Figure 8.5 – Setting an environment variable – step 4

Then, by referencing the implicit DATE format, the already defined value will be used:

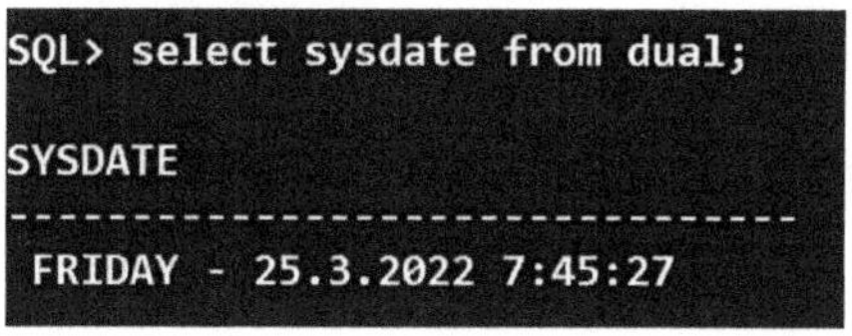

```
SQL> select sysdate from dual;

SYSDATE
----------------------------------
 FRIDAY - 25.3.2022 7:45:27
```

Figure 8.6 – Implicit conversion (DATE->character string) applying the defined format

NLS_DATE_FORMAT determines the output shape. However, what if the format uses a textual representation for individual elements, such as the name of the day or month in question? The format depends on the configuration of the NLS_DATE_LANGUAGE parameter covered in the next section.

NLS_DATE_LANGUAGE parameter

Another important parameter related to TO_DATE and TO_CHAR management is NLS_DATE_ LANGUAGE, which determines the textual language, mostly related to the day and month representations:

```
alter session set NLS_DATE_LANGUAGE='English';
select TO_CHAR(sysdate, 'FM DD (DAY).MONTH.YYYY')
  from dual;
--> 25 (FRIDAY).MARCH.2022
alter session set NLS_DATE_LANGUAGE='French';
select TO_CHAR(sysdate, 'FM DD (DAY).MONTH.YYYY')
  from dual;
--> 25 (VENDREDI).MARS.2022
```

The list of available languages can be found in the Oracle *Database Globalization Support Guide* here: https://docs.oracle.com/en/database/oracle/oracle-database/21/ nlspg/setting-up-globalization-support-environment.html.

However, even without studying the given documentation, the user can intuitively set the correct value, since the available specification options cover all commonly used language types.

Although the Gregorian calendar is generally used as the default option, different regions of the world often have their own rules and representations, and various calendar systems can be used here. The next section, therefore, covers the NLS_CALENDAR parameter.

NLS_CALENDAR parameter

The NLS_CALENDAR parameter represents the type of calendar used. The default option is Gregorian:

```
alter session set NLS_CALENDAR='Gregorian';
```

Besides that, the parameter can also take these values: Arabic Hijrah, English Hijrah, Japanese Imperial, Persian, ROC Official (Republic of China), and Thai Buddha.

The *Islamic calendar* (also known as *Hijrah*) is based on the moon. It consists of 12 months, the beginning of each of which is dependent on the visibility of the moon at the end of the previous month. Once the moon is sighted, the new month commences. Consequently, compared to the Gregorian calendar, which consists of 365 or 366 days (for leap years), the Hijrah calendar consists of 354 or 355 days.

The *Persian calendar* is used in Iran and Afghanistan. It is based on astronomical observations, not generally applied rules such as those used in the Gregorian calendar. It also consists of 12 months, where the first six months have 31 days and months 7 to 11 have 30 days. The last month (*Esfand*) generally has 29 days, and for leap years 30 days are present.

The 1st of January 2022 in the Gregorian calendar is represented as 1400 in the Islamic calendar and 1443 in the Persian calendar. Selecting different values here impacts the positioning of the day inside the calendar as well as the overall calendar representation. It might seem at first glance that one particular value does not significantly impact the processing, but the opposite is true. Namely, this impacts the other formats and NLS parameters, as demonstrated in the following code snippet.

Let's set the NLS_DATE_FORMAT in the first phase. However, by changing the NLS_CALENDAR value, the given NLS_DATE_FORMAT is changed as well, whereas its value is implicitly embedded by the selected calendar type:

```
alter session set NLS_DATE_FORMAT='DD.MM.YYYY HH:MI:SS';
alter session set NLS_CALENDAR='Arabic Hijrah';
select sysdate from dual;
--> 03 May 2022
```

Moreover, when dealing with calendars, the definition of a week inevitably has to be taken into consideration. What is the first day of the week? Is it Sunday or Monday? How does this impact the solution? The answer will be provided in the following section.

NLS_TERRITORY parameter

NLS_TERRITORY delimits the name of the given territory as used by the system, client, or the session. It determines the numbering conventions for days and weeks. As you may know, some countries consider Sunday as the *last day* of the week, while others prefer to *start* the week from Sunday.

The week starts on Sunday in the American region, whereas in Europe, the week starts on Monday. This can be set by specifying the NLS_TERRITORY parameter:

```
alter session set NLS_TERRITORY='America';
select TO_CHAR(sysdate, 'D') from dual;
--> 6
alter session set NLS_TERRITORY='Belgium';
select TO_CHAR(sysdate, 'D') from dual;
--> 5
```

At first glance, this may be obvious and unambiguous, but it must be borne in mind that if we want to find out what the day is, we may get incorrect values by changing the session characteristics. Also, comparison by textual day representation (the name of the weekday) may not produce the desired effect, as the result depends on the language configured in each case.

We demonstrate the depth of this problem with the following example (refer to *Figure 8.7*). Let's create a simple function to check whether the date is a workday or not. We suppose the following mapping – *Monday = 01*, *Tuesday = 02*, and so on:

```
create or replace function IS_WORKDAY (p_date DATE)
  return integer
is
  week_nr integer;
begin
  week_nr:=TO_CHAR(p_date, 'D');
  if week_nr between 1 and 5
     then return 1;
   elsif week_nr in (6,7)
     then return 0;
   else return -1;
  end if;
end;
/
```

The complete code of the IS_WORKDAY function can be downloaded from the GitHub repo of this chapter.

Calling the function works well if the mapping prerequisite is passed:

```
select sysdate, TO_CHAR(sysdate, 'DAY'),
        case IS_WORKDAY(sysdate) when 1 then 'IS WORKDAY'
                                  when 0 then 'IS WEEKEND'
                                  else 'UNKNOWN'
        end
   from dual;
--> 25.3.2022 11:34:44     FRIDAY    IS WORKDAY
```

Figure 8.7 shows the results in calendar form. March 25, 2022, represents a Friday. However, what about the order of the days in a given week? Does *Friday* refer to the fifth or sixth day of the week? What impacts and consequences could this have?

Figure 8.7 – Calendar displaying February 2022

Let's change the NLS_TERRITORY parameter to the *American region*. How would it affect the results? Well, the provided value for *Friday* will now be 6 instead of 5 as it was originally. As a result, Friday would be considered the *weekend*, as in the following snippet, which is incorrect:

```
alter session set NLS_TERRITORY='America';
select sysdate, TO_CHAR(sysdate+1, 'DAY'),
        case IS_WORKDAY(sysdate+1)  when 1 then 'IS WORKDAY'
                                     when 0 then 'IS WEEKEND'
                                     else 'UNKNOWN'
        end
   from dual;
--> 25.3.2022 11:34:44     FRIDAY    IS WEEKEND
```

Please be aware of this strict specification. Without proper management, incorrect data and decision-making can occur, as we've seen.

Even comparing textual representation does not provide sufficient power. Namely, although the provided value is independent of the NLS_TERRITORY parameter, some issues still can be present. Specifically, the value depends on the set language, defined by the NLS_DATE_LANGUAGE parameter. To highlight the problem, let's assume that the *English language* is used. The following function processes the position of the day in each week using textual representations:

```
create or replace function IS_WORKDAY(p_date DATE)
  return integer
is
  week_desc varchar(20);
begin
  week_desc:=trim(TO_CHAR(p_date, 'DAY'));
  if week_desc in ('MONDAY', 'TUESDAY', 'WEDNESDAY',
                   'THURSDAY', 'FRIDAY')
    then return 1;
  elsif week_desc in ('SATURDAY', 'SUNDAY')
    then return 0;
  else return -1;
  end if;
end;
/
```

As stated, if the NLS_DATE_LANGUAGE parameter is set to English, the overall processing works well and provides correct results:

```
alter session set NLS_DATE_LANGUAGE='English';
select sysdate, TO_CHAR(sysdate, 'DAY'),
       case IS_WORKDAY(sysdate) when 1 then 'IS WORKDAY'
                                when 0 then 'IS WEEKEND'
                                else 'UNKNOWN'
       end
  from dual;
--> 25.3.2022 11:34:44    FRIDAY    IS WORKDAY
```

The problem, however, occurs with language changes. Inside the function body, the comparison is made using English. However, if the language is changed, the mapping cannot be done, causing navigation to the ELSE clause of the conditional processing. As a result, an UNKNOWN value is returned:

```
alter session set NLS_DATE_LANGUAGE='French';
select sysdate, TO_CHAR(sysdate, 'DAY'),
        case IS_WORKDAY(sysdate) when 1 then 'IS WORKDAY'
                                 when 0 then 'IS WEEKEND'
                                 else 'UNKNOWN'
        end
  from dual;
--> 16.12.2021 16:05:33    VENDREDI     UNKNOWN
```

As is evident, the mapping was not done, because the comparison inside case did not find a suitable branch (unless the else branch covers the rest of the criteria).

In the preceding chapters, NLS parameters were applied to the session using the alter session command. This can, however, have negative consequences by influencing the whole session. Assume the expansion of the existing information system or application. Well, changing the behavior and format of the language on the session level can impact solutions already coded in existing applications, leading to incorrect outputs or exceptions being raised. Simply put, existing code might rely on the specific value of a parameter, thus it cannot be changed without revising and rebuilding the code.

The NLS_DATE_FORMAT parameter can be embedded in the TO_CHAR function as the third parameter (nls_parameter) to make it work in the specific context of a given statement only. Therefore, the session configuration itself will not be affected. However, NLS_TERRITORY cannot be applied in the TO_CHAR function, so in the next section, a workaround is introduced.

Embedding NLS parameters in the TO_CHAR function

To create a complex solution, the third parameter of TO_CHAR comes onto the scene. The language defined at the system, client, or session level needs to be relevant for applications that use durable database connections – but not only for them. Changing the date language can negatively impact already existing modules. If the user is not aware of the potential consequences, such an approach can be risky. Therefore, changing the processed language for the particular statement is more useful and robust. The third parameter (nls_parameter) of the TO_DATE and TO_CHAR conversion functions deals with the processed language:

```
TO_CHAR(<value> [, <format> [, <nls_parameter>]]
```

The main advantage of setting nls_parameter for the particular statement is robustness. It does not impact any other statements and strategies and only operates inside the proposed IS_WORKDAY function. Thus, correct results for whether a given day is a workday or not (based on the number of that day in the week) can always be obtained. Note that the value is in string format and encapsulated by apostrophes:

```
select TO_CHAR(sysdate, 'D', 'NLS_DATE_LANGUAGE=American')
  from dual;
```

The defined function is extended as follows to highlight the processed date language inside the TO_CHAR function call:

```
create or replace function IS_WORKDAY(p_date DATE)
 return integer
is
 week_desc varchar(20);
begin
 week_desc:=trim(TO_CHAR(p_date,
                         'DAY',
                         'NLS_DATE_LANGUAGE=American'));
  if week_desc in ('MONDAY', 'TUESDAY', 'WEDNESDAY',
                   'THURSDAY', 'FRIDAY')
      then return 1;
  elsif week_desc in ('SATURDAY', 'SUNDAY')
      then return 0;
  else return -1;
 end if;
end;
/
```

In this case, the solution is resistant to NLS_DATE_LANGUAGE parameter value changes at the system or session levels:

```
alter session set NLS_DATE_LANGUAGE='English';
select sysdate, TO_CHAR(sysdate, 'DAY'),
       case IS_WORKDAY(sysdate) when 1 then 'IS WORKDAY'
                                when 0 then 'IS WEEKEND'
                                else 'UNKNOWN'
       end
  from dual;
--> 25.3.2022 16:29:49    FRIDAY      IS WORKDAY
```

```
alter session set NLS_DATE_LANGUAGE='Slovak';
select sysdate, TO_CHAR(sysdate+1, 'DAY'),
       case IS_WORKDAY(sysdate) when 1 then 'IS WORKDAY'
                                when 0 then 'IS WEEKEND'
                                else 'UNKNOWN'
       end
  from dual;
--> 25.3.2022 16:29:49    PIATOK      IS WORKDAY
```

The comparison was done using full day names in the IS_WORKDAY function above. Let's return to the original aim of dealing with the weekday number. It is impossible to specify the territory in TO_DATE or TO_CHAR function calls. Only the NLS_DATE_LANGUAGE format can be used. Thus, the evaluation and transformation must be done on the full name level, followed by a transformation into the corresponding numerical value:

```
create or replace function IS_WEEK_DAY(p_date DATE)
 return integer
is
 week_desc varchar(20);
begin
 week_desc:=trim(TO_CHAR(p_date,
                        'DAY',
                        'NLS_DATE_LANGUAGE=American'));
 case week_desc
   when 'MONDAY'    then return 1;
   when 'TUESDAY'   then return 2;
   when 'WEDNESDAY' then return 3;
   when 'THURSDAY'  then return 4;
   when 'FRIDAY'    then return 5;
   when 'SATURDAY'  then return 6;
   when 'SUNDAY'    then return 7;
   else return -1;
 end case;
end;
/
```

The complete code of the IS_WEEK_DAY function can be downloaded from the GitHub repo of this chapter.

Now, get the name of the day and its sequence number inside the week. This provides the same results irrespective of NLS_TERRITORY or NLS_DATE_FORMAT:

```
alter session set NLS_TERRITORY='America';
alter session set NLS_DATE_FORMAT='DD.MM.YYYY HH24:MI:SS';
select sysdate, TO_CHAR(sysdate+1, 'DAY'),
       IS_WEEK_DAY(sysdate+1)
  from dual;
--> 25.3.2022 16:29:49    FRIDAY        5
alter session set NLS_TERRITORY='Belgium';
alter session set NLS_DATE_FORMAT='DD.MM.YYYY HH24:MI:SS';
select sysdate, TO_CHAR(sysdate+1, 'DAY'),
       IS_WEEK_DAY(sysdate+1)
  from dual;
--> 25.3.2022 16:29:49    FRIDAY        5
```

There are some things to consider regarding NLS_TERRITORY and the NLS_DATE_FORMAT value representation. By changing the territory, the selected date format will also be automatically overwritten. Therefore, in the preceding example, setting the territory is followed by setting the date format. Let's set the NLS_DATE_LANGUAGE parameter on the session level:

```
alter session set NLS_DATE_FORMAT='DD.MM.YYYY HH24:MI:SS';
select sysdate from dual;
--> 25.3.2022 16:29:49
```

Note if you change the territory using the NLS_TERRITORY parameter value, NLS_DATE_FORMAT will implicitly be changed as well. Thus, even if you specify the NLS_DATE_FORMAT parameter value in the first phase, it is then overwritten by changing NLS_TERRITORY, as evident in the following code snippet:

```
alter session set NLS_DATE_FORMAT='DD.MM.YYYY HH24:MI:SS';
alter session set NLS_TERRITORY='America';
select sysdate from dual;
--> 25-MAR-22
```

For completeness, we present a summary of other NLS parameters in the next section.

Other NLS parameters

The currently configured NLS parameter values can be obtained by querying the NLS_SESSION_PARAMETERS data dictionary view. It consists of two attributes – the parameter name (**Parameter**) and the associated representation (**Value**):

Parameter	Value
NLS_LANGUAGE	SLOVAK
NLS_TERRITORY	SLOVAKIA
NLS_CURRENCY	Sk
NLS_ISO_CURRENCY	SLOVAKIA
NLS_NUMERIC_CHARACTERS	, .
NLS_CALENDAR	GREGORIAN
NLS_DATE_FORMAT	DD.MM.YYYY HH24:MI:SS
NLS_DATE_LANGUAGE	ENGLISH
NLS_SORT	SLOVAK
NLS_TIME_FORMAT	HH24:MI:SSXFF
NLS_TIMESTAMP_FORMAT	DD.MM.RR HH24:MI:SSXFF
NLS_TIME_TZ_FORMAT	HH24:MI:SSXFF TZR
NLS_TIMESTAMP_TZ_FORMAT	DD.MM.RR HH24:MI:SSXFF TZR

Table 8.3 – NLS parameter values

Particular values can also be obtained using the show parameter command followed by the name specification with wildcards on both sides. Thus, the system looks for any occurrence of the defined string anywhere within the parameter name list:

```
show parameter NLS_DATE
--> NAME                TYPE    VALUE
--> ----------------    ------  --------------------
--> nls_date_format     string  DD.MM.YYYY HH24:MI:SS
--> nls_date_language   string  English
```

The same results would be provided by querying the data dictionary – `nls_session_parameters`. It provides two attributes, as well – the name of the NLS parameter (`parameter`) and value:

```
select * from nls_session_parameters
 where parameter like '%NLS_DATE%';
```

For the TIMESTAMP values formatting, the following NLS parameter definitions are available:

- NLS_TIME_FORMAT= 'HH.MI.SSXFF AM';
- NLS_TIMESTAMP_FORMAT= 'DD-MON-RR HH.MI.SSXFF AM';
- NLS_TIME_TZ_FORMAT= 'HH.MI.SSXFF AM TZR';
- NLS_TIMESTAMP_TZ_FORMAT= 'DD-MON-RR HH.MI.SSXFF AM TZR';

NLS parameters are always set for the whole system in the instance initialization file or by using defaults. System-level parameter values can be obtained by querying the `nls_database_parameters` data dictionary view:

```
select parameter, value from nls_database_parameters;
```

System-level values are inherited on the session level and can be changed dynamically. A useful solution for maintaining the session uses the event trigger associated with the login process. Inside the trigger, a particular session-specific NLS format can be set by calling the SET_NLS procedure of the DBMS_SESSION package. It takes two parameters – the name of the NLS parameter and the value to be set:

```
CREATE OR REPLACE TRIGGER CHANGE_DATE_FORMAT
AFTER LOGON ON DATABASE
begin
 DBMS_SESSION.SET_NLS('NLS_DATE_FORMAT','YYYYMMDD');
end;
/
```

This can also be set using the `alter session` command but be aware of the consequences related to EXECUTE IMMEDIATE and calling the DDL statement:

```
CREATE OR REPLACE TRIGGER DATABASE_AFTER_LOGON
  AFTER LOGON ON DATABASE
begin
  EXECUTE IMMEDIATE 'alter session set NLS_DATE_FORMAT =
```

```
                        ''DD-MON-YYYY HH:MI:SS''';
    end DATABASE_AFTER_LOGON;
    /
```

NLS parameters form an important component of the Oracle database, applying local rules used in a particular territory. We have dealt with the most important parameters concerning date and time processing, mostly related to format, language, and territory specifics. This section provided a summary of other NLS parameters, delimiting the used currency in a region, decimal separators, sort principles, and so on. By understanding the complexity of the principles described in this chapter, you will be able to create robust solutions that can applied worldwide, taking into account regional rules and principles.

Summary

In this chapter, you learned about regional settings. Oracle Cloud provides you with multilingual databases adjustable for each user separately. Date and time values are affected by the NLS parameters. `NLS_DATE_FORMAT` determines the output format and implicit conversions. `NLS_DATE_LANGUAGE` determines the language of date and time values when textual representations are used. `NLS_TERRITORY` is responsible for the definition of the week and whether the first day of the week is Monday or Sunday. All these parameters can be set on the *session* level or inherited from the *system* itself. A specific approach to the `TO_CHAR` function, embedding the NLS parameter inside the call, was also explored. Using this, the session configuration will not be affected, with the original parameters remaining in force.

This chapter concludes *Part 3* of the book – *Modeling, Storing, and Managing Date and Time*, dealing with the principles of storage, construction, conversion, extraction, and management functions, as well as the use of NLS parameters to apply regional rules.

The next chapter, which also marks the start of *Part 4*, *Modeling Validity Intervals*, focuses on the timepoint definition, duration modeling, interval representations, and transformations. It also deals with modeling and identifying unlimited validity.

Questions

1. What is the scope of the `NLS_DATE_FORMAT` parameter applied to the whole system?

    A. SPFILE only

    B. Memory only

    C. SPFILE and memory

    D. Physical database only

2. Which of the following parameters determines the first day of the week?

    A. NLS_CALENDAR

    B. NLS_TERRITORY

    C. NLS_REGION

    D. NLS_WEEK

3. Which NLS parameter can be embedded in the TO_CHAR function?

    A. NLS_DATE_FORMAT

    B. NLS_DATE_LANGUAGE

    C. NLS_TERRITORY

    D. NLS_CALENDAR

4. What is the name of the data dictionary that holds NLS parameter values for the system?

    A. NLS_SESSION_PARAMETERS

    B. NLS_SYSTEM_PARAMETERS

    C. NLS_PARAMS

    D. NLS_DATABASE_PARAMETERS

Further reading

Consider the following titles to learn more about the topics covered in this chapter:

- *The Database Globalization Support Guide*, covering NLS parameters as part of the Oracle documentation, can be read using the following link: https://docs.oracle.com/en/database/oracle/oracle-database/21/nlspg/setting-up-globalization-support-environment.html.

- *The Globalization of Language in Oracle – National Language Support*, by *James Koopmann*, deals with the Oracle data dictionary views that hold NLS parameters for different levels (database, instance, and session levels). It also emphasizes the `NLS_LANG` parameter, which can cause problems during maintenance and export if the value is not properly set and referenced. It provides two export scenarios and related instances of exception raising: `https://www.databasejournal.com/oracle/the-globalization-of-language-in-oracle-national-language-support/`.

Part 4:
Modeling Validity Intervals

The aim of this part is to provide a complex description of validity modeling using intervals. Duration modeling generally requires two timepoints expressing a start and end point. However, what about unlimited validity? How can you model it? It is generally clear that if a new state is to be loaded into the system, the user does not need to know when its validity will end – or even whether it ever will. Thus, we will describe various techniques of unlimited validity modeling. Oracle Database provides the `Period` definition to carry out this modeling.

Temporal models will also be discussed, focusing on various granularity precisions. But that is not enough. Therefore, temporal dimensions will also be explored, such as **validity**, **transaction validity**, the **IPL model**, the **IPLT model**, **online** and **offline** application modes, and **future valid record** management. We will draw your attention to the fact that the existence of attributes expressing date and time alone does not necessarily mean that the system is temporally oriented.

This part includes the following chapters:

- *Chapter 9, Duration Modeling and Calculations*
- *Chapter 10, Interval Representation and Type Relationships*
- *Chapter 11, Temporal Database Concepts*
- *Chapter 12, Building Month Calendars Using SQL and PL/SQL*

Duration Modeling and Calculations

Data and time values represent the finest processing precision and do not reflect any duration; just one point in time is referenced. Using the DATE data type, a granularity of up to one second can be used. Based on the definition, the granularity precision can be of the order of nanoseconds when dealing with TIMESTAMP. Oracle does not provide a specific data type for dealing with date elements with no time element. Therefore, the meaning of the DATE data type value depends on precision, and time elements do not need to be taken into account. Typically, an employment contract is in days rather than hours. Consequently, dealing with more precise granularity frames, such as hours, minutes, or even seconds, is unnecessary. But they always need to be stored in the DATE value.

Similarly, an invoice due date is a day, the time is not essential. When we look at energy payments, they are usually made once a month, so we can use monthly granularity. Thus, although the physical representation is DATE, there is no smaller granularity data type in Oracle, except for the explicit integer management, which, brings additional user code demands to ensure security and reliability. As discussed in *Chapter 5*, date and time can be managed at various granularities. The granularity definition and management affect duration management as well.

This chapter deals with the principles of duration management and discusses usability and constraints in temporal systems. There can be various meanings for the duration. Typically, the duration can express the validity, but transaction time reference can also be used for data corrections (a row can be considered valid for a limited time frame, followed by the data correction). Transaction approval timepoint can be useful for long-term transactions. When using distributed databases or data guards, it is inevitable to evaluate the duration of the data row spread; therefore, such timepoint elements can be stored and evaluated for each node. Thus, all these date and time references can express the duration.

From a logical modeling point of view, time elapsed can be represented in database systems by a single time value (timepoint) or by a validity time interval (duration).

This chapter will cover the following main topics:

- Characterizing one point in time, along with its meaning and representation
- Modeling duration expressed by the begin point of the validity
- Modeling duration using begin and end limits, forming closed-closed or closed-open representations
- Expressing unlimited validity

The source code for this chapter can be found in the GitHub repository accessible via `https://github.com/PacktPublishing/Developing-Robust-Date-and-Time-Oriented-Applications-in-Oracle-Cloud/tree/main/chapter%2009`. You can also scan the following QR code:

What timepoint means and how to use it

The moment in time is characterized by only one value expressing validity. It can be used to model either *one timepoint* or the *start point of the duration*. If the timepoint represents only one value, it must be ensured that a new state is inserted for each moment, regardless of the change. For example, a new data image must be entered into the database every month if monthly granularity is used. The management of the system is controlled by a plan (calendar) of events, which occurs periodically. These systems are suitable for obtaining a slice (image) of data in precisely defined periods but without further analysis of the changes in individual attributes. This limitation is just related to storage demands. Data is stored periodically, irrespective of the real change. As a result, many duplicates can be present. Expressing the duration by the *end point* is uncommon as it can be unknown or not very precise. People usually don't know in advance when a state will expire, but they do know when it started to become valid. A typical example is a price change. If you go to the market, you can tell that the price has increased, but you cannot specify the next price increase exactly in advance, right?

The duration of the state is defined by the granularity of the system, denoted by `chronon`, which expresses an indivisible unit of time used in a system. Let's look at the model shown in *Figure 9.1*. Each temporal data tuple is characterized by the beginning point of the validity (typically expressed by the **BD** attribute). As is evident, each tuple is valid for one month; thus, each month, a new state is stored. The limitation of the model shown in *Figure 9.1* is consistency when changing the granularity.

By transferring the system to the more precise characteristics (per day, per hour, per minute, or even per second), undefined states arise. Therefore, the whole system should be transformed and reconstructed to limit inconsistency by making copies of historical states according to the newly defined granularity. Consequently, more and more duplicate tuples are present. Specifically, when moving from daily to hourly granularity, it is necessary to copy each historical state for each hour (which would require adding 23 new states). Moreover, even if the data is not changed, a new state is always stored for each month.

Figure 9.1 shows the data. Let's consider a simple table consisting of the object identifier, the starting point of the validity, and the data itself. BD expresses the starting point of the validity. Please note that *starting point* and *beginning point* of the validity are synonyms, and both terms are used in the literature.

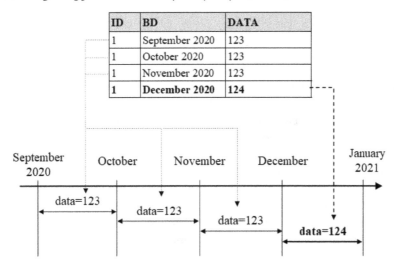

Figure 9.1 – Moment of time representation

The preceding principles and usage are system, source, and storage demand. Processed granularity is a critical factor influencing the performance of access and storage management. Data is always stored, regardless of the real change. Therefore, significant change identification can be complicated.

Changing the meaning of the timepoint representation makes achieving a significantly more effective solution possible. States (attribute values) do not need to change their values within a precisely defined period. Hence, the time value no longer expresses only one point but the beginning of the duration (mostly expressing the *validity* of the state). In this case, it uses the principles of unlimited validity of the current state; the approach of the **until further notice rule** is used (*Figure 9.2*). Namely, each consecutive state referencing a particular object terminates the validity of its predecessor by introducing a new state. Therefore, if no definition of the state is valid in the future, the current state is characterized by the highest value of the time reference or the last inserted record for the object.

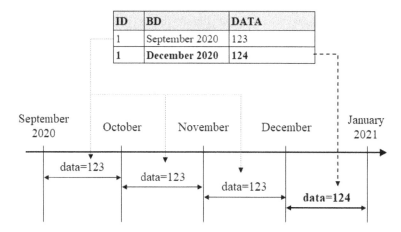

Figure 9.2 – Beginning point of the validity representation

When there is no upper limit to the duration and validity, and when the next state terminates its direct predecessor, it is crucial to know how to identify and obtain current valid states. This can be done by sorting individual states into partition groups formed by the objects. The principles of partition management are stated in the next section.

Getting the current state

To get the current state for each object, the RANK analytical function can be used to serve the result set for each object. The inner statement produces the data and the ranking, sorted based on the date value for each object (set by the PARTITION BY clause):

```
SELECT ID, BD, DATA
  FROM
    (SELECT tab.*,
            RANK() over(PARTITION BY ID ORDER BY BD desc)
                                               as rank
      FROM Tab
    )
   WHERE rank=1;
```

That statement produces the most up-to-date data version for each object (ID). This can also be done by using aggregations, as shown in the following code snippet:

```
SELECT *
  FROM Tab
    WHERE (ID, BD) in (SELECT ID, max(BD)
```

```
        FROM Tab
     GROUP BY ID);
```

It is not easy to determine which solution is better because it depends on various characteristics, mostly related to indexing. However, RANK is generally an analytical function optimized to number individual records that are part of the result set.

If valid future states can be present in the system, then a condition limiting the processing to the history and current states is added:

```
   WHERE BD<=sysdate
```

FETCH FIRST ROW ONLY will work correctly only if one object is handled. This syntax element does not have a partition clause, so the whole result set must be treated completely (irrespective of individual objects) like in the following snippet:

```
SELECT * FROM Tab
 ORDER BY BD
   FETCH FIRST 1 ROW ONLY;
```

The main advantage of beginning point management is associated with a significant reduction in the amount of data stored. Information value, however, remains the same. Moreover, such a system is resistant to changes in granularity and robustly identifies significant data value changes.

For each state, begin (BD) and end (ED) points of validity can be obtained. Merging each row with the next object state (based on the validity and timeline reference) produces the desired time limits. BD is obtained directly from the database. Each new state terminates its direct predecessor. Each ED is limited by the BD value of the next row. The following query transforms the data to the format covering the time interval from the beginning point to the end point of the validity. In this case, ED represents the first timepoint at which the referred state is not valid. Unlimited validity is, in this case, expressed by the NULL notation:

```
SELECT ID,
       BD,
       LEAD(BD) OVER(PARTITION BY ID ORDER BY BD) as ED,
       DATA
 FROM Tab;
 --> 1     01.01.2022 14:02:20   02.01.2022 14:02:20   data
 --> 1     02.01.2022 14:02:20   03.01.2022 14:02:20   data
 --> 1     03.01.2022 14:02:20   NULL                  data
 --> 2     11.12.2021 14:02:36   13.12.2021 14:02:36   data
 --> 2     13.12.2021 14:02:36   15.12.2021 14:02:36   data
```

In this case, unlimited validity is expressed by the NULL value for ED. In the *Modeling unlimited validity* section, we will discuss various options and limitations for this representation. Namely, the NULL value often does not reflect what we need to express and model.

Object granularity is used in the model principle described earlier in this chapter, meaning that all the data is retained (or copied from the existing state) in case of any update. However, what about a situation in which it is impossible to define a new valid state, but it is clear that the original state is no longer suitable? How do we model it if only BD is stated?

Well, the preceding model cannot manage such situations. The NULL value for individual data attributes does not provide sufficient power because they can have special notation and meaning and cannot be replaced generally. Therefore, the model architecture must also be extended to cover undefined states. Going deeper, it does not need to reflect the entire state. Thus, a special notation must exist for the undefined attribute value or the whole state. Such cases can have various causes, so it is important to categorize the reasons for further analysis. The next section deals with the duration expression.

Deploying duration models using timepoint borders

Although using timepoints in the system makes it possible to model the elapsed time duration, border identification can be demanding and too time-consuming; therefore, it is necessary to find the direct next state. The processing takes place in two phases. Firstly, the states where the validity begins after a defined timepoint (T) are identified, then the state with the lowest value of the beginning of validity is selected from this set. OBJ_REF expresses the referenced object identifier:

- First phase:

```
SELECT Tab.*, RANK() OVER(ORDER BY BD) as rank
  FROM Tab
    WHERE BD>T and ID=OBJ_REF;
```

- Second phase:

```
SELECT *
FROM
    (SELECT Tab.*, RANK() OVER(ORDER BY BD) as rank
      FROM Tab
        WHERE BD>T and ID=OBJ_REF
    )
WHERE rank=1;
```

Note that a solution where only the *end date of the validity* (*state expiration*) is modeled is uncommon. It would be impossible to identify the time of the first entry in the database, and a special flag would have to be introduced. However, the processing would be analogous to the already described principles of BD modeling and management.

In conclusion, for time duration management, it is preferable to use a time interval defined explicitly by two timepoints, which express the beginning and the end of the interval, to express the duration of a certain time range.

Generally, there are four types of time validity representations:

- `left open, right open interval (OO)`
- `left open, right closed interval (OC)`
- `left closed, right closed interval (CC)`
- `left closed, right open interval (CO)`

The following sections will explore the last two variants, characterized by left-closed intervals, because these types precisely define the beginning of validity, regardless of the granularity used.

Closed-closed representation

Representing a time period using CC representation is an intuitive type of duration modeling based on a conventional database environment. The validity interval is precisely limited. Both borders belong to the interval. *Figure 9.3* shows the data model for the `Employee` table. Date and time elements are expressed by the `Date_from` and `Date_to` attributes:

Employee			
⬦ Employee_id	Integer	NN	(PK)
Name	Varchar2(30)	NN	
Surname	Varchar2(30)	NN	
Date_from	Date	NN	
Date_to	Date		
Position	Varchar2(20)	NN	
Salary	Number(6,2)	NN	

Figure 9.3 – Employee structure

A typical example of a CC interval is the validity period, represented, for example, by an employment contract. For example, the employee with the `ID=1` personal number was hired on January 15, 2017, and the last day he still worked was December 31, 2020. As of January 1, 2021, he is no longer an employee of the company.

The formal mathematical notation is shown in *Figure 9.4*:

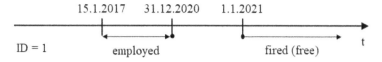

Figure 9.4 – Employment contract – CC representation

To get the list of the employees contracted during the T timepoint, the following statement can be used:

```
SELECT *
  FROM employee
    WHERE T between DATE_FROM and DATE_TO;
```

The main benefit of this CC interval approach is related to using the built-in between function. This function covers the numerical or date ranges and includes the borders of the < T1 ; T2 > interval. There is a natural assumption that T1 < T2. The method is *inclusive*.

When obtaining a list of employees in a defined time interval, it is necessary to consider the representation – the meaning of the time interval and the period coverage (*Figure 9.5*). If the user wants to get only those employees who were at the company during the entire defined period, < TBD ; TED >, the SELECT command would look like the following:

```
SELECT *
  FROM employee
    WHERE DATE_FROM<=TBD
      AND DATE_TO>=TED;
```

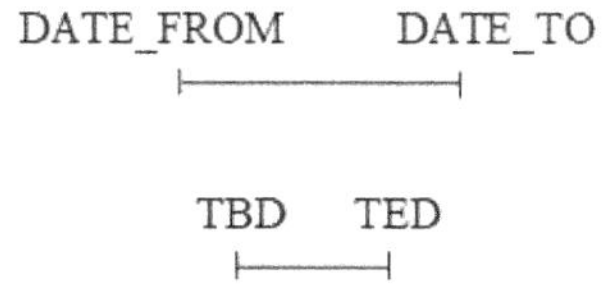

```
          DATE_FROM       DATE_TO
          ├───────────────────────┤

              TBD     TED
              ├───────────┤
```

Figure 9.5 – Interval positions – coverage – CC representation

If it were necessary to obtain those people who were employed for at least a while in a given time interval, < TBD ; TED >, the solution would be extended to focus on partial coverage. A logical interpretation is as follows:

```
SELECT *
  FROM employee
    WHERE <DATE_FROM, DATE_TO> intersects <TBD, TED>;
```

The physical representation is associated with the between function call. The intersection can be located either on the left or right side (*Figure 9.6*):

```
SELECT *
  FROM employee
    WHERE (TBD between DATE_FROM and DATE_TO)
          OR
          (TED between DATE_FROM and DATE_TO);
```

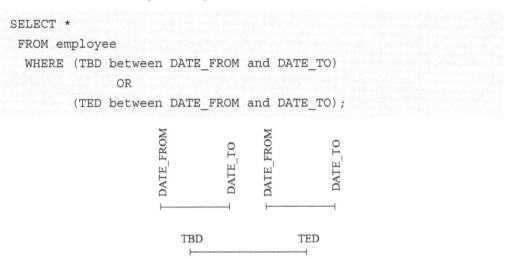

Figure 9.6 – Interval positions – overlaps – CC representation

By evaluating the records of employment contracts on the timeline, it may seem at first glance that there is an undefined state between the dates (refer to the gap between the states in *Figure 9.4*). Simply, the employment contracts of the *ID=1* employee do not directly follow each other. However, is it really true? Is an undefined state present? Well, the model uses a daily granularity. Thus, the time elements are not treated, and there is *no granularity* between the states. From the visual representation, without expressing the proper granularity of the processing, it does not need to be directly visible. To intensify the problem, note the dates February 28, 2015, and March 1, 2015. Is there an extra day? No, because this is the last day of February and the first day of March. However, be aware of leap years.

In the following table, the issue of leap years is clearly shown. It is necessary to keep this in mind for CC representations and daily granularity. In 2018, the last day of February was February 28. Thus, there is no day that the *ID=1* employee is not covered by a valid employment contract. However, what if a similar situation occurs in 2020, which is a leap year? Yes, February 29 would not be covered by the employment contract, and a gap would be present. The issue would not occur when using a CO interval. The following table showcases an example of CC temporal data management:

ID	BD	ED	DATA
1	1.1.2013	28.2.2018	123
1	1.3.2018	31.5.2012	456
2	1.1.2003	31.8.2020	789

Table 9.1 – Duration using leap year coverage

The data states and timeline reference for the ID=1 object are shown in *Figure 9.7*. From a graphical perspective, a gap between those states seems to be present; however, daily granularity is being used. On the other hand, a change in the granularity of the processing would cause a loss of consistency and, thus, incorrect results:

Figure 9.7 – Gap

There is no point in time between 28/02/2018 and 01/03/2018 if we use daily granularity. However, if we moved the solution to the time management level (for example, hourly granularity), the original record would express midnight (28/02/2018, 00:00:00), and another record would represent midnight of 01/03/2018, 00:00:00. But using the changed precision accuracy (transferring the processing from daily to hourly granularity), there would be 24 uncovered hours.

Closed-open representation

The second type of time interval is a CO representation. The right limit of the interval expresses the immediately following timepoint after the last time instant (granule) of validity of the given time interval. For example, < 1/1/2021 ; 1/1/2022 > expresses the validity during the whole year of 2021. Since we assume that the object can be defined at a given time by only one state (a person can have only one employment relationship with the company at a given time), the first possible date of the beginning of another state (another employment contract) may be 1/1/2022. Generally, each object in the temporal database is characterized by one valid state at any time (or no more than one).

The advantage of CO interval modeling is the easy identification of undefined states – states that are not covered by a valid record stored in the database. Furthermore, because the left boundary of the duration interval (BD) is identical to the right border (ED), if the intervals are directly connected, the complete coverage is easily identifiable by composing the state validity chain.

When changing the date and time processing granularity, it is unnecessary to modify the existing data in any way; undefined states do not arise during the transformation.

The command to get the state of an object at a defined point in time (T) can look like this:

```
SELECT *
  FROM employee
    WHERE T between DATE_FROM and DATE_TO-1;
```

Or it can look like this:

```
SELECT *
  FROM employee
    WHERE T >= DATE_FROM AND T < DATE_TO;
```

The -1 expression in the first code block subtracts the smallest managed piece of time. When using daily granularity, the 1 value expresses *one day*. By using hourly granularity, the particular smallest granule would be one hour, represented by the 1/24 value.

Now let's think about the coverage modeling. It is useful if we are looking for an overlap of states or when one state must be covered by another within the hierarchy modeling. For example, an employee can manage the device only if they have a valid employment contract in the given period.

Coverage

When using complete state coverage (*Figure 9.8*), the solution is as follows:

```
SELECT *
  FROM employee
    WHERE DATE_FROM <= TBD
      AND DATE_TO >= TED;
```

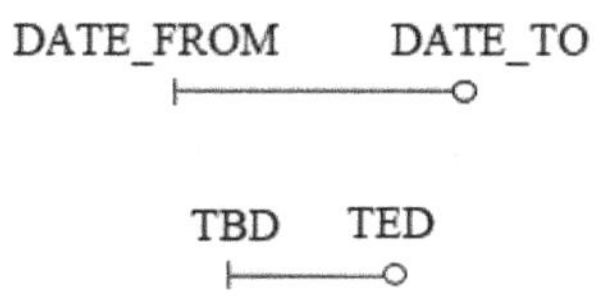

DATE_FROM   DATE_TO
|————————o

TBD   TED
|————o

Figure 9.8 – Interval positions – coverage – CO representation

Partial state coverage can be modeled and obtained using the logical representation of the CO interval associated with the `intersects` type (*Figure 9.9*):

```
SELECT *
  FROM employee
    WHERE <DATE_FROM, DATE_TO) intersects <TBD, TED);
```

The transformation of the `intersects` positional type is done by using two conditions evaluating the coverage of the duration, as stated in the following code block:

```
SELECT *
  FROM employee
    WHERE (TBD between DATE_FROM and DATE_TO-1)
              OR
          (TED between DATE_FROM and DATE_TO-1);
```

Alternatively, `between` can be replaced by two conditions:

```
SELECT *
  FROM employee
    WHERE (TBD >= DATE_FROM and TBD < DATE_TO)
              OR
          (TED > DATE_FROM and TED <= DATE_TO);
```

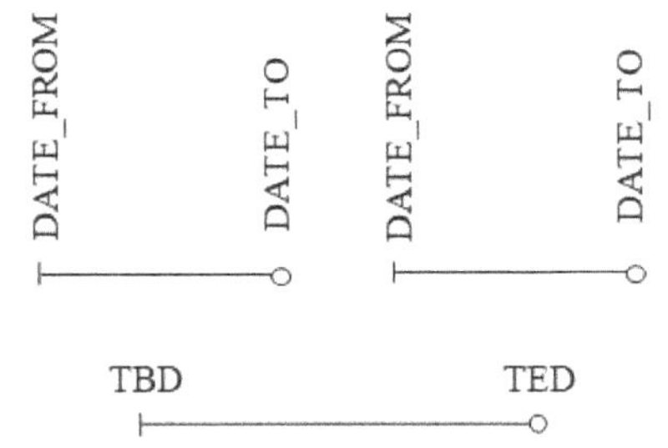

Figure 9.9 – Interval positions – overlaps – CO representation

The most common duration representations have been discussed. The next section explains how to transform them into each other.

Transformation of duration intervals

When creating and manipulating temporal databases, it is important to pay attention to the types of duration intervals because by confusing individual types, undefined periods of validity or periods during which two different states of one object are valid can be identified. Individual types of duration intervals can be transformed into each other while preserving the original time validity. *Table 9.2* to *Table 9.5* shows how to transform data from one type to another. The +1 term indicates the addition of one granule to the duration interval.

A CC interval (`<BD1, ED1>`) can be easily transformed into a CO interval (`<BD2, ED2)`). For the transformed interval to express identical values to the original ones, it must be true for the interval boundaries that `BD1 = BD2` and `ED1 + 1 = ED2`. For the opposite transformation, for example, from a CO interval (`<BD1, ED1)`) to a CC interval (`<BD2, ED2>`), the following rules apply so that both intervals express the same time validity: `BD1 = BD2` and `ED1 = ED2 + 1`.

The following table shows all the types of CC transformation. The first column takes two time intervals to be considered. The second column states the conditions that must apply to express the same value in different representations. The < symbol expresses the closed type, while) expresses the open type:

ORIGINAL VALUE	OUTPUT TYPE	PRECONDITIONS TO APPLY	
CC characteristics are defined by `<BD1, ED1>`	`<BD2, ED2>`	BD1=BD2	ED1=ED2
	`<BD2, ED2)`	BD1=BD2	ED1+1=ED2
	`(BD2, ED2>`	BD1=BD2+1	ED1=ED2
	`(BD2, ED2)`	BD1=BD2+1	ED1+1=ED2

Table 9.2 – Transformation rules from the CC duration interval to all other representations

The following table showcases the transformation rules based on the CO representation:

ORIGINAL VALUE	OUTPUT TYPE	PRECONDITIONS TO APPLY	
The CO characteristics defined by `<BD1, ED1)`	`<BD2, ED2>`	BD1=BD2	ED1=ED2+1
	`<BD2, ED2)`	BD1=BD2	ED1=ED2
	`(BD2, ED2>`	BD1=BD2+1	ED1=ED2+1
	`(BD2, ED2)`	BD1=BD2+1	ED1=ED2

Table 9.3 – Transformation rules from the CO duration interval to all other representations

The following table shows the transformation rules based on the OO representation:

ORIGINAL VALUE	OUTPUT TYPE	PRECONDITIONS TO APPLY	
The OO characteristics defined by (BD1, ED1)	`<BD2, ED2>`	`BD1+1=BD2`	`ED1=ED2+1`
	`<BD2, ED2)`	`BD1+1=BD2`	`ED1=ED2`
	`(BD2, ED2>`	`BD1=BD2`	`ED1=ED2+1`
	`(BD2, ED2)`	`BD1=BD2`	`ED1=ED2`

Table 9.4 – Transformation rules from the OO duration interval to all other representations

The following table shows the transformation rules based on the OC representation:

ORIGINAL VALUE	OUTPUT TYPE	PRECONDITIONS TO APPLY	
The OC characteristics defined by (BD1, ED>	`<BD2, ED2>`	`BD1+1=BD2`	`ED1=ED2`
	`<BD2, ED2)`	`BD1+1=BD2`	`ED1+1=ED2`
	`(BD2, ED2>`	`BD1=BD2`	`ED1=ED2`
	`(BD2, ED2)`	`BD1=BD2`	`ED1+1=ED2`

Table 9.5 – Transformation rules from OC duration interval to all other representations

If the duration interval were degraded to represent only one moment, then the transformation of the intervals would look like the one shown in the following table:

ORIGINAL VALUE	OUTPUT TYPE	PRECONDITIONS TO APPLY	
One timepoint (T)	`<BD, ED>`	`BD=T`	`ED=T`
	`<BD, ED)`	`BD=T`	`ED=T+1`
	`(BD, ED>`	`BD+1=T`	`ED=T`
	`(BD, ED)`	`BD+1=T`	`ED=T+1`

Table 9.6 – The transformation rules of one timepoint

With that, you are now familiar with timepoint and duration modeling, which comprises various representations and transformations. We examined the characteristics and evaluation principles of timepoint and duration modeling. In the next section, we will explain the issue related to gap creation if the granularity is changed. For a temporal model, such a transformation can be critical and demanding if the data transformation is necessary to limit undefined states.

Interval representation remarks

When using the CC interval representation, it is necessary to know additional information about the granularity of the processed data. Changing to a smaller granularity requires the transformation of the borders and the processing of undefined states. Existing data need to be updated to limit undefined states. The trend of current information systems and control applications is to increase accuracy and the amount and frequency of input data processing and evaluation. Therefore, a change in data granularity is often required during system deployment (or even on-demand dynamically). The process of changing and transforming the system is simple and fast when using an open-ended (CO) representation.

Conversely, if the interval is characterized as a closed representation from the right side of the interval (CC), the conversion can take a very long time. In addition, because the update statements expressed the transformation, it is also necessary to apply transaction locks to the data covered by the indexing. The relevant dataset is, therefore, unavailable for a long time.

The end point of the validity can be expressed by the NULL notation, as described in the previous sections. However, does it really mean unlimited validity? Honestly, the NULL value represents an *undefined value*, not unlimited validity. It is clear that some informational value is present there. Moreover, in general, the NULL value adds complexity to all SQL processing. Therefore, in the next section, there is a discussion about the meaning, representations, and modeling of unlimited validity.

Modeling unlimited validity

The reliability and security of the entire time-delimited system directly depend on the accuracy and quality of the data stored in it. The main emphasis is placed on the accuracy of individual values, undefined states, incorrect states, and validity limits.

As already mentioned, the validity of an object's state is mostly modeled using a logical and physical time interval definition. The implemented solution contains two timeline values expressing the beginning and end of the validity concerning the interval representation. The start of the validity is commonly precisely specified. Therefore, it is necessary to know exactly when the validity of the new state begins. The end point of the validity can be more complicated. If a synchronization layer is present in the system and the data is modified with precisely defined time periodicity, the validity of each state can be clearly defined. However, the problem occurs when the data change is asynchronous. In this case, it is not possible to limit the state's validity from the right side of the validity definition. It is simply impossible to determine for how long the particular state will be valid in advance. This problem is mostly associated with the currently valid states or states whose validity will begin in the future.

A typical example of this situation is a person's address. By default, it is not feasible to determine when, or if at all, the person will move. Another example is the change of job. If the employment is a fixed-term contract, then the limit is clear. However, if there is an employment contract for an indefinite period, it is unknown when the person will be fired or will leave. There are many other examples in the field of industry, transport, and sensor processing where it is impossible to limit the validity in advance. Simply, the current state applies until it is changed. Therefore, the question remains how to correctly express an unlimited, undefined validity at a given time.

To represent an undefined value, either because the user does not know the value, the value is not needed for the system, or the value is inaccurate, NULL is used in conventional systems (e.g., the wedding date of a person who is still single is expressed as a NULL value, or the graduation date of just enrolled student is undefined (NULL)). The main characteristic of the NULL value is complete ignorance of the value (although, in the case of sensory networks, it is sometimes possible to estimate or calculate the missing value based on business analytics). However, this is not the case with time constraints expressing the validity spectrum. Although we do not know the exact value of the expiry, it is clear that this event has not occurred yet. If it does occur, it will definitely be in the future. Therefore, for unlimited validity, the terms *until changed* or *until further notice*, are used. Therefore, it cannot be said that the value of the end of the interval is completely unknown. From a time point of view, it is evident that the state validity's right border will definitely be placed in the future.

Despite the unclear temporal meaning of the NULL value representation, in the first phases of temporality management models, a NULL value was also used for the time element. However, it was soon discovered that this representation was inappropriate in several ways. The first important way is the fact that the value is only partially unknown. In any case, the change in validity will certainly be in the future. The impossibility of storing an undefined value in standard indexes (B-tree, B+tree) is also a significant performance element since the NULL values cannot be compared using mathematical operations. Therefore, traversing the index structure by depending on the required smaller (traversing the tree to the left) or greater (traversing the tree to the right) value is impossible. Consequently, the index for the data location could not be used, meaning the whole dataset needs to be scanned sequentially, block by block. Thus, the performance of the data retrieval would degrade.

Last but not least, there is the issue of entity integrity. In temporal systems, the attributes expressing validity are part of the unique identifier of the entity – the *primary key* or *unique index*. The definition of the primary key (or unique constraint more generally) requires any part (element) to be precisely defined. Thus, it cannot contain a NULL value inside any primary key element. But since the expiration (end point of the validity) can be undefined and thus hold a NULL value, it cannot be part of the primary key, right? Thus, it would mean that the validity interval and its physical representation would have to be divided into special attributes with the need for separate comprehensive management of the states. Existing models, mainly based on object granularity, would be unusable and would need to be modified and transformed with undefined expiration values.

Since it is clear that the unbounded validity of the state expresses a timepoint in the future, the NULL value has been replaced by a sufficiently large time value in the future. However, the question remains as to what the term *sufficient value* means. Is it the year 2500? 3000? 5000? Or is the year 2100 enough? How can we distinguish between real validity in the future and the special expression of an undefined boundary and thus ensure a robust and correct solution?

The MaxValueTime notation was introduced in the *Managing Time in Relational Databases* book (Tom Johnston, Randall Weis, Morgan Kaufmann, 2010). It replaces the conventional principle of relational databases. It is based on the value associated with the maximum date (currently December 31, 9999). The difference lies mainly in the comparison and explicit definition of the date component. From a logical point of view, this definition uses the MaxValueTime notation. Although the physical implementation uses a date value, the directly specified value is not identical to the MaxValueTime notation, which we can demonstrate in the following example. As you can see, the physical representation is the same, but the logical view is not. Thanks to that, it is possible to distinguish individual values and store states whose validity is strictly defined in exactly the same manner as indicated using the MaxValueTime notation. The following code shows the principle of undefined value modeling, representation, and evaluation:

```
DECLARE
   date1 DATE:=MaxValueTime;
   date2 DATE:=TO_DATE('31.12.9999', 'DD.MM.YYYY');
BEGIN
  IF date1=date2 THEN
   DBMS_OUTPUT.PUT_LINE('values are the same...');
  ELSE
   DBMS_OUTPUT.PUT_LINE('values are different...');
  END IF;
END;
/
```

The preceding code produces values are different.... Although it is rather a logical concept, it is evident, that there should be a specific notation for replacing undefined values with a precise definition, allowing the system to compare and sort values.

Note that the NULL values cannot be mathematically compared by providing NULL in the evaluation.

A significant element of the already-mentioned MaxValueTime notation is binding to interval (relational) operations that can be used. The following condition would be used to find the currently valid states:

```
WHERE sysdate between BD and ED
```

To get the list of current states that have unlimited validity, the following `WHERE` clause can be used to identify them:

```
WHERE sysdate <= BD AND ED = MaxValueTime
```

A solution using a `NULL` value would look as stated in the next section. The condition should be extended by handling the `NULL` values or replacing the undefined value.

Getting current valid states

If the validity is expressed by the lower and upper limits (BD and ED) and unlimited validity is expressed by the `NULL` notation, the following code snippets are applicable.

The first solution shows three conditions interconnected by `or` and `and`. To express the currently valid state, BD must be smaller than (or equal to) `sysdate`. The ED value should be higher than `sysdate`, or an undefined value can be present:

```
WHERE sysdate <= BD and (ED >= sysdate or ED is null)
```

The second code takes the first condition to express the current state for the defined value of ED. The second condition structure references the situation in which the ED value is undefined:

```
WHERE (sysdate between BD and ED)
      OR (sysdate<=BD and ED is null)
```

Finally, this solution transforms the undefined value by using the `nvl` function call:

```
WHERE sysdate between BD and nvl(ED, sysdate)
```

In the next section, we will see how to apply consistent rules for managing unlimited validity.

List of current states that have unlimited validity

To get a list of the states with unlimited validity, ED must have an undefined value, or the `MaxValueTime` notation can be used:

```
WHERE sysdate <= BD and ED is NULL
```

Ensuring the replacement of the `NULL` values with the `MaxValueTime` notation can be done using a trigger that is executed before the `INSERT` or `UPDATE` operations:

```
CREATE OR REPLACE TRIGGER MAXVALTRIG
  BEFORE INSERT OR UPDATE ON Tab
FOR EACH ROW
```

```
BEGIN
  IF :new.ED IS NULL
    THEN :new.ED:=MaxValueTime;
  END IF;
END;
/
```

The default value has another solution. Oracle 12c introduced the DEFAULT ON NULL extension, which replaces the non-specified value, as well as the explicitly defined NULL value. This extension clause is part of the attribute constraint definition:

```
CREATE TABLE Tab
  (ID integer,
   BD DATE,
   ED DATE default on null MaxValueTime,
   data ...);
```

Date value management can also be done explicitly via user management. It is commonly covered by the object type or supervised by the package. In both cases, it is possible to distinguish between multiple date element sources, which are mostly related to processing delays, inconsistencies, and temporal corrections.

Validity can be modeled by the two limits – BD and ED – or each model can be transformed into that representation, as discussed earlier. However, how do we express the elapsed time? How do we calculate it? The answer is provided in the next section, reflecting the various granularity perspectives.

Managing duration – getting the elapsed time

Architectures approaching the duration clearly state that the most useful solution is based on modeling the begin and end point of the validity explicitly, represented physically by the DATE or TIMESTAMP data formats.

It is unnecessary to do any calculations to locate the states in the timeline. Moreover, undefined states do not need to be handled explicitly nor stored physically. Instead, several representations of the time interval exist, supervising the closed or open characteristics from the left and right parts of the duration set.

The elapsed time between two points is commonly represented by days, months, or years. Let's look at the DATE values first. The difference between two DATE values produces a numeric value expressing the number of days between those points. The result value can be decimal, whereas the time elements are also taken into consideration. If the first timepoint has a lower value than the second value, then the result holds a negative value. You can use simple mathematical subtraction to get the difference in days between the two timepoints. If the time spectrum is truncated (or the same), an integer value is provided:

```
select TO_DATE('15.1.2022', 'DD.MM.YYYY')
          - TO_DATE('10.1.2022', 'DD.MM.YYYY')
   from dual;
```

> **Time spectrum**
>
> The term **time spectrum** is commonly used in temporal theory. This is because it expresses all *time elements*. Thus, the term time spectrum can be used instead of stating *hour, minute,* and *second* element values. This is valid for DATE values, while TIMESTAMP values are enhanced by the finer precision *second fractions*.

This code produces the value, 5, meaning 5 days have elapsed between the 2 timepoints. The preceding code uses CO characteristics, but the transformation using the smallest value can be used. Generally, a decimal value is produced. Let's consider the following example with different values for the time:

```
select TO_DATE('15.1.2022 15:12:03',
                  'DD.MM.YYYY HH24:MI:SS')
          - TO_DATE('10.1.2022 5:18:26',
                       'DD.MM.YYYY HH24:MI:SS') as days
   from dual;
--> 5,412
```

The appropriate multiplication of the accuracy is used to provide the elapsed time with more precise granularity:

- 24 for hours:

    ```
    select (TO_DATE('15.1.2022 15:12:03',
                       'DD.MM.YYYY HH24:MI:SS')
              - TO_DATE('10.1.2022 5:18:26',
                          'DD.MM.YYYY HH24:MI:SS'))*24
              as hours
       from dual;
       --> 129,894
    ```

- *1440* for minutes (24*60):

    ```
    select (TO_DATE('15.1.2022 15:12:03',
                       'DD.MM.YYYY HH24:MI:SS')
    ```

```
              - TO_DATE('10.1.2022 5:18:26',
                      'DD.MM.YYYY HH24:MI:SS'))*24*60
           as minutes
    from dual;
    --> 7793,617
```

- *86,400* for seconds (24*60*60):

```
    select (TO_DATE('15.1.2022 15:12:03',
                 'DD.MM.YYYY HH24:MI:SS')
         - TO_DATE('10.1.2022 5:18:26',
                   'DD.MM.YYYY HH24:MI:SS'))*24*60*60
           as seconds
    from dual;
    --> 467617,000
```

A subtraction operation provides us with the number of days between two timepoints, also taking leap years into account:

```
select TO_DATE('1.2.2022', 'DD.MM.YYYY')
        - TO_DATE('1.1.2022', 'DD.MM.YYYY')
 from dual;
 --> 31
select TO_DATE('1.3.2022', 'DD.MM.YYYY')
        - TO_DATE('1.2.2022', 'DD.MM.YYYY')
 from dual;
 --> 28
select TO_DATE('1.3.2024', 'DD.MM.YYYY')
        - TO_DATE('1.2.2024', 'DD.MM.YYYY')
 from dual;
 --> 29
```

To get the number of months between two DATE values, we can use the MONTH_BETWEEN function.

Common methods available in database systems (using the Standard package) related to duration management use just one granularity element. The result can be either the number of days (provided by the subtraction operation), the number of months (using MONTHS_BETWEEN), or the number of years. Typically, a more useful and user-friendly solution is provided by combining multiple elements. A value of 2 years, 1 month, 2 days, 4 hours, 10 minutes, and 11 seconds is far more useful than 764,174 days.

It is difficult, or even impossible, to obtain a more complex structure from an ordinary numerical expression. Take, for example, the number of elapsed months. It cannot be obtained from the numerical value if the starting (reference) point is not present. The next section aims to analyze and discuss a more robust solution, which must be coded by the user. We will take you through the process, the result of which will be in textual format, providing values for each date and time element.

Sophisticated solution for getting the duration

To provide a more sophisticated output, let's create a function that takes two DATE values (P_DATE1 and P_DATE2). First, process the common date elements – day, month, and year. When dealing with multiple elements, always end with the finest precision.

To get the number of elapsed years, calculate the number of months between two timepoints and divide by 12. The integer part of the response denotes the number of elapsed years. The remainder is taken for further processing:

```
v_year_count:=TRUNC(MONTHS_BETWEEN(v_date2,v_date1) / 12);
```

The V_DATE1 and V_DATE2 values are copied from the function call input parameters. They are not taken directly. Therefore, we use separate local variables. The reason is duration consistency. It is assumed that P_DATE1 is less than or equal to P_DATE2. If so, particular input parameter values are copied to the local variables. If not, they are exchanged with each other.

From the user's perspective, the easiest way is to shift the beginning point by adding the number of years obtained from the preceding code snippet. Thanks to this, the month calculation can be done directly. Furthermore, it is evident that the overlap across the years can be ignored – simply, the duration cannot now be longer than one year.

For this purpose, the V_UPDATED_DATE local variable is defined:

```
v_updatedDate:=ADD_MONTHS(v_date1, v_year_count * 12);
```

The duration in months can be obtained by the calculation of the MONTHS_BETWEEN function call. Similarly, the integer value is taken using the TRUNC function:

```
v_month_count:=TRUNC(MONTHS_BETWEEN(v_date2,
                                    v_updatedDate));
```

The V_UpdatedDate local variable is calculated again. Now, all the month elements are added using the ADD_MONTHS function call (the years have already been processed):

```
v_updatedDate:=ADD_MONTHS(v_updatedDate, v_month_count);
```

The difference between the shifted-left border (V_UPDATEDDATE) and the original end point (V_DATE2) denotes the day granularity perspective, removing the impact of the month and year. Thus, only the remainder of the preceding operations is taken into account:

```
v_day_count:= TRUNC(v_date2 - v_updatedDate);
```

In the preceding code, only the integer part is taken, which is done using the TRUNC function.

The decimal part of the day can be transformed into the time elements, represented by the number of hours, minutes, and seconds. For example, decimal part 0.5 expresses half of a day, exactly 12 hours. 1 day has 86,400 seconds (60x60x24), so the day fragment is transformed into seconds elapsed (v_second_difference). For example, 12 hours are expressed by 43,200 seconds. From that value, the number of hours can be extracted. 1 hour is 3,600 seconds. Then, taking the remainder, minutes can be calculated (1 minute lasts 60 seconds). And finally, the result of extracting the hour and minute values expresses the second element. The principles of the time value calculation are shown in the following code snippet by introducing three variables, one for each time element:

```
v_hour_count:=TRUNC(v_second_difference/3600);
v_minute_count:=TRUNC((v_second_difference
                        - v_hour_count*3600)/60);
v_second_count:=TRUNC(v_second_difference
                        - v_hour_count*3600
                        - v_minute_count*60);
```

When dealing with duration management, a separate local variable was created for each precision segment. Finally, all obtained values are grouped in the required format and provided to the user as follows:

```
return case v_shift when true then '-'
                    when false then '+'
        else '' end
    || v_year_count || ' years, '
    || v_month_count || ' months, '
    || v_day_count || ' days, '
    || v_hour_count || ' hours, '
    || v_minute_count || ' minutes, '
    || v_second_count || ' seconds. ';
```

Note that the V_SHIFT variable is used to denote the order of parameters. As stated, there is an assumption: PDATE1 <= PDATE2. If this is not the case, these values are interchanged, saving information about it in the V_SHIFT Boolean variable.

The complete solution of the function definition is provided in the following code snippets:

- The header and variable declaration:

```
create or replace function GET_DIFFERENCE_DATE
                            (pDate1 DATE, pDate2 DATE)
 return varchar
is
 v_shift boolean:=false;
 v_date1 DATE;
 v_date2 DATE;
 v_updatedDate DATE;
 v_year_count integer;
 v_month_count integer;
 v_day_count integer;
 v_second_difference integer;
 v_hour_count integer;
 v_minute_count integer;
 v_second_count integer;
```

- Interchanging values to ensure consistency:

```
begin
 if pDate1 > pDate2
   then v_date1:=pDate2;
        v_date2:=pDate1;
        v_shift:=true;
   else
    v_date1:=pDate1;
    v_date2:=pdate2;
 end if;
```

- Calculating elapsed time using various granularity levels – years, months, and days:

```
v_year_count:=
       TRUNC(MONTHS_BETWEEN(v_date2,v_date1) / 12);
v_updatedDate:=ADD_MONTHS(v_date1, v_year_count * 12);
v_month_count:=
       TRUNC(MONTHS_BETWEEN(v_date2,v_updatedDate));
```

```
    v_updatedDate:=ADD_MONTHS(v_updatedDate,
                              v_month_count);
    v_day_count:= TRUNC(v_date2 - v_updatedDate);
```

- Calculating elapsed time using various granularity levels – hours, minutes, and seconds:

```
    v_second_difference:=(v_date2
                        - ADD_MONTHS(v_date1,
                                 12*v_year_count
                              + v_month_count)
                        - v_day_count)*24*60*60;
    v_hour_count:=TRUNC(v_second_difference/3600);
    v_minute_count:=TRUNC((v_second_difference
                          - v_hour_count*3600)/60);
    v_second_count:=TRUNC(v_second_difference
                          - v_hour_count*3600
                          - v_minute_count*60);
```

- Getting the output:

```
        return case v_shift when true then '-'
                      when false then '+'
                      else ''
            end
                || v_year_count || ' years, '
                || v_month_count || ' months, '
                || v_day_count || ' days, '
                || v_hour_count || ' hours, '
                || v_minute_count || ' minutes, '
                || v_second_count || ' seconds. ';
    end;
    /
```

The source code for the GET_DIFFERENCE_DATE function can be obtained from the GitHub repository of this chapter.

Calling the proposed function results in the following output:

```
select GET_DIFFERENCE_DATE(
   TO_DATE('10.1.2020 15:10:22', 'DD.MM.YYYY HH24:MI:SS'),
```

```
    TO_DATE('29.4.2023 11:00:11','DD.MM.YYYY HH24:MI:SS'))
            as difference
        from dual;
--> +3 years, 3 months, 18 days, 19 hours, 49 minutes,
    49 seconds.
select GET_DIFFERENCE_DATE(
    TO_DATE('8.4.2016 15:10:22', 'DD.MM.YYYY HH24:MI:SS'),
    TO_DATE('6.3.2014 11:00:11','DD.MM.YYYY HH24:MI:SS'))
            as difference
        from dual;
--> -2 years, 1 months, 2 days, 4 hours, 10 minutes,
    11 seconds.
```

The proposed function provides a more complex solution with multiple levels of precision. It works for the DATE values. However, what happens if we use the TIMESTAMP values as the input? The next section gives you the answer.

Remarks – DATE versus TIMESTAMP

Dealing with the date and time elements in the way we did in the previous section is valid for the DATE data types. TIMESTAMP management works differently. Subtraction of the TIMESTAMP values does not provide the number of days between them. Instead, an INTERVAL DAY TO SECOND data type value is obtained:

```
select TO_TIMESTAMP('8.4.2016 15:10:22',
                     'DD.MM.YYYY HH24:MI:SS')
        - TO_DATE('6.3.2014 11:00:11',
                   'DD.MM.YYYY HH24:MI:SS')
    from dual;
--> +764 04:10:11.000000
```

Hour, minute, and second elements can be easily obtained from the INTERVAL data type, for example, by using the EXTRACT functionality:

```
select EXTRACT(
          hour from (TO_TIMESTAMP('8.4.2016 15:10:22',
                                   'DD.MM.YYYY HH24:MI:SS')
                 - TO_DATE('6.3.2014 11:00:11',
```

```
                               'DD.MM.YYYY HH24:MI:SS')))
     from dual;
```

Day, month, and year precision management can seem more complicated. Subtracting the TIMESTAMP values results in the number of elapsed days. The snippet stated at the beginning of this section returns the number of days as 764. Can the number of elapsed months and years be extracted from that value? How many days correspond to one month? 28, 29, 30, or 31? Similarly, what about the number of days in the year? Is it 365 or 366? Yeah, it is impossible to get the right answer just from the number of days. It is necessary to use the beginning point of the validity, the left border.

The proposed function, GET_DIFFERENCE_DATE, has this functionality, but the second fractions are ignored. It uses the fact that the TIMESTAMP and DATE values can be interchanged by calling a conversion function (TO_TIMESTAMP or TO_DATE or implicit conversion can be used). As a result, second fractions are ignored by converting the value to the DATE data type. Then, the processing is similar to what we have already discussed:

```
select GET_DIFFERENCE_DATE(
                TO_TIMESTAMP('8.4.2016 15:10:22.12',
                             'DD.MM.YYYY HH24:MI:SS.FF'),
                TO_TIMESTAMP('6.3.2014 11:00:11.17',
                             'DD.MM.YYYY HH24:MI:SS.FF'))
          as difference
     from dual;
--> -2 years, 1 months, 2 days, 4 hours, 10 minutes,
    11 seconds.
```

The function takes two values, which are evaluated and processed.

The task is then to maintain the fractions as well. Is this possible? How must the implementation change? Let's focus on the management of seconds. As stated, when the TIMESTAMP values are treated as the DATE values, fractions are ignored. If we add this level of precision, original values may need correction. Namely, consider the second fraction elements of two TIMESTAMP values and find the difference between them. The result can be negative. In that case, it is necessary to refer to the upper granularity level, minutes, which must be taken into consideration.

Let's consider the following example, which deals with two timepoints. For the purposes of the example, we'll only look at seconds and fractions:

```
T1 = 21.625 (SS.FF)
T2 = 12.802 (SS.FF)
```

If we subtract these values in the DATE domain, the result would be 9, because fractions are ignored. Going deeper, considering the second fractions as well, the difference is a bit smaller, 8.823. Generally, it can influence the minute, even hour, or date management:

```
T3 = 5:21.045 (MI:SS.FF)
T4 = 3:21.890 (MI:SS.FF)
```

Taking the preceding values, the subtraction of T3 and T4 results in 1 minute and 59.155 seconds. Thus, the processing and calculation should always start with the finest precision, whereas it can influence the other elements. For example, second fractions can influence seconds, and seconds can influence minutes, and so on.

Therefore, to calculate the elapse between two TIMESTAMP values, the original DATE approach can be directly used to get a reliable solution, but the second fraction management should precede it. Thus, these two TIMESTAMP values to be considered are taken. In the first phase, the fractional part for each value is extracted, and these two provided values are compared. If the subtraction of the extracted second fractions of those two TIMESTAMP values is negative, one second must be added for the core evaluation. Then, the *corrected* TIMESTAMP values can be treated as the DATE types directly. In our case, the TO_CHAR function is used to get the fractions (the second parameter value is FF). Alternatively, EXTRACT can be used. In that case, however, second and fractions are obtained, so the fraction must be differentiated. The V_FRACTION1 and V_FRACTION2 variables hold the fractions for the specified timepoints:

```
v_fraction1:=TO_CHAR(pTimestamp1, 'FF');
v_fraction2:=TO_CHAR(pTimestamp2, 'FF');
```

If the subtraction is negative, one second is added. Thus, if nanosecond precision for TIMESTAMP is used, one second is defined by the 1000000000 value:

```
if v_fraction2 < v_fraction1
  then
    v_fraction2:=v_fraction2 + 1000000000;
      -- adding one second;
    v_date2:=v_date2 - Interval '1' second;
  end if;
```

Then, the processing can be directly related to the DATE management, where the principles are the same. The following code shows the complete function body, highlighting the differences between DATE and TIMESTAMP management. Note that after fraction management, the original input values (pTimestamp1 and pTimestamp2) are treated as the DATE values due to the MONTHS_BETWEEN function call, as well as the mathematical DATE difference, which, in comparison with the TIMESTAMP points, provides the number of elapsed days. In the return section, the fraction management extends the processing. The explicit conversion converts the result of the mathematical subtraction to the character format. The changed or added code snippets are in bold:

- Header and variable declaration:

```
create or replace function GET_DIFFERENCE_TIMESTAMP
                    (pTimestamp1 TIMESTAMP,
                     pTimestamp2 TIMESTAMP)
 return varchar
is
 v_shift boolean:=false;
 v_date1 DATE;
 v_date2 DATE;
 v_fraction1 integer;
 v_fraction2 integer;
 v_updatedDate DATE;
 v_year_count integer;
 v_month_count integer;
 v_day_count integer;
 v_second_difference integer;
 v_hour_count integer;
 v_minute_count integer;
 v_second_count number(5,2);
```

- Interchanging values to ensure consistency:

```
begin
if pTimestamp1 > pTimestamp2
    then v_date1:=pTimestamp2;
         v_date2:=pTimestamp1;
         v_shift:=true;
         v_fraction1:=TO_CHAR(pTimestamp2, 'FF');
         v_fraction2:=TO_CHAR(pTimestamp1, 'FF');
    else
         v_date1:=pTimestamp1;
         v_date2:=pTimestamp2;
         v_fraction1:=TO_CHAR(pTimestamp1, 'FF');
         v_fraction2:=TO_CHAR(pTimestamp2, 'FF');
    end if;
```

- Getting and evaluating fractions:

```
if v_fraction2 < v_fraction1
 then
  v_fraction2:=v_fraction2 + 1000000000;
                             -- adding one second
  v_date2:=v_date2 - Interval '1' second;
end if;
```

- Calculating elapsed time using various granularity levels – years, months, and days:

```
v_year_count:=
        trunc(MONTHS_BETWEEN(v_date2,v_date1) / 12);
v_updatedDate:=
        ADD_MONTHS(v_date1, v_year_count * 12);
v_month_count:=
        TRUNC(MONTHS_BETWEEN(v_date2,v_updatedDate));
v_updatedDate:=ADD_MONTHS(v_updatedDate,
                          v_month_count);
v_day_count:= TRUNC(v_date2 - v_updatedDate);
```

- Calculating elapsed time using various granularity levels – hours, minutes, and seconds:

```
v_second_difference:=(v_date2
                       - ADD_MONTHS(v_date1,
                                    12*v_year_count
                                    + v_month_count)
                       - v_day_
count)*24*60*60;
v_hour_count:=TRUNC(v_second_difference/3600);
v_minute_count:=TRUNC((v_second_difference
                       - v_hour_count*3600)/60);
v_second_count:=TRUNC(v_second_difference
                      - v_hour_count*3600
                      - v_minute_count*60);
```

- Getting the output:

```
    return case v_shift when true then '-'
                       when false then '+'
                  else ''
    end
              || v_year_count || ' years, '
              || v_month_count || ' months, '
              || v_day_count || ' days, '
              || v_hour_count || ' hours, '
              || v_minute_count || ' minutes, '
              || v_second_count ||'.'
         || TO_CHAR(v_fraction2-v_fraction1)
         || ' seconds. ';
    end;
/
```

The source code for the GET_DIFFERENCE_TIMESTAMP function can be obtained from the GitHub repository of this chapter.

Summary

In this chapter, we dealt with duration management. The discussion started with one timepoint modeling expressing the validity. This solution is strongly limited by the efficiency of data storage and subsequent data processing because many duplicate tuples can be present. However, by changing its meaning, the duration can be modeled. Namely, the timepoint can represent the beginning point of the validity. Thanks to that, each new state of the object automatically limits the validity of the direct predecessor. Valid future states can also be covered. There is, however, one strong limitation related to completely undefined states, which cannot be modeled, and special notation must be introduced.

Therefore, a general solution is provided by two timepoints representing the begin and end points of the validity duration frame. They can have various representations based on whether the begin and end points of the duration interval are part of the validity or not. The most often used models are *closed-open* and *closed-closed* representations. The principle used is strict and must always be applied to the data attributes in question. However, as described, all models are directly transformable using one extra timepoint.

This chapter also provided a discussion of unlimited validity modeling. Typically, developers tend to use the NULL value to express the unlimited right border of the duration interval (ED). However, the NULL value does not have a specific meaning, nor is any information present. In duration modeling, it is obvious that ED has not occurred yet but will be present in the future. Therefore, modeling using MaxValueTime is more robust.

The next chapter deals with duration models more deeply by examining the interval representations and positional relationships used in the temporal models. Duration does not need to only model the *validity* frame; any duration interval can generally be used, modeled, and represented.

Questions

1. What is the main limitation of using only the beginning point of the validity coverage?

    A. It's impossible to model completely undefined states

    B. It's impossible to model valid future states

    C. Inefficiency caused by the duplicate tuples

    D. Significant costs to identify current valid states

2. Which model can change the granularity by limiting the gaps?

    A. Closed-closed

    B. Closed-open

    C. Open-closed

    D. All of the above

3. What is the output of the calculation ED - BD if both values are the DATE data types?

    A. Time elapsed in seconds

    B. Time elapsed in minutes

    C. Time elapsed in days

    D. Time elapsed in years

4. What is the output of the calculation ED - BD if both values are the TIMESTAMP data types?

    A. Time elapsed in seconds

    B. Time elapsed in days

    C. Value of the Interval Day to Second data type

    D. Value of the Interval Year to Month data type

Further reading

- *Developing Time-Oriented Database Applications in SQL* by Richard T. Snodgrass. It provides you with best practices for integrating past and current data in your applications.

- *Time Granularities in Databases, Data Mining, and Temporal Reasoning* by Claudio Bettini, Sushil Jajodia, and X. Sean Wang. It presents a technical framework for many issues in temporal databases, mostly related to duration, formats, and granularity levels.

10
Interval Representation and Type Relationships

So far, we have dealt with time points and durations expressed via closed-closed or closed-open characteristics. We have shown how to model unbounded validity. The focus has been on the transformation rules across the representations. This chapter deals with the duration interval positions and interaction principles between multiple representations. These interaction characteristics are widely used in the temporal environment by examining the evolution of the object's states and changes. The emphasis must be on collision detection and undefined state identification. As we will describe in the first part of this chapter, interval interaction types form a tree structure summarizing possible interval positions and interactions.

Then, the second part of this chapter will highlight the temporal validity, represented by the PERIOD data structure. You will get a complex understanding of modeling principles, representations, usage, and associated packages for manipulating the data image. PERIOD definition and management are also supervised by the data dictionary views and other methods by which the defined table structures can be represented. By the end of this chapter, you will be familiar with the ENABLE_AT_VALID_TIME method of the DBMS_FLASHBACK_ARCHIVE package, which provides the image of the data at a defined timepoint.

In a nutshell, this chapter will cover the following main topics:

- Relationships between time interval representations
- Using positional relations in temporal space
- Modeling temporal validity using the PERIOD structure, allowing you to extend the DBMS_FLASHBACK_ARCHIVE package to manage temporality

The source code for this chapter can be found in the GitHub repository accessible via this link: `https://github.com/PacktPublishing/Developing-Robust-Date-and-Time-Oriented-Applications-in-Oracle-Cloud/tree/960e892ada2c4ff3247f4967ca9b3d866e0f1c4f/chapter%2010`. You can also scan the following QR code:

Relationships between time interval representations

Relationships between time intervals form an important part of processing in time-delimited environments. Individual states of objects must be defined positionally, emphasizing possible overlaps of time intervals so that the object is always defined correctly in time. Each object needs to be covered by just one valid state at any time (based on the assumption that the undefined states are explicitly defined by the user). Therefore, it is necessary to identify the different types of time and interval relationships. This section describes the existing taxonomy of relationships according to Tom Johnston and Randall Weis in their book *Managing Time in Relational Databases: How to Design, Update and Query Temporal Data*. We will point out the potential interpretations and the importance of specific types in time-oriented databases. Temporal solutions based on positional representation were defined in 1983 and became the basis of the temporal concept in the process of creating a temporal norm. However, this standard has not been adopted as a whole; only some parts have been applied. Therefore, time-based relationships only define a logical framework. The physical implementation and representation are left to the programmer.

> **Important note**
> The relationships between time intervals are mostly enclosed by parentheses to distinguish the text from technical language, so we will follow this rule in this publication as well.

Throughout this chapter, we will use the **closed-open** (**CO**) interval representation introduced in *Chapter 9*. We leave any correlations to other models (CC, OC, or OO `representations`) to you by reflecting on the transformation principles discussed in *Chapter 9*. Each interval is characterized by the start point of the validity (BD) and end point of the validity (ED). There is a natural assumption of consistency, so $BD < ED$.

The types of representations and their individual relationships between time intervals form a binary tree, as shown in the following diagram:

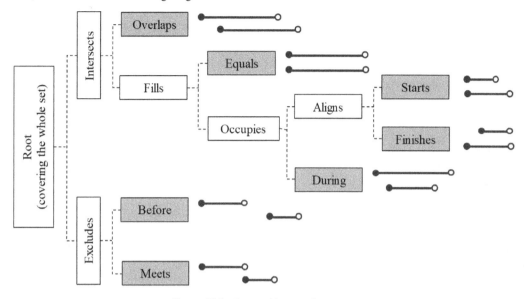

Figure 10.1 – Interval interaction types

> **Important note**
>
> The preceding diagram has been created using Tom Johnston's and Randall Weis's representation from their book *Managing Time in Relational Databases: How to Design, Update and Query Temporal Data.*

The tree's root represents all types of interval relationships. At the first level, we can divide them in terms of overlap. In principle, the specified intervals can overlap ([intersects]) or are separated from each other, so they do not have a single point in time. In this case, an [excludes] relationship is defined. If the intervals have at least one common time moment (they are in an [intersects] relationship), we can divide them again into two categories – either defined by the total coverage ([fills]) or by a partial overlap. If they are in a position of partial overlap, they have neither the same start nor endpoint, but there is at least one common point (either BD1 < BD2 and ED1 >= BD2 and ED1 <> ED2 or BD2 < BD1 and simultaneously ED2 >= BD1).

Conversely, in the case of complete coverage, four types can be identified at the leaf level of the tree. They are as follows:

- The [equals] relationship, where the intervals are completely identical (BD1 = BD2 and ED1 = ED2)

- The [starts] relationship, which covers intervals that have a common beginning but do not have a common end (BD1 = BD2 and ED1 <> ED2)

- The [finishes] relationship, where intervals have a common end but do not have a common beginning (BD1 <> BD2 and ED1 = ED2)

- A special type of full coverage is the [during] relationship, in which one time interval is a subset of the second time interval (either BD1 < BD2 and ED1 > ED2) or the intervals are positionally opposite, namely BD2 < BD1 and ED2 > ED1

In principle, for the [during] relationship, any part of the interval (sub-interval) is fully covered by the second relationship. The [starts] and [finishes] relationship intervals are covered at the hierarchy level by the [aligns] type – the left part for [starts] and the right part for [finishes].

If the intervals have no common points in time, they are in the [excludes] position. Two cases can arise here; in the first case, they follow each other directly (that is, they have a [meet] relationship (ED1 = BD2 or ED2 = BD1)). The second case covers the situation where there is at least one timepoint between them expressing the [before] relationship (ED1 < BD2 or ED2 < BD1).

What about a situation where a time interval has only one time point (T)? How do we model positional relationships? Concerning the previous characteristics and the binary tree, the [overlaps] and [equals] relationships can be excluded from processing as they can never occur. If the time interval has only one timepoint, then the following categories can be identified:

- The moment in time has the same point as the beginning of the interval – the [starts] type

- The moment in time has the same point as the end of the interval – the [finishes] type

- The time moment is positionally within the time interval (BD < T and T <= ED) – the [during] type

This is shown in the following diagram:

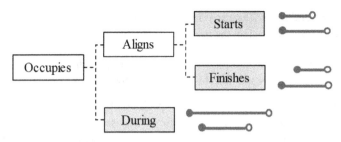

Figure 10.2 – Interval interaction types – time point

> **Important note**
>
> The preceding diagram has been created using Tom Johnston's and Randall Weis's representation from their book *Managing Time in Relational Databases: How to Design, Update and Query Temporal Data.*

If the time moment does not belong to a defined interval, it is covered by the [excludes] relationship, either the [before] or [meets] types.

If both intervals contain exactly one point in time, the [equals] relationship can be used if they are the same. The [starts] and [finishes] relationships require exclusivity (both relationships cannot be valid), so they cannot be used. Thus, their supertypes ([aligns], [occupies], [fills], and [intersects]) do not apply as well. If the values are not identical ([excludes]), this can reflect a [before] relationship or a [meets] relationship if they follow immediately after each other.

This section has summarized all options characterizing duration interval interactions and transformations of individual duration interval representations between each other. Their usage in the temporal space is discussed in the next section, which highlights the multi-temporal model. In principle, any temporal attributes can be used.

Using positional relations in temporal space

A temporal database is characterized by extending the object identifier by the time dimensions introduced in the *Exploring temporal dimensions* section in *Chapter 11*. Typically, the object cannot be directly identified by the ID. Time validity must be applied as well. Thus, there are multiple states for each object. However, each data object tuple characterizes the object's values at a particular time frame. It must be ensured that each state is defined by no more than one valid state at any time. The **uni-temporal model** uses a *one-time dimension*, mostly related to validity. The temporal integrity covers the reliability of the system. Supervision is done by the transaction manager using the positional relationships in the temporal space. The aim is to ensure that each object is characterized by one state, allowing users to sort the states based on the timeline. To do that, the following relationships are used:

- The [intersects] relationship is used when defining validity by inserting a new state of an object. As stated earlier, each object is represented by one valid state.

- The [fills] relationship applies in the composition of objects and states, especially in the **temporal is-a (ISA) hierarchy**, where ancestors and descendants are present. Ancestors are used for generalization (acting as an object type root), while ancestors store specific values based on the object type. Thus, to get complex object state information, each descendant must be completely covered by the ancestor. This type of relationship is also important in data modeling when strict coverage is required within referential integrity.

- The [excludes] relationship determines the uniqueness of the definition of states at any point in time.

- Using the [before] relationship, we can distinguish individual states and sort them in time.

- The [meets] relationship is intended to identify the direct successor of the state. If the object does not have a valid state in a certain time interval, the solution is moved to the [before] relationship control level. With the resulting set, it is then necessary to use relational criteria to find the direct ancestor – the nearest correctly defined relationship.

The extension of the validity definition creates a **bitemporal system**, which allows you to correct already-made states and preserve information about their original definitions. *Transaction validity* is defined for these purposes – the time duration for which the stored data was considered correct. We use the [during] relationship to identify fixes and data corrections. For the current version of the data, transaction validity is unlimited. The order of modifications is defined by the [meets] relationship. When planning and correcting future changes, it is advisable to use the [during] and [starts] relationships.

Transaction validity cannot contain undefined time intervals. It expresses the period during which the data is considered to be correct. Even the fact that, at some point in time, we cannot identify a valid state is important and limited by transaction validity.

After the transition to more complex temporal systems, it is important to ensure the integration of the whole temporal model for any time spectrum with any meaning. Specifically, we cannot limit ourselves to validity alone. Emphasis must also be placed on individual versions of objects, their corrections over time, and undefined states. The **multi-temporal model** provides a general solution. Multiple time frames can be associated with one state. Some of the most common temporal representations and references are listed here:

- T1: The arrival of a new record in the system clarifying input stream strength

- T2: Placing the record in the input queue

- T3: Taking the record for an integrity check

- T4: Time of considering the value (the new state) to be significant (insignificant changes below the limit of accuracy can be ignored while maintaining the validity of the original state)

- T5: Inserting a new tuple in the master system

- T6: Spreading information to individual nodes and slaves

- T7: Ending the transaction by approving the changes

The PERIOD data type has been introduced in Oracle Database as another option for expressing the duration, which is managed automatically. The next section introduces the PERIOD data type and the PERIOD identification methods.

Modeling temporal validity using Oracle Database's embedded PERIOD

The original concept of the PERIOD data type was specified in the temporal paradigm definition, but it was not, however, completely accepted. The principle was based on providing an extra data type covering the duration, internally modeled by the beginning and end points of the state validity. It was based on direct mapping and comparison; thus, the positional relationships were assumed to be widespread. Thanks to that, sorting the states over the evolution was easy, and there was an emphasis on the consistency of the duration and the definition – the beginning point was placed before the end point. It was primarily designed for validity management, but any time frame could be used generally. Since the standardization process did not approve the concept of the PERIOD data type, many research and development prospects were either refused or at least not further developed. As a result, individual developers had to specify the principles and manage rules independently. Thus, each system handles the principles differently.

Temporal validity is a robust approach offered by Oracle Database for tracking changes. It is based on splitting and highlighting current valid data, but historical or future states can also be modeled and stored in a core table. Temporal validity takes the active flag or the inactive flag for each row. Some references also use the valid and invalid flags, but it should be emphasized that invalid data does not represent data and does not pass the data modeling constraints. States lose their validity because they are replaced by newer states or will become valid later (loaded as future plans). Thanks to that, the whole evolution can be part of the main table. Therefore, each data image or snapshot is always related to the specified time frame, either by the time point or the time range. Thus, a simple temporal system is created, commonly formed by the validity time frame, covered by the PERIOD specification, which is physically modeled by two time points. For example, validity is denoted by the contract when dealing with employment. It defines the effective date range. Namely, the validity of the employment is related to insurance, access to the company information systems, labs, technologies, and so on. Thus, many activities and security statements are associated with a validity range (valid time – VT). Also, transaction time (TT) is present in temporal systems.

Reflect on the preceding example about employment. What time attribute limits the validity of the employment contract? In principle, it's the hire date. However, data is not loaded into the database exactly at the time of the validity entry point, is it? Therefore, the transaction time is a system-managed DATE or TIMESTAMP value representing the point of the insertion. Based on the settings, transaction time can mean *insert* operation time, transaction approval, and so on.

By using the PERIOD data type embedded by Oracle Database, temporal concepts can be easily defined. It is, however, important to get familiar with the principles and techniques to identify its usage practically. The next section leads you through the modeling techniques, followed by the rules to identify temporal validity.

Concepts of temporal validity

A valid time period consists physically of two Date and Time columns in the table definition. They can be specified explicitly, or the columns can be created automatically by making the table temporal implicitly. Moreover, temporality does not need to be specified during table creation. It can be added at any time by altering the table (with the ALTER TABLE command). Let's create a table called EMP, covering the temporal validity modeled by the date_from and date_to attributes. An employment contract can be unlimited. Therefore, the date_to attribute value can hold a NULL value, or the MaxValueTime concept can be used. The highlighted row represents the temporal validity definition using the PERIOD concept:

```
Create table EMP
  (employee_id integer not null,
   name varchar(20) not null,
   surname varchar(20) not null,
   date_from date not null,
   date_to date,
   position varchar(20) not null,
   salary number(6,2) not null,
   PERIOD FOR VALIDITY(date_from,date_to)
  );
```

The PERIOD definition forms the temporal constraint. In the preceding case, it is delimited by two physical attributes (date_from and date_to), but generally, particular borders can be created implicitly as internal attributes (which are not directly visible and referenceable). In that case, the following name notation is used: border names consist of the PERIOD name (in the preceding case, validity) followed by the underscore (_) and start or end. The following code block shows an example of the table with the PERIOD definition. Attributes forming the PERIOD definition are marked as hidden implicitly. The PERIOD definition is hidden as well:

```
Create table Emp
 (employee_id integer not null,
  name varchar(20) not null,
  surname varchar(20) not null,
  position varchar(20) not null,
  salary number(6,2) not null,
  PERIOD FOR validity
 );
```

The following select statement referencing a {USER | ALL | DBA}_TAB_COLS data dictionary view can be used to get the table definition extended by the hidden columns and internal definitions:

```
select COLUMN_NAME, DATA_TYPE, COLUMN_ID,
       SEGMENT_COLUMN_ID, INTERNAL_COLUMN_ID,
       HIDDEN_COLUMN
 from USER_TAB_COLS
  where TABLE_NAME='EMP';
```

The output of the preceding statement is shown in the following snippet:

```
--> COLUMN_NAME        DATA_TYPE
-->    COLUMN_ID   SEGMENT_COLUMN_ID   INTERNAL_COLUMN_ID
-->       HIDDEN_COLUMN
--> VALIDITY_START     TIMESTAMP(6) WITH TIME ZONE
-->            1                         1
-->      YES
--> VALIDITY_END       TIMESTAMP(6) WITH TIME ZONE
-->            2                         2
-->      YES
--> VALIDITY           NUMBER
-->                              3
-->      YES
--> EMPLOYEE_ID        NUMBER
--> 1              3                     4
-->      NO
--> NAME               VARCHAR2
```

```
--> 2                   4                       5
-->        NO
--> SURNAME             VARCHAR2
--> 3                   5                       6
-->        NO
--> POSITION            VARCHAR2
--> 4                   6                       7
-->        NO
--> SALARY              NUMBER
--> 5                   7                       8
-->        NO
```

We have listed column_id, segment_column_id, and internal_column_id to highlight the differences between individual attributes. Period management is covered by the hidden columns – start point (validity_start), end point (validity_end), and the period itself (validity), which is represented by the NUMBER data type. Their column_id values are not specified. Physically, storage requirements are covered by the start and end point (segment_column_id) for the period (validity is NULL). And finally, all attributes are covered by the internal_column_id unique value.

The temporal reference in terms of duration consistency is checked using the PERIOD definition for the table. Thus, the begin point value must be positioned on the timeline before the end point. In other words, a particular duration definition should last at least one time point. The definition must be reliable as well. This is ensured by the internal check constraint created automatically for the period. The following query lists the constraints, related to the EMP table, but we are not covering the NOT NULL constraints. The search_condition and search_condition_vc values are the same, but there is a difference in their data types. The search_condition column is LONG, whereas search_condition_vc is VARCHAR (vc is the abbreviation for VARCHAR):

```
select constraint_name, constraint_type, search_condition
  from user_constraints
   where table_nane='EMP'
     and search_condition_vc not like '%NOT NULL%';
```

The data management is the same as with a common conventional relational table. Let's insert five rows, as follows:

- Insert one row with limited employment contract validity:

```
insert into EMP(employee_id, name, surname,
                date_from,
```

```
                              date_to,
                         position, salary)
          values(1, 'Michal', 'Kvet',
                 TO_DATE('1.1.2010', 'DD.MM.YYYY'),
                 TO_DATE('1.1.2020', 'DD.MM.YYYY'),
                 'Researcher', 1000);
```

- Insert another row with unlimited employment contract validity (the same person):

```
    insert into EMP(employee_id, name, surname,
                         date_from,
                         date_to,
                         position, salary)
          values(1, 'Michal', 'Kvet',
                 TO_DATE('1.1.2020', 'DD.MM.YYYY'),
                 null,
                 'University teacher', 1200);
```

- Insert a new employee with unlimited employment contract validity:

```
    insert into EMP(employee_id, name, surname,
                         date_from,
                         date_to,
                         position, salary)
          values(2, 'Karol', 'Matiasko',
                 TO_DATE('1.1.2001', 'DD.MM.YYYY'),
                 null,
                 'Manager', 1500);
```

- Insert one extra employee with two unlimited employment contracts. Thus, two rows need to be inserted:

```
    insert into EMP
      values(3, 'Stefan', 'Toth',
                 TO_DATE('1.1.2005', 'DD.MM.YYYY'),
                 null, 'Developer', 1300);
    insert into EMP
      values(3, 'Stefan', 'Toth',
```

```
    TO_DATE('1.1.2006', 'DD.MM.YYYY'),
    null, 'Tester', 700);
```

All these statements inserting data into the EMP table are valid. However, there can be an overlap of the validity periods, which is relevant for employment contracts – one person can have multiple positions simultaneously. On the other hand, when dealing just with the beginning point of the validity or sensor-based networks, the developer must ensure consistency and reliability by checking and limiting overlaps. For example, there cannot be multiple states relating to one sensor (such as temperature or pressure) at one referenced time point.

The validity definition modeled by PERIOD only ensures the correctness of the time borders in the timeline, namely date_from < date_to, and the duration must cover at least one time point based on the granularity used. Generally, there can be undefined validity in terms of storing NULL for date_to. However, an undefined date_from value can be placed as well.

There are two statements listed in the following section. Notice the highlighted time borders, mapped to the date_from and date_to attributes. The first statement is invalid because the first timepoint is greater than the second. The second statement is also refused because it does not last for any time:

```
insert into EMP
    values(4, 'Emil', 'Krsak',
           TO_DATE('1.1.2005', 'DD.MM.YYYY'),
           TO_DATE('1.1.2002', 'DD.MM.YYYY'),
           'Development Manager', 1400);
--> ORA-02290: check constraint KVET3.VALIDITYcon violated
insert into EMP
    values(5, 'Vitaly', 'Levashenko',
           TO_DATE('1.1.2005', 'DD.MM.YYYY'),
           TO_DATE('1.1.2005', 'DD.MM.YYYY'),
           'Development Manager', 1400);
--> ORA-02290: check constraint KVET3.VALIDITYcon violated
```

Let's focus on the validity duration. It has been mentioned that check constraint ensures that the duration is at least one time point. What about precision? Is the granularity a day, an hour, a minute, or even a second? Validity duration always works with the finest granularity associated with the referenced borders. Thus, if the DATE data type is used for the PERIOD definition, the finest granularity is second. As a result, the following statement is accepted, even if the duration is just one second:

```
insert into EMP
    values(6, 'Elena', 'Zaitseva',
           TO_DATE('1.1.2005 00:00:00',
```

```
                        'DD.MM.YYYY HH24:MI:SS'),
           TO_DATE('1.1.2005 00:00:01',
                        'DD.MM.YYYY HH24:MI:SS'),
            'Development Manager', 1400);
```

When dealing with the TIMESTAMP elements, the precision is fractions of a second, up to nanoseconds. In the following example, microsecond precision is used. When there is a difference of one microsecond, a particular row is accepted. The table we are creating is related to sensor management. It gets the ID of the sensor, the value, and the validity period in timestamp format:

```
create table sensor_tab
  (sensor_id integer not null,
   value integer,
   ts_from TIMESTAMP(6),
   ts_to TIMESTAMP(6),
   PERIOD FOR duration(ts_from,ts_to)
  );
```

The insert statement uses the TIMESTAMP constructor:

```
insert into sensor_tab
   values(1,5,
          TO_TIMESTAMP('12.1.2022 14:12:22.643891',
                        'DD.MM.YYYY HH24:MI:SS.FF'),
          TO_TIMESTAMP('12.1.2022 14:12:22.643892',
                        'DD.MM.YYYY HH24:MI:SS.FF'));
```

The precision of the timestamp is six decimal places. What about smaller granularity? Will it be ignored in terms of truncation or rounding? Let's take these two statements. The difference is in the seventh decimal place. In the first statement, the difference is 400 nanoseconds:

```
insert into sensor_tab
  values(2,3,
          TO_TIMESTAMP('12.1.2022 14:12:22.6438910',
                        'DD.MM.YYYY HH24:MI:SS.FF'),
          TO_TIMESTAMP('12.1.2022 14:12:22.6438914',
                        'DD.MM.YYYY HH24:MI:SS.FF'));
```

The insertion of the preceding row to be inserted is refused due to the PERIOD check constraint related to the duration. By replacing 4 with 5, processing can be done whereby it is rounded up; thus one microsecond has elapsed, as shown in the next statement:

```
insert into sensor_tab
 values(3,8,
        TO_TIMESTAMP('12.1.2022 14:12:22.6438910',
                     'DD.MM.YYYY HH24:MI:SS.FF'),
        TO_TIMESTAMP('12.1.2022 14:12:22.6438915',
                     'DD.MM.YYYY HH24:MI:SS.FF'));
--> 1 row inserted.
```

Then, select the data for the third sensor:

```
select * from sensor_tab where sensor_id=3;
--> SENSOR_ID:   3
--> VALUE:       8
--> TS_FROM:     12.01.22 14:12:22,643891000
--> TS_TO:       12.01.22 14:12:22,643892000
```

Let's go back to the original role of employee management through the validity of the contract (validity period). Individual queries can reference time frames explicitly like this:

```
select *
 from Emp
  where date_from <=sysdate
     and NVL(date_to, sysdate)>=sysdate;
```

Here is an alternative:

```
select *
 from EMP
  where date_from <=TO_DATE('1.1.2022', 'DD.MM.YYYY')
     and date_to IS NULL;
```

That alternative provides the following results:

```
--> EMPLOYEE_ID  NAME        SURNAME
-->    DATE_FROM              DATE_TO
-->         POSITION                    SALARY
--> 1             Michal      Kvet
```

```
-->    01.01.2020 00:00:00
-->           University teacher            1200
--> 2              Karol       Matiasko
-->    01.01.2001 00:00:00
-->           Manager                       1500
--> 3              Stefan      Toth
-->    01.01.2005 00:00:00
-->           Developer                     1300
--> 3              Stefan      Toth
-->    01.01.2006 00:00:00
-->           Tester                        700
```

From the management point of view described so far, there is no significant difference in terms of performance and data management. Check constraints can still be managed explicitly. As is evident from the employee table definition (EMP), based on the business rules, contract validity should use the day granularity. However, system management of the duration's consistency is maintained automatically at the second level. Thus, developers must take care of the minimal duration using the check constraints and triggers.

The main advantage of the management of PERIOD in terms of validity management is temporality. Tables enclosed by the PERIOD definition have one or more dimensions of user-defined time. Each has a start and end point. Queries applying the FLASHBACK queries can easily reference the validity frame by forming the ASOF version queries (getting the data image as it existed at the defined TIMESTAMP value).

Utilizing data images using the DBMS_FLASHBACK_ARCHIVE package

Using methods from the DBMS_FLASHBACK_ARCHIVE package, you can directly obtain valid data at a specified date and time or duration. Basically, it is not necessary to specify time ranges for each query. Instead, just the session parameter is set for reference. This is done by the ENABLE_AT_VALID_TIME procedure enabling session-level visibility control by the FLASHBACK query:

```
DBMS_FLASHBACK_ARCHIVE.ENABLE_AT_VALID_TIME (
    level        IN    VARCHAR2,
    query_time   IN    TIMESTAMP DEFAULT SYSTIMESTAMP);
```

There are three available levels – ALL, CURRENT, and ASOF:

- The ALL option sets the temporal visibility to the whole spectrum (this is the default option)

- The CURRENT option focuses on the currently valid data within the validity period at the session level

- The ASOF option takes the second parameter (QUERY_TIME) referencing the timepoint of the validity that should be covered:

```
DBMS_FLASHBACK_ARCHIVE.ENABLE_AT_VALID_TIME
   ( { ALL | CURRENT | ASOF,QUERY_TIME } )
```

Let's consider the following examples demonstrating these principles. First, get the whole dataset:

```
select * from EMP;
```

The preceding select statement produces the following data. There are four employees in total, identified by employee_id. Employees 1 and 3 have two contracts, while employees identified by IDs 2 and 6 have just one contract. The output is as follows:

```
--> EMPLOYEE_ID          NAME            SURNAME
-->   DATE_FROM                  DATE_TO
-->        POSITION                         SALARY
--> 1                    Michal      Kvet
-->   01.01.2010 00:00:00   01.01.2020 00:00:00
-->        Researcher                       1000
--> 1                    Michal      Kvet
-->   01.01.2020 00:00:00
-->        University teacher               1200
--> 2                      Karol     Matiasko
-->   01.01.2001 00:00:00
-->        Manager                          1500
--> 3                      Stefan    Toth
-->   01.01.2005 00:00:00
-->        Developer                        1300
--> 3                      Stefan    Toth
-->   01.01.2006 00:00:00
-->        Tester                            700
--> 6                      Elena     Zaitseva
```

```
-->    01.01.2005 00:00:00   01.01.2005 00:00:01
-->         Development Manager              1400
```

Now, let's apply various DBMS_FLASHBACK_ARCHIVE levels and observe the output of the command. We will use the CURRENT, ALL, and ASOF keywords:

- ENABLE_AT_VALID_TIME('CURRENT'):

  ```
  EXECUTE         DBMS_FLASHBACK_ARCHIVE.ENABLE_AT_VALID_
  TIME('CURRENT');
  select * from EMP;
  --> EMPLOYEE_ID       NAME         SURNAME
  -->   DATE_FROM           DATE_TO
  -->        POSITION              SALARY
  --> 1             Michal       Kvet
  -->   01.01.2020 00:00:00
  -->        University teacher     1200
  --> 2             Karol        Matiasko
  -->    01.01.2001 00:00:00
  -->        Manager                1500
  --> 3             Stefan       Toth
  --> 01.01.2005 00:00:00
  -->        Developer              1300
  --> 3             Stefan       Toth
  -->   01.01.2006 00:00:00
  -->        Tester                 700
  ```

- ENABLE_AT_VALID_TIME('ALL'):

  ```
  EXECUTE DBMS_FLASHBACK_ARCHIVE.ENABLE_AT_VALID_
  TIME('ALL');
  select * from EMP;
  --> EMPLOYEE_ID       NAME         SURNAME
  -->   DATE_FROM           DATE_TO
  -->        POSITION                  SALARY
  --> 1             Michal       Kvet
  -->    01.01.2010 00:00:00  01.01.2020 00:00:00
  -->        Researcher                1000
  --> 1             Michal       Kvet
  ```

```
-->   01.01.2020 00:00:00
-->        University teacher          1200
--> 2                 Karol    Matiasko
-->   01.01.2001 00:00:00
-->        Manager                     1500
--> 3                 Stefan   Toth
-->   01.01.2005 00:00:00
-->        Developer                   1300
--> 3                 Stefan   Toth •
-->   01.01.2006 00:00:00
-->        Tester                       700
--> 6                 Elena    Zaitseva
-->   01.01.2005 00:00:00  01.01.2005 00:00:01
-->        Development Manager         1400
```

- ENABLE_AT_VALID_TIME('ASOF', <TIMESTAMP>):

   EXECUTE
   DBMS_FLASHBACK_ARCHIVE.ENABLE_AT_VALID_TIME('ASOF',
   **   TO_DATE('1-DEC-2020 00:00:00',**
   **   'DD-MON-YYYY HH24:MI:SS'));**

   ```
   select * from EMP;
   --> EMPLOYEE_ID        NAME          SURNAME
   -->   DATE_FROM              DATE_TO
   -->        POSITION                     SALARY
   --> 1                 Michal   Kvet
   -->   01.01.2020 00:00:00
   -->        University teacher          1200
   --> 2                 Karol    Matiasko
   -->   01.01.2001 00:00:00
   -->        Manager                     1500
   --> 3                 Stefan   Toth
   -->   01.01.2005 00:00:00
   -->        Developer                   1300
   --> 3                 Stefan   Toth
   -->   01.01.2006 00:00:00
   -->        Tester                       700
   ```

The PERIOD management supervised by the DBMS_FLASHBACK_ARCHIVE package uses closed-open characteristics; thus date_to expresses the first time point (based on the granularity precision used), which is not covered by the validity.

The user must have been granted the privileges to execute methods of the DBMS_FLASHBACK_ARCHIVE package. They can be granted using the following command by specifying the username of the grantee:

```
GRANT execute ON DBMS_FLASHBACK_ARCHIVE to <username>;
```

Now you know how to model and use a PERIOD structure in a table definition. However, have you thought about how to identify such a structure in an existing structure? The next section will show you.

Identifying temporality

Time-delimited management of data tuples is an important part of life cycle data management. The conventional paradigm of the relational database system is based on storing currently valid data. Thus, the update operation is used to change an object's state. Simply, original values are replaced with new ones. After the process, only transaction logs maintain the executed operation reference by storing change vectors. Temporal management encapsulates the data with the validity time frame. Therefore, multiple states bordered by the validity period, commonly modeled by the start and end points, can define each object. It is not easy to observe whether the table is temporal or not. A PERIOD definition inside the table can provide an answer of temporality usage. However, how do we identify the PERIOD data type? Looking at the *data dictionary views*, there is no IS_TEMPORAL attribute or any variant. One possible solution is provided by the **data definition language** (DDL) definition of the table. A full data definition can be obtained by calling the GET_DDL function of the DBMS_METADATA package. From this character string, PERIOD can be extracted. The GET_DDL function takes two parameters – the structure type and the object name:

```
select DBMS_METADATA.GET_DDL('TABLE', 'EMP') from dual;
```

The GET_DDL function result provides not only the statement used for the creation of the object but also other physical storage parameters, as listed in the following code block, which gets the structure of the EMP table:

```
CREATE TABLE "KVET3"."EMP"
   ( "EMPLOYEE_ID" NUMBER(*,0) NOT NULL ENABLE,
     "NAME" VARCHAR2(20) NOT NULL ENABLE,
     "SURNAME" VARCHAR2(20) NOT NULL ENABLE,
     "DATE_FROM" DATE NOT NULL ENABLE,
     "DATE_TO" DATE,
     "POSITION" VARCHAR2(20) NOT NULL ENABLE,
```

```
    "SALARY" NUMBER(6,2) NOT NULL ENABLE
 ) SEGMENT CREATION IMMEDIATE
PCTFREE 10 PCTUSED 40 INITRANS 1 MAXTRANS 255
NOCOMPRESS LOGGING
STORAGE(INITIAL 65536 NEXT 1048576
        MINEXTENTS 1 MAXEXTENTS 2147483645
        PCTINCREASE 0 FREELISTS 1 FREELIST GROUPS 1
        BUFFER_POOL DEFAULT FLASH_CACHE DEFAULT
        CELL_FLASH_CACHE DEFAULT)
TABLESPACE "SMALL_TBLSPC"
ALTER TABLE "KVET3"."EMP"
      ADD PERIOD FOR "VALIDITY"("DATE_FROM","DATE_TO")
```

Thus, the condition for the evaluation, whether the keywords of the validity period are present or not, can be placed.

The following block can be used to get a list of the temporal tables. First, the cursor is defined, which lists the tables from the *data dictionary*. The output is then tested to see whether the temporal validity clauses are there or not by getting the DDL. The next block can provide a solution. A cursor can be either explicitly or implicitly defined and managed. One of the possible solutions using a FOR cycle is as follows:

```
declare
 ddl_code clob;
 is_temporal integer;
begin
 for i in (select trim(table_name) as tab_name
           from user_tables)
  loop
   select DBMS_METADATA.GET_DDL('TABLE', i.tab_name)
         into ddl_code
    from dual;
   select count(*) into is_temporal
    from dual
     where ddl_code like '%PERIOD FOR%';
   if is_temporal>0 then
    DBMS_OUTPUT.PUT_LINE(i.tab_name);
```

```
    end if;
  end loop;
end;
/
```

The block first extracts a list of the table names. Secondly, the DDL script is generated using the GET_DDL function of the DBMS_METADATA package. It produces a **Character Large OBject (CLOB)**, which is extracted to evaluate the presence of the PERIOD FOR keywords. If they are present, the relevant table is listed. Although the preceding code does not guarantee temporality (because the PERIOD definition does not generally ensure it), it provides at least a basic insight for deeper evaluation.

Another perspective is related to the *hidden columns*. The presence of hidden columns does not, however, ensure temporality. Thus, a particular reference is just a primitive temporality indicator:

```
select column_name, hidden_column
  from user_tab_cols
    where table_name='EMP';
--> COLUMN_NAME          HIDDEN_COLUMN
--> VALIDITY             YES
--> EMPLOYEE_ID          NO
--> NAME                 NO
--> SURNAME              NO
--> DATE_FROM            NO
--> DATE_TO              NO
--> POSITION             NO
--> SALARY               NO
```

It cannot be totally ensured that the table is temporal using any of the preceding methods. But on the other hand, it is evident that the database system must provide a robust interface for management. Otherwise, the DBMS_FLASHBACK_ARCHIVE package could be used, and reliability would be compromised. Note that without the PERIOD definition, the DBMS_FLASHBACK_ARCHIVE package does not apply the following rules. The principles of historical data image definition using the ENABLE_AT_VALID_TIME method of the DBMS_FLASHBACK_ARCHIVE package are described in the following paragraphs.

Let's create a copy of the EMP table with no period reference. Then, fill the table with all the data from the original table and apply the ENABLE_AT_VALID_TIME method from the DBMS_FLASHBACK_ARCHIVE package.

Create a table called EMP2 with the following structure:

```
Create table EMP2
 (employee_id integer not null,
  name varchar(20) not null,
  surname varchar(20) not null,
  date_from DATE not null,
  date_to DATE,
  position varchar(20) not null,
  salary number(6,2) not null
 );
```

Apply ENABLE_AT_VALID_TIME('ALL') and insert the content of the existing EMP table to it:

```
EXECUTE DBMS_FLASHBACK_ARCHIVE.ENABLE_AT_VALID_TIME('ALL');

insert into EMP2 select * from EMP;
```

Select the number of tuples in the EMP and EMP2 tables:

```
select count(*) from EMP2;        --> 6
select count(*) from EMP;         --> 6
```

Apply ENABLE_AT_VALID_TIME('CURRENT'):

```
EXECUTE
    DBMS_FLASHBACK_ARCHIVE.ENABLE_AT_VALID_TIME('CURRENT');
```

Select the number of tuples in the EMP and EMP2 tables:

```
select count(*) from EMP2;        --> 6
select count(*) from EMP;         --> 4
```

Without PERIOD management, *a table is not treated as temporal*! Looking deeper into the internal structures, temporal identification can be made via the SYS_FBA_PERIOD view. It does not use the USER, ALL, or DBA prefixes. It instead consists of the following structure:

```
desc SYS_FBA_PERIOD;
--> Name            Null?       Type
--> OBJ#            NOT NULL    NUMBER
--> PERIODNAME      NOT NULL    VARCHAR2(255)
--> FLAGS                       NUMBER
```

```
--> PERIODSTART    NOT NULL    VARCHAR2(255)
--> PERIODEND      NOT NULL    VARCHAR2(255)
--> SPARE                      NUMBER
```

SYS_FBA_PERIOD provides several attributes. From the temporal perspective, the most relevant are PERIODNAME (for dealing with the name of the validity period) and PERIODSTART and PERIODEND covering the duration borders. The owner of the SYS_FBA_PERIOD view is the SYS user.

Be aware there is no TABLE_NAME reference or an OWNER reference. Instead, the identification is made by an internal numerical identifier, obj#. Its value can be obtained from the USER, ALL, or DBA_OBJECTS data dictionary views. So, the period validity identification query can be defined as follows:

```
select periodname, periodstart, periodend
  from sys.SYS_FBA_PERIOD
    where obj# IN
      ( select object_id
          from user_objects
            where object_name = 'EMP'
      );
--> PERIODNAME        PERIODSTART      PERIODEND
--> VALIDITY          DATE_FROM        DATE_TO
```

Alternatively, a view can be defined by providing the validity definition, as well as frame borders:

```
create or replace view validity_period_view as
    select name table_name,
           periodname period_name,
           periodstart period_start,
           periodend period_end
      from sys.sys_fba_period join
           sys.obj$ using(obj#)
        where owner# = userenv('SCHEMAID');
```

Grant the select privileges to all users:

```
grant select on validity_period_view to public;
```

Optionally, you may create `public synonym` for it:

```
create or replace public synonym validity_period_view
  for kvet3.validity_period_view;
```

Final remarks

The mechanism of duration management using a `PERIOD` definition is associated with the time reference. In contrast to pure date and time management, it automatically checks the consistency and duration to ensure that at least one time point belongs to it. The `DBMS_FLASHBACK_ARCHIVE` package can provide the limits of the time validity reference. Although there is robust time management, a particular table cannot be directly flagged as temporal from a database theory point of view. This is where the problem of approving the concept of temporality becomes visible. Namely, the temporality definition ensures that one object is covered just by one valid state at any time. However, the `PERIOD` definition does not guarantee it. Let's go back to the preceding example. Each employee is defined by a unique identifier (`employee_id`). This attribute cannot be directly shifted to the temporal environment. As is evident from the evaluated data, multiple employment contracts referencing the same `employee_id` attribute value can be valid in parallel. This is not, however, valid according to the technical temporality definition. Thus, it is always important to state the principles used and the applicable temporal integrity.

When dealing with employees, there is no significant problem. One employee identifier references just one person physically, who can be referenced by multiple contracts. These contracts can be valid simultaneously as well. Conversely, when dealing with a measured dimension, it is clear that we cannot define multiple states at once. For example, car components cannot be given more than one color at any time. A sensor provides only one temperature value. A button's state is either on or off. Naturally, these object values can be changed, but the real change can always be timeline referenced. As a result, the object itself cannot be denoted just by the identifier, but the validity or other time management spectrum (such as transaction reference) should be emphasized. Therefore, the validity time frame is covered by the identifier. That definition means additional constraints must be applied, mostly related to unlimited definition management, which needs to be replaced by another notation. In contrast, the `NULL` value cannot be part of the primary key in the definition.

Summary

This chapter dealt with interval representation and type relationships. In the first part, transformation principles across time intervals were stated and demonstrated by adding the smallest time element to the interval. Then, the focus shifted to the positional and temporal relationships by discussing interval interaction types. These positions are crucial for state monitoring in a timeline.

Oracle Database provides a `PERIOD` structure that applies to the table definition expressing the duration forming temporal validity. This chapter dealt with its concepts, modeling, and definition. By reading this chapter, you will have become familiar with using the data dictionary to evaluate and identify the PERIOD structure for a particular table that already exists. The PERIOD structure is associated with the ENABLE_AT_VALID_TIME function of the DBMS_FLASHBACK_ARCHIVE package. By using that function, temporal principles can be applied. It allows you to highlight current valid states, look at all states, or get a snapshot at a defined timepoint.

The next chapter discusses temporal database concepts from an architectural point of view. The core structure, which has already been discussed, is an *object-level* temporal architecture. If the granularity is finer, *attribute-oriented* or *synchronization group* management models can be used. Besides the temporal architectures, temporal dimensions are explored in the next chapter.

Questions

1. Let's define the following duration with a closed-closed representation: < 1st of January 2022 ; 28th of February 2022 >. Which option expresses an analogous solution using a closed-open model?

    A. < 1st of January 2022 ; 28th of February 2022)

    B. < 1st of January 2022 ; 29th of February 2022)

    C. < 1st of January 2022 ; 1st of March 2022)

    D. (31th of December 2021 ; 1st of March 2022)

2. Which relationship characterizes two interval interactions with no common point?

    A. not_fills only

    B. occupies only

    C. The whole excludes category

    D. meets only

3. Which interval interaction type is used in temporal management to monitor the temporal state evolution with no undefined states?

    A. before

    B. after

    C. meets

    D. fills

4. If the `PERIOD` specification is defined by `PERIOD FOR VALIDITY`, what are the names of the temporal attributes?

    A. `PERIOD_START, PERIOD_END`

    B. `VALIDITY_START, VALIDITY_END`

    C. `BD, ED`

    D. An exception is raised because the names are not explicitly specified

5. Which option takes current valid data from the period?

    A. `DBMS_FLASHBACK_ARCHIVE.enable_at_valid_time('CURRENT')`

    B. `DBMS_FLASHBACK_ARCHIVE.enable_at_valid_time('ALL_VALID')`

    C. `DBMS_FLASHBACK_ARCHIVE.enable_at_valid_time('ASOF', 'CURRENT')`

    D. `GET_CURRENT_PERIOD`

Further reading

- *Managing Time in Relational Databases: How to Design, Update and Query Temporal Data* by Tom Johnston and Randall Weis. It provides a practical guide to data modeling and query management of date and time delimited tuples.

- *Reference documentation of the DBMS_FLASHBACK_ARCHIVE package*: `https://docs.oracle.com/en/database/oracle/oracle-database/21/arpls/DBMS_FLASHBACK_ARCHIVE.html#GUID-5A76DE8C-F834-4CF2-8940-1FFF2C2177EC`.

11

Temporal Database Concepts

This chapter will provide you with complex information related to temporal modeling and physical data model representations. You will be introduced to the historical evolution of databases originating from conventional (non-temporal) models by using additional extensions, following which the processing level – object, attribute, and group granularity – will be discussed. The object level is signified by the primary key extension, while attribute granularity associates the temporal sphere for each attribute separately. Therefore, a universal solution is built on group granularity, where temporal groups can be composed dynamically based on real workloads. Thanks to that, the system can automatically apply new rules and build on and structure existing groups. However, always keep in mind that the principle of the temporal table is based on the opportunity to hold multiple states for each object, delimited by the validity time frame.

We will also look at several aspects of temporal requirements. By accepting the proposed rules, performance-effective and robust solutions can be developed. The temporal definition is not only about the state validity itself but also several other dimensions. All the rules, principles, and models mentioned in this chapter are practically oriented and practically described.

To put it succinctly, this chapter will cover the following main topics:

- Conventional structure extension by forming a **current-history** and **header-temporal** model
- An **object-oriented** temporal approach
- **Uni-temporal** and **bi-temporal** solution principles
- Requirements for temporal systems to ensure performance
- **Temporal dimensions**
- **Attribute-oriented** granularity
- **Group-level** temporal architecture – logical and physical representations
- Conventional tables with time-delimited attributes

The origin and evolution of temporal models

The need to preserve individual states and limit them over time arose with the advent of the first database systems in the 1960s. Several ideas were brought forward for managing and modeling these types of data, with an emphasis on validity, but most of them were only treated theoretically. The main limitation was the technical support costs in terms of the hardware components, limiting the possibility of the further expansion of technical equipment.

A more significant shift occurred during the boom of relational database systems. They are based on entities and relationships, creating a data model separate from the physical hardware. The main features are integrity, individual rows being identified by primary keys, as well as the precise structure supervised by transaction support. Moreover, individual structures can be optimized using the process of data normalization to maintain efficiency and data quality. The data access itself is provided by instance processes and the optimizer, which selects individual operations based on relational algebra. This made it possible to ensure that up-to-date data was managed efficiently, quickly, and securely. In addition, huge data amounts could be managed by applying the techniques of data optimization, model optimization, partitioning, and data resource distribution. However, resources were still limited. Only current valid states were retained, making these conventional database systems.

Therefore, the pressure and requirements to create a state management solution increased and became more significant over time. The management and preservation of current valid states only were not sufficient. As a result, there were several attempts to create a temporal paradigm, but the pre-existing principles usually blocked them. Specifically, the main obstacle to any new solution was transactional support, which stores change vectors. Thus, by optimizing the structure of the transaction logs, historical data could be reconstructed. However, it must be noted that this process was too demanding and was not very reliable if the transaction logs were rewritten or were simply removed or missing.

Moreover, transaction logs also store a lot of other information dealing with transactions, not just the original and new state. Individual logs had to be parsed and evaluated step by step, row by row, greatly limiting performance. Thus, sooner or later, the technique was rejected due to the inefficiency of the processing. The data stream was later optimized and is currently used by **Flashback technologies**, which we will discuss more deeply in *Chapter 13*. It is clear that such an approach cannot be considered temporal, as future valid states cannot be observed. It is used rather as a tool for data recovery during a short time window to allow users to refuse operations after processing, even at the committed stage.

Although temporal databases were not a universally proven solution initially, there was a lot of research done into strategies to tackle temporal paradigm complexity. Unfortunately, most temporal solutions were optimized only for the specific information system and were not widespread or made general, forcing developers to manage data independently.

There were two feasible conventional systems created that should be mentioned. The **current-history** architecture splits management into two layers, extending traditional (conventional) principles. The original, current valid data is managed by the first layer. Historical data is stored in a separate table created for each conventional table. Future valid data can be stored and evaluated as well by applying the database jobs to transfer the future valid data to the current states at a defined time point. *Figure 11.1* shows the architecture of the solution. **TAB1** and **TAB2** are conventional and do not manage validity time frames at all. Non-current data is then in the **Temporal_data1** and **Temporal_data2** tables. One obstacle is referential integrity, which only applies to current valid data. It is impossible to maintain coverage of historical and future data in terms of validity (mostly in terms of the impossibility of storing the state without proper reference to another table object's state).

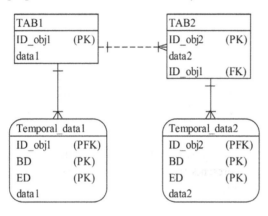

Figure 11.1 – The current-history concept

The preceding architecture's disadvantage is that it stores data in two separate layers. Current valid data is treated separately from outdated data. Suppose the current valid data needs to be time-bordered. In that case, particular tuples are duplicated in both layers, which on the one hand violates normalization principles, and also imposes additional costs and requirements in terms of integrity and overall consistency.

Another approach is the **header-temporal** model, which holds all the data in a temporal layer (**Temporal_data1** and **Temporal_data2**). Conventional data does not hold values for all attributes; just the identifiers and referential integrity are covered (*Figure 11.2*). Validity correspondence must, however, be treated explicitly as well.

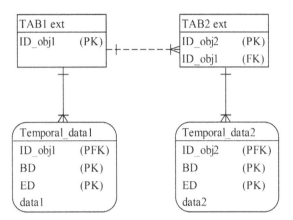

Figure 11.2 – The header-temporal concept

The third concept is based on using **nested tables**. In this case, each table is treated conventionally, but it has additional object type attributes expressing a nested table, in which outdated states are stored. However, this approach is not powerful. Nested tables cannot be indexed directly, and data retrieval and evaluation are too demanding.

Looking at the models, each tuple is enclosed by the validity time frame forming the whole object state. Thus, each database row represents one state – the whole object. Thus, let's move on to describe the principles of the object-oriented temporal approach in depth.

The object-oriented temporal approach

It is clear that despite several improvements to historical databases, referenced database concepts are unsustainable for practical use in the long run. In order to make the right decision and reflect data changes, it is not enough to store and evaluate only current valid data; historical data also needs to be defined and operated. Over the decades, several attempts have been made to create a temporal model. The object-oriented temporal approach was the basis of this. Moving conventional processing to an object-level temporal model is based on the **primary key**, which uniquely identifies any object in a conventional database. A conventional system uses just one object version expressing the current valid state, while a temporal model stores the whole evolution. Thus, to manage the object, several versions must be identified and processed. To do so, the original primary key is extended to reference the version. It is mostly modeled by the validity time frame, but generally, any temporal meaning can be used. Those additional attributes forming the temporal references are then used as keys to order the states of the object by time.

Figure 11.3 shows the available table definitions from a temporal point of view. Common conventional tables (the first model in *Figure 11.3*) do not process individual versions. Thus, by executing update operations, the original state is replaced by new data and the original states are no longer present in the system. While the changes are always enclosed by the transactions, the change vector part of the transaction logs can temporarily provide original states. However, identifying evolution using transaction logging is insufficient performance-wise.

The second approach in *Figure 11.3* uses one temporal dimension and is called the **uni-temporal model**. The primary key is composite, defined by the object identifier itself and the time interval duration – BD (the beginning of validity – beginning point, sometimes referred to as the start of validity, which is synonymic) and ED (the end of validity).

The third model in *Figure 11.3* was created by simplifying the previous model by referencing only the beginning of the duration. The principle is based on the assumption that each consecutive state borders the validity of the previous state in a timeline. Therefore, this lowers the storage demands. Namely, the ED value is not stored physically. Instead, it can be dynamically calculated. On the other hand, identifying the current valid state is more demanding and time-consuming.

Please note that the duration border shown in the third model in *Figure 11.3* is logical rather than physical. ED can be unlimited, but how can you model and express it? One of the solutions is to use the NULL value, characterizing unlimited validity. However, this has two drawbacks. The NULL value cannot be part of the primary key, forcing the system to exclude the ED attribute from the primary key definition (which is possible in principle, whereas the {ID, BD} pair is unique within the definition). The second limitation is based on the meaning. With the NULL value, it is not obvious that it refers to the future. Well, if the state is still valid, it is naturally clear that the new change has not occurred yet.

The second solution for unlimited validity modeling replaces an undefined value with another expression, usually a sufficiently large time representation (31-12-9999 as the current maximum time that can be modeled in database system). In this case, this particular value is also logical rather than physical. It does not really mean that the validity will end in the year 9999. The aforementioned solution clearly states that there is no new state. However, before attempting to store the new object state, the ED value of the original state has to be changed (the original unlimited validity must be replaced by the real value).

The expiration date (the validity endpoint) is often undefined from a logical perspective, but it is not physically possible to use a NULL value precisely because this attribute is part of the primary key.

In principle, this is technically possible because the {ID, BD} pair itself must have a unique constraint definition, as the time intervals for a particular object must not overlap. ID in this representation means a logical concept of the object identifier. Physically, it can be modeled by multiple attributes (which can be composite) based on the requirements of the application domain. On the other hand, the time frame thus defined is not correct from a logical perspective. Where two attributes mostly model the validity interval, it must be highlighted that they are inseparable and interdependent to secure consistency and prevent overlaps, collisions, undefined states, and so on.

All the temporal characteristics that have been discussed in this section so far were based on an irregular duration interval validity period. A new state could occur at any time. The fourth model in *Figure 11.3* uses predefined synchronization clocks that specify the occurrence of the next change. For example, the change is recorded once a day at a specific time. In that case, it would be unnecessary to store the beginning of the validity (BD), because the values can be calculated dynamically on demand. Instead, version sequence numbers can be used.

The uni-temporal models available are shown in *Figure 11.3*. Primary key attributes are marked in gray. BD defines the start point of the validity and ED represents the endpoint of the validity. SEQ_ID is used in the last model to make the versions of the object sortable. For simplicity, the data table content is modeled and represented using the **DATA** expression.

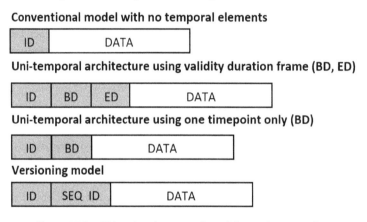

Figure 11.3 – Object-level temporal models – uni-temporality

The object state does not need to be temporally defined only by the validity. The limitation of the uni-temporal system is associated with the impossibility to make data corrections to existing systems. Thus, an additional temporal layer must be created to serve not only the validity time frame but also the transaction reference. The **bi-temporal** model defined in the next section handles the temporal model extension by managing two time dimensions. Generally, the **multi-temporal** model can be used to cover any number of temporal dimensions.

The temporal aspect of management

In the previous section, we referred to uni-temporal systems. The time spectrum was expressed using either a time interval (BD, ED) or a single time point (BD). This allows you to store individual states and their positional arrangement over time. Uni-temporal systems are usually characterized by the validity of the record, where overlaps are not allowed. The temporal paradigm rule requires the creation of a consistent state in which each attribute is assigned exactly one value.

In such a system, it is impossible to correct the existing states. It would be necessary to physically overwrite them, which could cause problems with the consistency and usability of the whole system, whereas the states before modification could already be used for reports, analyses, and so on. Repeatedly calling these functionalities with the same parameters would achieve different results. However, the reasons for the changes would not be directly identifiable, as individual corrections would not be present in the system.

Therefore, other time elements are added by forming a bi-temporal approach. The time references are expressed by two time representations, usually expressing validity and transaction definition. The bi-temporal model also allows us to store corrections of existing states and create individual versions for each state. Like validity, the transaction definition expressed over time must not overlap for specific states of the same object. A representation is depicted in the following figure:

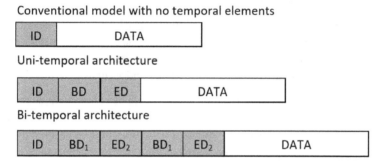

Figure 11.4 – Object-level temporal models – multi-temporality

Thus, each state is delimited by the validity time frame (BD_1, BD_2), allowing you to store corrections differentiated by the transaction validity (BD_2, ED_2). Furthermore, the current version has an unlimited transaction validity reference, meaning that data correction is not supposed to be done later. If it is necessary to correct the existing state, the original state transaction reference is automatically ended by creating a new state (commonly supervised by the *trigger*).

You are hopefully now familiar with the object-level temporal model and have understood the principles of temporal modeling. However, what are the requirements of the temporal system? Which aspects must be covered to build a solution with effective performance and ensure robustness and reliability? The next section deals with these temporal requirements.

Temporal system requirements

The requirements for temporal systems were first summarized in the book *Managing Time in Relational Databases* by Tom Johnston and Randall Weis. They presented the first two aspects – usability and performance. Based on our own research, we'll extend the definition to the rest of the aspects. The main reason for the extension is to ensure complexity and describe in detail the temporal approach used:

- **Usability**: This focuses on the ease of use of methods for users. The aim is to provide a robust and easy-to-use solution, regardless of whether we access currently valid records, historical data (images of objects that were valid in the past), or records that will become valid in the future, but information about these states is already stored in the database. This aspect also focuses on the possibility to monitor changes, perform statistical evaluation, and create a platform by providing input data for decision-making.

- **Performance**: This is based on the accuracy of the required data and the speed of access to it, emphasizing the time spectrum. The time costs to access outdated values should not be significantly higher compared to the current valid states, regardless of the data structure used. Simply, access to a data image at any defined point in time should be feasible and accurate in terms of processing time.

- **Transparency**: From a user's perspective, the solution should be independent, so the requirements and methods should be the same for any architecture and granularity used. Transformation and access to individual pieces of data in different structures must be ensured by the processes of the instance and the temporal database system. So, if the user wants to obtain an object's state at a defined time point, the result must be precisely shaped regardless of the internal data modeling and representation (the level of *objects*, *attributes*, or *temporal group* sets).

- **Data structure**: This is a requirement that focuses on using optimized data structures for a specific application, so that duplicate tuples do not occur and the system covers all changes at any level of accuracy and granularity. It is also necessary to emphasize the need to normalize the data model to which the time perspective must be applied (normalization of the data model at the conventional, or untimed, level is insufficient because an enormous number of duplicates may be generated or the temporal aspect of individual data types may be missing). In addition, you must be able to change the granularity or management of existing attributes in the solution (changing types, temporal orientation, the addition of other attributes, and so on). The system must be able to automatically adapt to changes in the data model, as the system stores the states of objects in a long time spectrum. The representation and structure of each data model is again modeled using a temporal approach limiting its validity.

- **Relevance**: The criterion of relevance is met by identifying significant changes and is used mainly in cases of data processing from sensors, for example, in industry, transport, or the field of processing, evaluating, and storing medical data. Each sensor has a defined measurement accuracy, which the user can extend to identify a significant change. Not every attribute change based on a sensor reading is important and must be saved. It should also be possible to locate the areas that need to be monitored at a given time instead of whole environment monitoring. At the given time, only data from some sensors is processed, emphasizing their position, forming a spatiotemporal solution.

- **Transaction support**: A core element of relational systems is transaction support, moving the database from one consistent state to another consistent state. The original conventional transaction properties – **Atomicity, Consistency, Isolation, and Durability (ACID)** – are passed. Additionally, the speed of disaster recovery is also emphasized in the temporal world, so changes are grouped based on the object reference and are cumulatively written from memory to the physical database in order to minimize processing time and technical demands. Emphasis is also placed on *isolation*. Data changes are very dynamic and can be associated with multiple transactions manipulating the same object in a versioning manner. Therefore, it is necessary to properly define techniques to recover transaction versions at any depth.

- **Limited temporal usability**: This is related to the various data management techniques described as follows:

  - *Historical states* can lose their importance and relevance by either having tuples removed by passing the criterion or being moved to an archive repository (mostly modeled by the aggregated data warehouse).

  - *Future planned states* can be updated, corrected, or even canceled before becoming current and valid.

  - **Modeling undefined and incomplete states**: This refers to representing (partially) undefined and incorrect states, either using a default value, a NULL value, or a generalization using functionality and applying a pointer to an object maintained in the instance's memory.

- **Access optimization**: This focuses on access using index structures and relevant block identification by limiting the restriction of NULL values that are not covered by *B-tree* indexes. Thanks to that, sequential table scanning is replaced by block pointers. Note that sequential scanning brings many drawbacks, such as fragmentation, free blocks, or **High Water Mark (HWM)** references.

- **Correctness**: This covers almost the whole issue of data integrity – state overlaps, state ISA hierarchy, and referential integrity coverage. It also focuses on the issue of correcting existing states, either in the form of direct state replacement or individual version management.

- **Security**: This involves dealing with the data distribution, partitioning, backups, encryption, recovery, **Automatic Storage Management (ASM)** (a feature that simplifies data storage management by controlling ASM Disk Groups, covered by the Oracle system), load balancing within a distributed solution, and **Data Guard**. In the case of temporal management, we use either the 2-N, 3-N, or N-N approach. The first character represents the minimum number of nodes on which each record must be located in order to be declared as permanently and securely stored in the system. The second character represents the total number of nodes (replicas) of the system. This technique is used if the node failure frequency is relatively high (for example, caused by unreliable network communication).

- **Attribute temporality**: This is a way of modeling and distinguishing static attributes. If the **STRICT** approach is used, the characteristics of the attribute values cannot be changed. The **RELAXED** approach is associated with code lists. Existing values must not be changed, but the domain itself can be extended by new elements.

The temporal model is commonly associated with validity, but generally, several dimensions can be covered. So far, we have mainly focused on validity and transaction support in a bi-temporal approach. The next section deals with temporal dimensions, which can be covered by modeling and representation.

Exploring temporal dimensions

Time can have various meanings and representations in temporal database systems. It does not have to be just about validity; the time when the `insert` statement was actually executed, processed, or approved can be referenced. There is no complex formulation and categorization of individual time aspects in current systems and the literature overview. Modeling and interpretation are mostly left to the developers or system admins. Therefore, we will summarize the main principles and reference models in the upcoming subsections. You will understand that it is not enough to refer only to validity; we need to create a platform for recording state corrections over time or solving synchronization across multiple nodes. For example, we may have a state that we consider to be valid for a certain period of time. However, it is possible that later we discover that this state was not actually correct and needs to be corrected. Or perhaps, the valid state recorded in the databases may not match the actual start time of the event. These are some of the real-world issues that you might be faced with and this section is aimed to give you the knowledge needed to address them. Let's get started.

Validity

Validity expresses the time duration (various definitions can be used) for which a particular row is physically valid. For example, an employment contract is defined by the validity frame regulating its applicability, employee access to the systems, and so on.

Transaction validity

Transaction validity, sometimes referred to as **reliability**, denotes the time duration for which the state is considered correct. Thus, individual states delimited by the validity can have multiple versions characterizing corrections. The physical `update` statement is replaced by the logical `insert` operation managing the new version. Transaction validity is modeled either by the start point of the validity or two time elements, BD and ED. In principle, the right border (ED) is unlimited for the current version. It is assumed that the version will never be changed and is correct.

To be able to sort these versions (corrections) and identify the order, the definition of transaction validity is introduced (analogous to the original bi-temporal approach). The physical representation is identical to the validity modeling itself; by default, the transaction validity is expressed using only one attribute, which characterizes the physical timestamp of inserting the relevant record (or status correction) into the database. In this case, each new version limits the validity of the previous one.

There has been a tendency to replace transactional validity with a simplified version characterized by a number sequence in the past. However, we do not consider this solution to be appropriate due to these two major problems:

- The correction of the state of the object cannot be located in time. To find out when the new version has been physically registered would be impossible. Moreover, it causes problems with individual functions and analytics.

- It would be necessary to obtain the last serial number before inserting the new version, either by using a trigger or by calling a user-defined method that returns the object's version number. However, in an environment of parallelism, collisions arise, necessitating the use of strict locking.

The IPL model

In systems that manage sensor data, making entries to the database itself is preceded by pre-processing and filtering the incoming data. There may be situations in which the input data flow is too large and therefore data cannot be processed immediately. To solve the problem, the **Input, Preprocessing, Load** (IPL) model can be used, highlighting the data flow in a temporal reference. It uses a three-dimensional temporal architecture covering these steps:

1. **I – The input operation**: Each record entering the system is assigned a timestamp and is given to the input queue (**First In, First Out (FIFO)**).

2. **P – Preprocessing/filtration**: Individual records are gradually selected from the input queue, checking their accuracy and reliability and linking to the existing states by identifying significant changes and the correctness and integrity of the provided data using the *Epsilon approach*. If the system evaluates that the data should be stored in the database, the algorithm will proceed to the next step (the **Load** operation). Otherwise, the processing ends, and the particular record is removed from the input queue and is not further evaluated. The processing continues by taking the next element from the input queue.

3. **L – Load**: In this step, the process of storing a new record in the database takes place.

Figure 11.5 shows the architecture of the IPL model. The input stream is routed to the input queue. Individual tuples and values are then preprocessed and evaluated to distinguish whether a particular value is relevant and needs to be stored in the database or should be refused (for example, if there's no significant change compared to the pre-existing condition of the state that was identified).

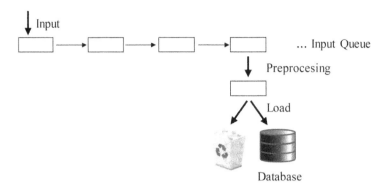

Figure 11.5 – Input queue in the IPL model

The IPL model uses a time spectrum with three time values – the entry of the record into the system (TI) and the beginning of the evaluation or filtration (TP). The third time dimension is the time the record was entered into the database (TL).

The IPLT model

The **Input, Preprocessing, Load, Transaction Coverage** (**IPLT**) model extends the IPL model by adding a fourth dimension – time information about the termination (approval) of the transaction. The data entered into the database can be part of relatively complex and long-lasting transactions. Therefore, a special modeling technique is proposed, characterized by the timestamp of the direct insert operation within the transaction (TL) and the time of its confirmation (TT).

In this context, we would like to note that the transaction may be longer than the validity of the state itself. Thus, there is a situation where the records are no longer valid at the time of insertion and confirmation (an anti-date issue).

Future valid record management

A fully temporal system focuses on all time spheres. It processes records valid in the past, those that are currently valid, and also those states that will be valid in the future. Commonly, states that will be valid in the future are entered into the database earlier, with the certainty that the system will provide automated transformation and projection into currently valid states at a given time. In such a model, three time dimensions can be identified:

- **Dimension 1**: The insertion of the state into the system (*reliability*).
- **Dimension 2**: The beginning of the validity (*validity*).

- **Dimension 3**: The third dimension is specific to systems that structurally separate future states and currently valid records. It expresses the time of the transformation of the future state into the current one (**Future Response Time (FRT)**). Ideally, the time value of the second and third dimensions should be the same. However, in reality, they may be different. Because of the large amount of data that needs to be processed, delays can occur that may be significant and need to be recognized and recorded. This dimension also evaluates the performance reliability of the entire system, allowing for optimization. If the difference between the dimensions is large, it points to a bottleneck in the whole system – transaction logs and their processing by the transaction log management process (operated by the **Log Writer (LGWR)** background process).

Note that the meaning of "*significant diversity of the second and third dimensions*" is application-dependent. It can even reflect the number of microseconds or nanoseconds.

Online and offline application modes

Many information systems require a direct and permanent connection to a data source. All changes are encapsulated within transactions. After successful completion, data becomes permanent but, at the same time, immediately accessible to all users.

Mobility and accessibility anywhere are increasingly required of information systems. This trend has become even more important with the advent of intelligent information systems and mobile devices. Although most of us have reliable internet access as well as mobile connectivity, there are still large areas where mobile internet is unavailable or too slow or unstable, and ultimately unreliable and therefore unusable as a result. Accordingly, concepts, approaches, and systems with asynchronous processing and state updates have been developed. This means that data is stored locally for each client, and synchronization with the central database system is performed on request or automatically by identifying the connection availability.

When using an offline-mode data processing architecture, it is necessary to distinguish the logical and physical time from the perspective of the central data node. Since the record is initially stored locally, the transaction time expresses a logical representation – the point in time at which the record was inserted into the local database, or the time interval during which the record was considered correct for a specific client. In the case of synchronization with the central database repository, the physical transaction time (the time when the record was entered into the central data register) is used for the main database system, where the data is then shared by all clients. From an update and synchronization point of view, two situations can occur:

- Synchronization is successful, so the logical transaction time (expressing client processing) and the physical transaction time (the time of synchronization with the server) are stored.

- Collisions are detected in the synchronization process. In this case, it is necessary to resolve conflicts using the transaction rules (shortening, replanning, or refusing state), or transaction priority can be applied to support more important nodes.

The architecture of the synchronization model is shown in *Figure 11.6*. There is a central system integrating data from individual nodes.

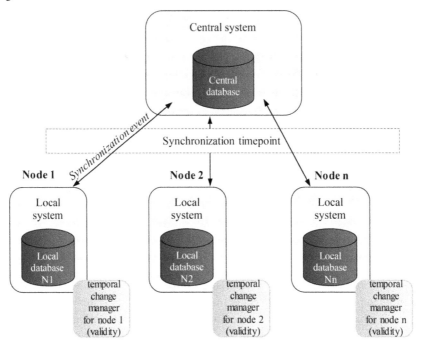

Figure 11.6 – Data node synchronization architecture

Replicas

Generalizing these approaches makes it possible to create a temporal solution for replica management in a distributed database system environment. In this sense, we define validity. The transaction time expresses the immediate timestamp of reaching and processing the specified dataset on a particular node. From this perspective, it is therefore possible to distinguish between the local and global transaction time:

- The **local transaction time** is stored on a specific node of the distributed database system and represents the data replica input to that node.

- The security policy of the system characterizes the **global transaction time**. If the user requests two replicas of each piece of data in the system, the global transaction time represents the greater value of the local transaction times. If the user requires more than two replicas, they can still define a **global partial transaction time**, which expresses the time at which the replicas were processed by two nodes (or more depending on the organization's security policy and implemented temporal system). However, at that time, particular data can be marked as attainable even in the event of a node failure.

Note that the processing itself can optionally be extended with pre-processing and filtering modules. However, these operations are usually performed before replicas are created and data is distributed.

The next section contains some important remarks about time dimensions, discussing physical and logical times, as well as processed granularity.

Final remarks on the temporal dimensions

The preceding sections show several possibilities for the expression of time dimensions. It is always important to correctly model time but also to accurately express the meaning and characteristics of the individual time domains used for a particular system. Only by reaching both requirements can effective and reliable solutions be achieved. Place an emphasis on all time periods, whether in a *logical* (the validity of states, corrections, and versions) or *physical* (such as pre-processing, evaluation delays, and so on) manner.

So far, we have considered that the object itself is completely bounded by the validity, that is, the entire record shares a common validity value. Similar principles can, however, be applied to other time-domain expressions. Note that you may often encounter time dimensions and time domains in the literature. These terms are synonymous.

Object granularity is the easiest manageable model. Each state is formed precisely from one row in the database. On the other hand, if the frequency of change is not the same for all attributes, many duplicate values can be present, resulting in reduced storage efficiency and limiting real change identification possibilities. Although various temporal dimensions can be used, there is still a limitation associated with the object-level solution. The attribute-oriented approach takes the opposite perspective – each attribute is enclosed by the temporal dimension. Thus, the object state image is made up of the status of individual attributes, which are separately monitored over time.

The next section introduces the attribute-oriented temporal model by identifying static, conventional, and temporal attributes, which can be modeled and processed together when maintaining performance.

The attribute-oriented approach

An object-level temporal solution is a relevant solution if the changes are synchronized and occur simultaneously. Thus, frequency, as well as the completeness of the change, must inevitably be synchronized too. Otherwise, duplicate values will be present, resulting in performance degradation and rising disc storage costs. This is caused by the primary key extension dealing with the time elements expressing various characteristics, mostly related to the validity frame. The opposite of object-level temporal granularity is associated with the attribute itself. Therefore, each attribute is treated separately, providing robustness with an emphasis on the different granularity and frequency of the changes. As a result, a particular system can cover any attribute in the core table:

- **Static attributes** are not changed at all. Once they are initialized, there is no possibility of change. The existing set can be extended by adding new elements (if approved and accepted).

- **Conventional attributes** can evolve over time but covering and evaluating non-current data is unnecessary. Thus, the change is done by physical replacement or original state with no history management.

- **Temporal attributes** need to be monitored over time.

Attribute-oriented granularity manages each attribute separately, delimited by the time frame. It does not use physical architecture like the object-level granularity. Rather, the complex architecture operated by temporal management is used. It uses three layers (current valid state, temporal management, and outdated states). *Current valid states* (the first layer) are formed by the object reference with no change compared to the conventional approach. Thanks to that, existing solutions can run without the necessity to rebuild or rewrite the code. The core part of the processing is formed of *temporal management* (the second layer), which is operated by the **Temporal Manager** background process monitoring individual changes and registering them to the `temporal_table` module. It is responsible for the future valid data transformation based on the validity time frame. The third layer, *outdated states*, deals with the history, as well as *future plans*, which can be treated for each attribute separately by creating a new table (a pure attribute-level system) or by reflecting the data type categories, limiting the number of internal structures (an extended attribute-level system). The architecture is shown in *Figure 11.7*.

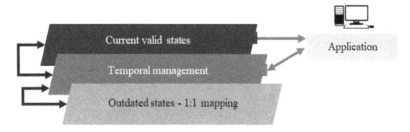

Figure 11.7 – Attribute-oriented architecture

The `temporal_table` structure consists of the identifier of the change (`id_change`). Individual changes to the particular object can be linked and sorted using the reference to the last change stored in the `id_previous_change` attribute. Besides, it is necessary to identify the source table (`id_tab`), source object (`id_row`), and changed attribute (`id_column`). The original data value is referenceable using the `id_orig` value, which can hold an undefined value if there is no previous state. As the name of the table suggests, there is also a temporal reference defining the start point of the validity (`bd`).

The principle of data flow is shown in *Figure 11.8*. In the `Tab1` and `Tab2` tables, the current valid states are present. Thus, if there is an attempt to change the state, individual attribute changes are extracted and stored in the `Temporal_table` table. Before replacing the original value in the `Tab1` or `Tab2` table, it is automatically stored in the tables consisting of the outdated states.

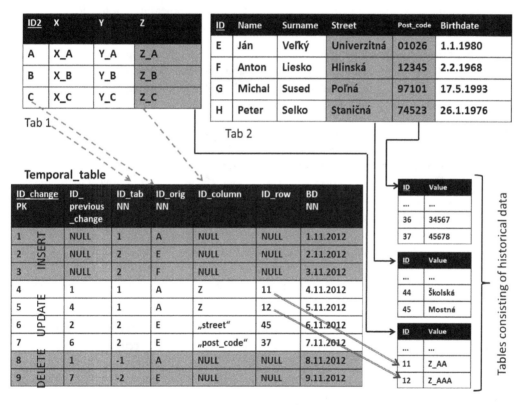

Figure 11.8 – Data flow

The attribute-oriented temporal approach uses 1:1 mapping for the outdated attribute states. Thus, for each temporally monitored attribute, one extra segment in the third layer is created. Extended attribute management uses base data types. Thanks to that, the amount of data to be processed and referenced is decreased and optimized.

The extended attribute-oriented temporal approach

The extended attribute-oriented temporal approach uses a three-layer architecture as well. The main difference compared to the original attribute-oriented granularity solution is based on the third layer, which categorizes individual referenced values based on the *data type categories*. Thus, the number of tables is fixed (5-6 tables will often suffice, depending on the characteristics and structure of the stored temporal value) and does not relate to the number of temporally processed data attributes or tables. Each data type category (such as character strings, numerical data, and date and time values) has its own table holding outdated data references. To improve the performance perspective and storage demands, several lengths of character strings can be referenced. But to hold the outdated data references, it is not necessary to split the data defined for the real data types such as `Varchar(20)` and `Varchar(30)`. Just developing one category is enough, characterized by the `Varchar(max)` basic type. Similarly, there are no separate approaches for integer, real, or other numerical representations. Instead, the general `Number` type category is used to cover all numerical types.

The content of the third layer of the extended attribute-oriented temporal approach consists of one table for each data type category. Physically, it consists of the identifier (`ID`) obtained by `sequence` and `trigger` and the referenced value itself (`value`). To optimize the storage and limit duplicate tuples, there can be an additional temporal background process merging the data tuples stored there.

Besides the third layer dealing with outdated data references, some changes can be found in the main temporal system of the second layer. Namely, the executed statement type is explicitly stated. It does not refer just to the `insert`, `update`, and `delete` operations. Specifically, what does the `delete` operation represent? Does it really express the physical data reference removal? Well, it is commonly still necessary to store information about the existence of an object in the past. Moreover, the **volatility** aspect and other regulations for holding history must be carefully considered. However, such an operation is still physically operated as a `delete` statement. And finally, the `delete` operation can often relate to the transfer of data to other repositories, such as **data warehouses**, **marts**, or **archive repositories**. In addition, this transfer can be enclosed by aggregation, anonymization, and other processing techniques. The parameters of the shift are stored in the temporal table (the second layer of the extended attribute-oriented temporal approach).

Figure 11.9 shows the architecture of the extended attribute-oriented temporal approach and the data flow is expressed by the arrows. As is evident, current valid states can be directly obtained, allowing the interconnection of the database layer with the existing conventional applications. Requests getting historical or future valid data access the temporal layer. Internally, the temporal layer interconnects all layers. Any data change is recorded in the `extended_temporal_table` table.

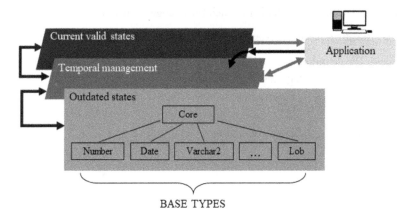

Figure 11.9 – Attribute-oriented architecture – base types

The core part of temporal management is defined by the `Extended_temporal_table` table, consisting of the following attribute set:

- `id_change`: The primary key obtained by the sequence and trigger.

- `id_previous_change`: This references the previous state of the particular object forming the state chain.

- `statement_type`: This determines the executed command's type:

  - `I = Insert`: A new state has been added, either by identifying a new object completely or by getting the new state for the existing object.

  - `U = Update`: This makes a correction to the already existing state.

  - `D = Delete`: This removes the data states from the database, either physically or by moving the data to the archive repository or another analytical model, supervised by data transformation, aggregation, and so on.

  - `P = Purge`: The removal of historical states in terms of their transfer to external archival repositories, protected by an optional transformation module (such as anonymization and aggregation), or movement to data warehouses, lakes, or marts.

  - `X = Invalidate`: This creates an undefined state of the object by moving the original state to a separate storage holding only invalid object states.

  - `V = Validate`: A restore operation for the particular state or object itself. Note that a valid state must pass all the constraints.

- `id_tab`: The identifier of the table in which the change occurred.

- `id_orig`: The attribute that carries information about the identifier of the changed row.

- `data_type`: The definition of the referenced basic data type for the changed temporal attribute. The following data type categories can be identified:

  - `C = Char / Varchar2`

  - `N = numerical values` (*real*, *integer*, ...)

  - `D = Date`

  - `T = Timestamp`

  - `L = Lob`

  - `X =` XML, and so on

- `id_column`: This references the temporal attribute, on which the change was made.

- `id_row`: This references the original value in the third layer.

- `bd`: The start point of the validity.

These attribute sets are depicted in the following table:

Extended_temporal_table		
id_change	Integer	NN (PK)
id_previous_change	Integer	
statement_type	Char(1)	NN
id_tab	Integer	NN
id_orig	Integer	NN
data_type	Char(1)	
id_column	Integer	
id_row	Integer	
bd	Date	NN

Figure 11.10 – An extended temporal table model

The object-level temporal approach provides processing at the level of objects. On the contrary, attribute-oriented granularity separates the processing into individual attributes. Both systems can have significant disadvantages and limitations associated with performance if the model does not fit the requirements of the environment. The group-level temporal system provides an inter-layer solution and ensures efficiency for any granularity.

The group-level temporal system

Temporal system input data is usually produced asynchronously. An attribute-oriented system provides a suitable solution. On the other hand, it is often possible to encounter cases of synchronization, or partial synchronization of individual attribute changes. Subset data (from the same source) is provided synchronously in batches. In this case, the physical change is divided into individual attributes, which are processed separately for each attribute by writing a new record to the temporal layer. For these purposes, an extension of the attribute-oriented approach to the management of synchronization groups (temporal groups) was created in 2017. It creates an architecturally new solution, expands the possibilities of temporal systems, and ensures efficiency, robustness, reliability, and performance in an environment of partial synchronization of temporal state changes.

Figure 11.11 shows the data flow. For simplicity, we consider only current and historical states.

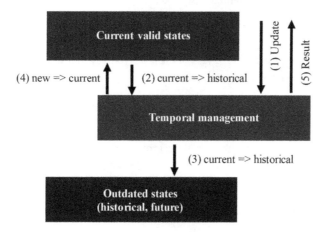

Figure 11.11 – Attribute-oriented temporal model – data flow

If a new state comes to the system, a message is provided to the temporal layer (process **1**) that ensures the change of the current state to the historical state from the current state layer (process **2**) and the historical state (process **3**). After successful execution, the new state is written to the current layer (process **4**), and processing is terminated by informing the input source about the successful update (process **5**). If future states were processed, the procedure would be extended to move the new state to the outdated data layer directly. At the same time, an automatic update process (*job*) would be planned, which would send a message about the newly created state in the defined state. Then, the procedure would be the same as shown in *Figure 11.11*.

Besides the temporal dimensions, temporal systems can also store spatial and positional data references. In this case, we are talking about a spatiotemporal system, which often follows moving elements – usually vehicles – determining their position over time, either synchronously by period or asynchronously in the event of a significant change. Communication technologies are based on an ad hoc network architecture, where individual network nodes are connected and disconnected from the data center dynamically. The data is provided in batches depending on the possibility of communication itself, the changes obtained, as well as the signal strength. The time during which nodes can provide data is relatively short and it is therefore imperative to optimize the data flow to maximize the information value. Although an attribute-oriented approach is used as the core definition, the solution is significantly extended to include synchronization layers (synchronization group detection and management). Thus, we will refer to the concept as a hybrid temporal system, which, through synchronization, passes from attribute granularity to object granularity (if all attribute changes are synchronized, we get to an object-oriented temporal system).

The proposed system extends the attribute-oriented approach by three layers. The first extended layer deals with automatic group synchronization. The second layer deals with group creation, its management, and deletion operations. The task of the last layer is synchronization itself. Two essential parameters affect the behavior of synchronization processes:

- The number of synchronized changes, after which a new group will be created. Once created, each group is then managed and monitored.

- The number of out-of-sync changes until the group is dropped. In principle, the group can be dropped immediately if the new attribute values do not arrive simultaneously. Still, the costs of building and restructuring the group as such must be emphasized.

It is also necessary to consider the possibility of outages, temporary delays in values, delays in evaluating reliability, and so on.

Furthermore, immediate cancellation would significantly increase the cost of relocating and splitting the group. Therefore, assessing the situation more comprehensively is appropriate, either in terms of the number of changes or a defined time frame. As we will show later, group building and deleting can be managed manually or automatically by the temporal system.

Figure 11.12 shows the architecture of the hybrid approach formed by the temporal synchronization group management.

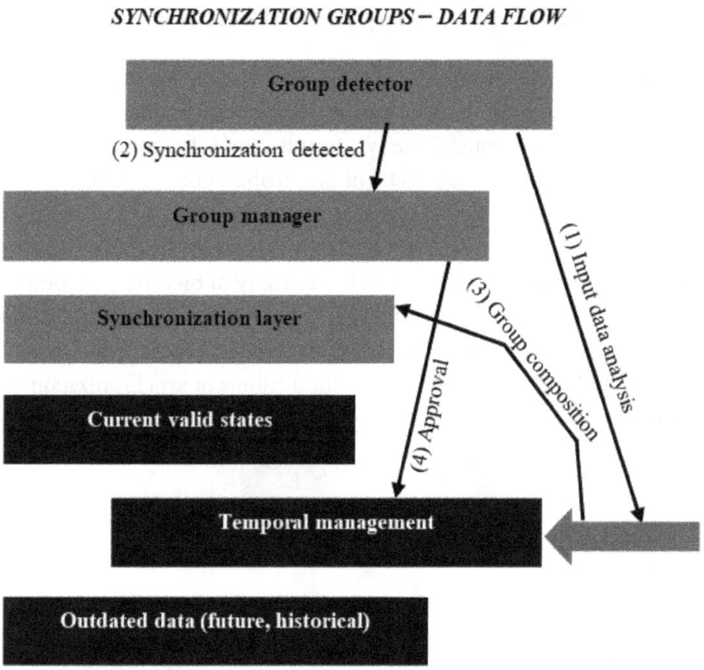

SYNCHRONIZATION GROUPS – DATA FLOW

Figure 11.12 – Synchronization group management – data flow

Only the first layer of detection has the *Select* privilege over the main temporal structure. Its task is to detect, evaluate, and reflect on group processing techniques using the `create`, `alter`, and `drop` commands in the second layer – the group manager. The third layer supports the synchronization itself. The physical architecture is managed by its own **Temporal Synchronization Manager** (**TSyncMan**) process, which is of the master type, and its supporting processes, **Temporal Synchronization Workers** (**TSyncWorkn**).

The existing attribute-oriented approach forms the core layer of data processing, whether in direct form or pre-processed by synchronization. The synchronization group can be detected either automatically – by the first layer (the *detector*) – or manually by the user. The processed data is sent to the second layer (the *manager*), which requests the creation of a group (a new group can consist of individual attributes, but also of existing groups, which it expands). After creating the group, the synchronization manager informs the first layer about the success of the operation. Subsequently, the attributes and groups are combined. Their processing is moved to the synchronization layer (the third layer), which communicates with the temporal database for the whole spectrum – *current*, *historical*, and *future* states. The task of the synchronization layer is to manage the input values and create a *batch* (the data does not have to arrive at the same time; each group has a defined *Epsilon value*, which expresses the maximum allowable time window in which the data is still considered time-synchronized, forming a fuzzy approach).

The logical sphere is clear now, isn't it? Let's look at the physical representation of the temporal group.

Temporal group – the physical representation

When the temporal system is created, each attribute is initially processed separately. Thus, the system does not contain a synchronization group. After running the instance synchronization processes, the changes are detected and a group is created if the synchronization is found to be consistently available and not accidental. The group is composed directly of the attributes or by extending an existing group (*Figure 11.13*). From the perspective of the temporal core layer, the `data_val` representation can be either one physical attribute, or a set can be used, generally represented by a non-empty group. At the physical level, the model is characterized by the **ISA hierarchy** (a hierarchy decomposing data into logical sub-components formed by multiple tables). The root is the `data_val` notation, which is created by aggregating existing synchronized modules (`attribute/group`). *Figure 11.13* shows the composition of the `data_val` reference, delimited by the attribute or synchronization group. The term **ISA** is an abbreviation of *IS A*, e.g., student *IS A* person, a car *IS A* vehicle, by forming the hierarchy.

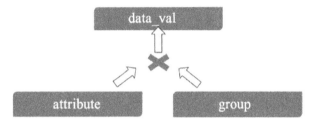

Figure 11.13 – The data_val data hierarchy

Each temporal group is detected and associated based on time synchronization. Its validity is limited by the duration interval modeled by the `date_from` and `date_to` attributes. The endpoint of the validity is commonly undefined (or expressed by the `MaxValueTime` notation or by its enhancements). Theoretically, it would be possible to simplify the model and store only the beginning of the validity and the attribute with a bivalent value (Boolean) characterizing whether the group is still valid or not. However, this concept is not reliable because, after the cancellation of the group, it is divided into individual attributes or subgroups. The physical data representation is shown in *Figure 11.14*.

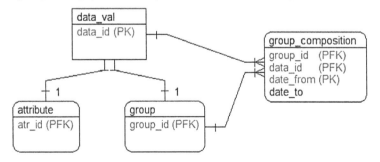

Figure 11.14 – data_val representation in the data model

As we have shown, there are three basic types of temporal models in terms of granularity – *object*-level, *attribute*-level, and *group*-level systems. The group-level temporal model provides a universal solution because it can cover any data change frequency by maintaining performance and optimizing storage demands. Moreover, each group is enclosed by the validity frame, so the group can accommodate evolving data characteristics.

Having DATE or TIMESTAMP data type attributes in a table, however, does not actually directly mean that the temporal table is defined. In the next section, we will discuss a conventional table with time-delimited attributes.

Conventional table with time-delimited attributes

The fact that a table consists of DATE or TIMESTAMP data type attributes does not automatically mean a temporal table has been composed. The main difference between conventional and temporal tables is the primary key definition and the ability to store multiple versions. The primary key of a conventional table consists of the object identifier (it can be composite). By adding temporal attributes to be part of the primary key, a temporal table can be created. To explain issue, let's create a table dealing with personal data (PERSON_TAB) by referring to the employment contract table (EMPLOYEE_TAB). Despite the fact that the employee table contains time durations as represented by the start_date and end_date attributes, we cannot talk about a temporal model for several reasons. The first problem is that the time interval does not denote the validity of the record itself; it just signifies the length of the employment contract. Another important factor is the definition of the **primary key**, which contains only an employee identifier (emp_id) without time limits. Therefore, if the contract is terminated but the same person is hired again as an employee later, it is necessary to assign them a new identifier; the original one cannot be used later on, so the employment history will be lost. Similarly, if the employee's salary or job position is changed, the system will no longer be able to store the original values. Consequently, complex state information about the past will be lost. Finally, this kind of table cannot process and hold states that are valid in the future. For example, it is impossible to store information about a specific employee whose salary will be increased by 20% from next year.

A conventional table with time-delimited attributes is shown in *Figure 11.15*, pointing to the EMPLOYEE_TAB table, which is connected to the PERSON_TAB table using a non-identifying relationship with a one-to-many cardinality type.

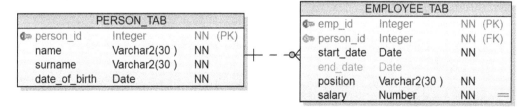

Figure 11.15 – Data model – conventional EMPLOYEE_TAB table with time-delimited attributes

The principle of a temporal table is based on the opportunity to hold multiple states for each object, delimited by a validity time frame. Common attributes (even time-delimited attributes) that are not part of the tuple identifier cannot serve that requirement generally due to them producing a conventional table only. To get complex insights, always consider whether multiple object versions can be stored for an object. If so, a temporal table is present. If not, only a conventional approach has been applied.

Summary

This chapter forms a crucial part of the physical implementation of temporality. You navigated through the *object-level temporal architecture* based on the primary key extension. *Attribute granularity* is formed completely differently. Instead of storing the whole state, only changed attributes are listed. The general solution applied between the attribute and object level is defined by *temporal groups*, which are composed dynamically and delimited by the group validity. Synchronization groups can be detected and processed as one attribute, decreasing the demands and requirements of the temporal reference layer. In principle, the attribute itself, as well as the whole object, can be covered by a group.

Each temporal state deals with the object reference, as well as the time dimensions, mostly expressing validity. The *uni-temporal solution* uses one dimension for the temporality while *bi-temporal models* use two, commonly expressing *validity* and *transaction references*. In this chapter, we defined the referenceable dimensions supported by various models and implementations, such as the *IPL* and *IPLT* model, *future valid* record management, *offline* mode, and *replicas*.

Most current systems use date values for operations. In the application layer, the user commonly does not need to specify the date value directly (explicitly), but it can be selected from the month calendar wizard. The next chapter deals with the *month calendar* definition using *SQL* and *PL/SQL*. Although a development environment usually has pre-prepared components to this end, it is always useful to go to the roots and understand the principles.

Questions

1. Choose the correct statement related to the *header-temporal* concept:

   A. It cannot manage references using primary and foreign keys.

   B. The header consists of the current valid data; the temporal layer is used just for the history.

   C. Future valid data cannot be managed by it.

   D. The header layer is used for the references; attribute values are stored separately in a temporal manner.

2. Which model does not use temporal management?

    A. A conventional model

    B. A uni-temporal model

    C. A bi-temporal model

    D. A fixed temporal model

3. Historical states can lose their importance and relevance by being removed from the main system or moved to an aggregated data warehouse repository. Which aspect characterizes these options?

    A. Relevance

    B. Transaction support

    C. Limited temporal usability

    D. Correctness

4. How many dimensions are covered by the IPL model?

    A. 1

    B. 2

    C. 3

    D. 4

5. Which system takes the validity for each column separately to optimize the storage demands if the frequency rate of the changes for individual attributes is not the same?

    A. The object-level temporal model

    B. The attribute-oriented temporal model

    C. The uni-temporal model

    D. The IPL model

6. A temporal group definition uses data_val. Choose the statement that characterizes its structure:

    A. data_val is composed of the attribute or existing temporal groups

    B. data_val can be composed of attributes only

    C. data_val cannot be applied to a uni-temporal model

    D. data_val cannot be used for unlimited validity

Further reading

- *Concept of temporal data retrieval: Undefined value management* by Michal Kvet, Štefan Toth, and Emil Kršák discusses the complexity of the temporal data retrieval process in the temporal environment, focusing on undefined value management: `https://onlinelibrary.wiley.com/doi/epdf/10.1002/cpe.5399`

- *The Complexity of the Data Retrieval Process Using the Proposed Index Extension* by Michal Kvet and Jozef Papán deals with an index extension to ensure the performance of the `select` statement. The performance optimization techniques are based on limiting migration rows, post-indexing, data pointer reflectors, and priority management by introducing architecture enhancements. It is published in the *IEEE Access* journal and is open-access: `https://ieeexplore.ieee.org/document/9763539`

12

Building Month Calendars Using SQL and PL/SQL

Each date value consists of the month and year value references. Therefore, it is often necessary to create reports referencing the defined granularity. For example, the evaluation of employee attendance, a project summary, or resource consumption is commonly associated with month precision. Moreover, individual activities can be bordered by the left and right borders of the duration frame. Thus, it can be necessary to get the visual form of the month in terms of individual day's serial numbers, as well as weekday references. In the past, this had to be hardcoded. Nowadays, mostly pre-prepared visual components forming the calendar can be used.

Nevertheless, in our opinion, it is good to summarize the principles of building calendars and focus on weekday references with an emphasis on the calendar being defined by PL/SQL code, followed by the direct SQL statement approach. The weekdays can be placed either as columns or can be transposed to rows.

In this chapter, we're going to cover the following main topics:

- PL/SQL calendar definition
- SQL calendar definition
- Name day management

The source code for this chapter can be found in the GitHub repository accessible via this link: https://github.com/PacktPublishing/Developing-Robust-Date-and-Time-Oriented-Applications-in-Oracle-Cloud/tree/960e892ada2c4ff3247f4967ca9b3d866e0f1c4f/chapter%2012. Alternatively, you may scan the following QR code:

Defining a calendar using PL/SQL

You might've been in a situation where you need to agree on meeting dates, synchronize activities and events, or write down deadlines for individual tasks and activities. When you receive an invitation to a conference, meeting, or just an online call, you usually open Google Calendar or your smartphone calendar app and check your availability. What do you primarily see? A calendar that's daily or monthly organized. If you click on a specific day, hour references are loaded.

However, many times, you don't need to know the exact time of the activity. You will get a task from your boss that needs to be completed by the end of the week. But they do not need to know exactly what time you will do it. It's enough to ensure that the job is done and not exactly *when* it was finished. Similarly, when you are asked to prepare a vacation schedule, which precision will be used? Yes, daily granularity. In the same way, for an employment contract, we store the signing date, date of application, and when the validity is to be terminated. Evidently, we store this data on a daily basis but we don't need more accuracy. Therefore, it is necessary to use the *day* as the basic unit of value processing and its placement in the monthly calendar. As such, we will show you how to create such a calendar using SQL and PL/SQL.

Firstly, let's take a look at the month calendar, according to which individual weekdays will be denoted in rows (Monday to Sunday). In order to save the dates of individual activities in the projects, typically, daily accuracy is sufficient. The required result format for February 2022 is shown in the following figure:

```
MON :        7  14  21  28
TUE :    1   8  15  22
WED :    2   9  16  23
THU :    3  10  17  24
FRI :    4  11  18  25
SAT :    5  12  19  26
SUN :    6  13  20  27
```

Figure 12.1 – Calendar – February 2022

A PL/SQL block can generate such output. The composition of the monthly calendar is a staged process. Firstly, individual weekdays need to be referenced. Secondly, the offset (padding) should be used to build consistent output. In the following part, we will describe the proposed and implemented solution. In the declaration section, five variables are defined (`composed_date`, `cur_date`, `first_month_date`, `output_text`, and `week_days`). The `composed_date` variable holds the last day of the preceding month. It is used as a left border for the processing:

```
composed_date:=TRUNC(TO_DATE('&MONTH.&YEAR', 'MM.YYYY'),
                'MM')-1;
```

The `first_month_date` variable represents the weekday of the first day of the month in a numerical format.

The outer loop deals with the weekday processing, executing the body seven times, and forming the header of the lines. The inner `while` loop ensures the processing of all days of the month, separately for each week.

The weekday representation abbreviations are stored in the nested `week_days` table and used to calculate the value of the `cur_date` local variable. Therefore, the format is crucial in the reference stream:

```
cur_date:=TO_CHAR(NEXT_DAY(composed_date, week_days(i)),
                'DD');
```

By using `NEXT_DAY`, one week shift is done. It is done in a cycle, while the individual days still belong to the given month. It is processed seven times, for each day of the week separately (specified by the `week_days` variable). The declared nested table defines abbreviations for the days' correspondence.

The currently processed day is treated by the `cur_date` variable, protected by the cycle for each weekday. After the processing, a particular value is incremented by one week. Thus, if it does not overlap the defined month, the next loop iteration is processed. Vice versa, shifting the `cur_date` value out of the designated month automatically limits the evaluation and the loop execution ends:

```
while cur_date <= TO_CHAR(LAST_DAY(composed_date+1), 'DD'))
```

Finally, the result set is written to the console using the `PUT_LINE` method of the `DBMS_OUTPUT` package. The `if` condition inside the loop cycle ensures the correct position of the particular day's numeric value (value offset):

```
if first_month_date >= i
    then output_text:=output_text || '      ';
end if;
```

If the `if` condition stated before was not specified or omitted, for *February 2022*, the first position of *Monday* would not be empty, forming misleading output:

```
MON :   7  14  21  28
TUE :   1   8  15  22
WED :   2   9  16  23
THU :   3  10  17  24
FRI :   4  11  18  25
SAT :   5  12  19  26
SUN :   6  13  20  27
```

Figure 12.2 – Calendar – February 2022 – padding problem

The preceding code is language-specific, related to the `week_days` nested table, but can be filled based on the particular session parameter to ensure a proper and independent solution. Similarly, the `first_month_date` variable would be influenced by the NLS session parameters.

It is assumed that the following parameters are set for the session:

```
alter session set NLS_TERRITORY='Belgium';
alter session set NLS_DATE_FORMAT='DD.MM.YYYY HH24:MI:SS';
alter session set NLS_DATE_LANGUAGE='English';
```

It is assumed that the European region is used, so the first day of the week refers to Monday (compared to the regions in America where the week starts on Sunday). Thus, `NLS_TERRITORY` can refer to any European region – in our case, we used Belgium.

The complete code of the PL/SQL calendar management can be downloaded using this web link:

```
https://github.com/PacktPublishing/Developing-Robust-
Date-and-Time-Oriented-Applications-in-Oracle-Cloud/blob/
960e892ada2c4ff3247f4967ca9b3d866e0f1c4f/chapter%2012/PLSQL%20
calendar.sql
```

Alternatively, you can use this QR code:

The implementation of the universal solution independent of NLS parameter settings is left to you.

The transposed matrix forms the opposite solution, so the columns instead of rows reference the weekdays. Local variables consist of `composed_date`, and the first (`first_month_date`) and the last day (`last_month_date`) of the particular month are obtained in the same manner as in the preceding code. Firstly, the header is composed of the nested table separated by spaces:

```
for i in week_days.first .. week_days.last
  loop
     DBMS_OUTPUT.PUT(week_days(i) ||' ');
  end loop;
```

The values are sent only to the buffer (using the PUT method), and then flushed by the PUT_LINE method of the DBMS_OUTPUT package. The starting position (`cur_date`) is set to a value of 1 (the first day of the month). The text to be used as the output (`output_text`) takes the number of the first day (`first_month_date`) and prepares the format offset:

```
output_text:=lpad(' ', 5*(first_month_date));
```

Finally, the second loop processes individual days of the month by separating the values into the lines for each week:

```
loop
      output_text:=output_text || rpad(cur_date, 5);
      if mod(cur_date+first_month_date,7)=0
         then DBMS_OUTPUT.PUT_LINE(output_text);
                   output_text:='';
      end if;

        if cur_date>last_month_date
            then exit;
        end if;
      cur_date:=cur_date+1;
  end loop;
```

For February 2022, the following output is provided:

MON	TUE	WED	THU	FRI	SAT	SUN
	1	2	3	4	5	6
7	8	9	10	11	12	13
14	15	16	17	18	19	20
21	22	23	24	25	26	27

Figure 12.3 – Calendar – PL/SQL output

The complete code of the PL/SQL calendar management can be downloaded from the GitHub repo of this chapter.

PL/SQL is a procedural language, so the structure and format can be easily defined. In the next section, an analogous solution to build the month calendar is provided but defined by SQL. The solution discussed can treat individual weekdays either as rows or columns. Both approaches are introduced and discussed.

Modeling a calendar using SQL

Although it is possible to hardcode a calendar using a PL/SQL block, it can look a bit complicated at first sight.

Now, let's look at the monthly calendar definition using SQL. For simplicity, we will use the month extracted from the sysdate value, but generally, it can be applied to any specified DATE value. Note that the processing can be optimized by grouping multiple select statements together. However, for explanatory purposes, step-by-step processing is more convenient. Therefore, individual steps are nested. Each step takes the data from the preceding processing.

Step 1 – Getting the day numbers for each day in the month

The goal of this step is to get the same number of rows as the number of days in the month. To do that, the dual table is referenced, which holds just one row. Therefore, the CONNECT BY LEVEL clause is used with the value expressing the numerical representation of the difference between the last day of the month and the first day of the month. The resulting value is incremented by 1 to cover all days:

```
select level as day_val
  from dual
  CONNECT BY LEVEL <=LAST_DAY(sysdate)-TRUNC(sysdate, 'MM')+1
```

Step 2 – Grouping the data based on the weekday

This step aims to group data based on the weekdays. It is done by using the LISTAGG aggregate function, forming the rows of the calendar. The group is defined by the *ISO week*:

```
select LISTAGG(lpad(day_val, 4))
         WITHIN GROUP (order by day_val)
                                        as calendar_row
  from
  (select level as day_val
    from dual
     CONNECT BY LEVEL <=LAST_DAY(sysdate)-TRUNC(sysdate,
```

```
                                                  'MM')+1
     )
     group by to_char(TRUNC(sysdate, 'MM') + day_val -1,
                       'IW')
```

Step 3 – Padding the data

In general, the last week is always formatted to the left margin, and the first week is associated with the right margin. The rest of the rows completely fill the character string length so that they can remain original. The first highlighted CASE branch places emphasis on the first week. The proper position placement identification is made by getting the first value of the row. If it is lower than 7, it is clear that it is the right option. The LPAD function call covers the processing. Vice versa, the last week is covered by the second condition ensured by the RPAD function call. The value of 28 for the padding expresses the total length of the row.

Please note that the first and second condition does not need to strictly cover only the first or the last week, respectively. It is based on the assumption that LPAD or RPAD functions do not impact output if the input string length is the same as the total size (output format):

```
select case when SUBSTR(calendar_row,3,2)<7
                           then LPAD(calendar_row,28)
            when SUBSTR(calendar_row,3,2)>20
                 and SUBSTR(calendar_row,-1)=' '
                           then RPAD(calendar_row,28)
            else calendar_row
        end as text
  from
    (select LISTAGG(lpad(day_val, 4))
                  WITHIN GROUP (order by day_val)
                                      as calendar_row
      from
        (select level as day_val
          from dual
           CONNECT BY LEVEL <=LAST_DAY(sysdate)
                              - TRUNC(sysdate, 'MM')+1
        )
      group by to_char(trunc(sysdate, 'MM') + day_val -1,
                        'IW')
```

Step 4 – Getting the header and ordering the rows in the result set

The separate query gets the header, formed of individual weekday representations, by combining the text with the previously obtained calendar rows. Then, the rows are sorted to ensure relevance. Thus, the textual header is the same. Then, individual rows are placed in the output buffer to form the calendar output, delimited by the first value of a particular row.

The complete code for SQL calendar management can be downloaded from the GitHub repository of this chapter.

The output of the preceding discussion and statement definition is the calendar. Individual weekdays are treated as columns. For June 2022, the result set would be like the following:

```
 TEXT
MON  TUE  WED  THU  FRI  SAT  SUN
                1    2    3    4    5
  6    7    8    9   10   11   12
 13   14   15   16   17   18   19
 20   21   22   23   24   25   26
 27   28   29   30
```

Figure 12.4 – Calendar – SQL output

A general solution can be built by referencing any user-specified DATE value. Thus, instead of using sysdate, the DATE value is created by the user-provided parameter values.

Note that there is an assumption of an English date language representation:

```
alter session set NLS_DATE_LANGUAGE='English';
```

The different tasks can be associated with the requirement to transpose the result set so that the individual weekdays will be placed in rows instead of columns. The management is also composed of multiple nested queries for description purposes.

To achieve proper results, multiple steps must be carried out. Each step takes the data from the preceding process:

1. The first step is to get the list of all days included in the particular month, obtained by referencing the dual table. It uses the CONNECT BY LEVEL extension, forming the limitation of the first and last day of the month, which is obtained by calling the TRUNC and LAST_DAY functions:

```
select level as day_id
    from dual
```

```
                  CONNECT BY LEVEL <=LAST_DAY(sysdate)
                                    - TRUNC(sysdate, 'MM')+1
```

2. Then, the date is composed, followed by weekday extraction in a numerical format:

```
select TO_CHAR(TRUNC(sysdate, 'MM')+day_id-1, 'D')
                                   day_of_week,
        day_id
          from
            (select level as day_id
              from dual
                CONNECT BY LEVEL <=LAST_DAY(sysdate)
                                  - TRUNC(sysdate,
                                     'MM')+1

            )
```

3. In the third step, individual values are aggregated using LISTAGG, formatted by the LPAD function. The group associated with the aggregate function deals with individual weekdays:

```
select day_of_week,
        LISTAGG(lpad(day_id,4)) WITHIN GROUP
            (order by day_id) as text
  from
  (select TO_CHAR(TRUNC(sysdate, 'MM')+day_id-1, 'D')
                                      day_of_week,
          day_id
      from
        (select level as day_id
          from dual
          CONNECT BY LEVEL <= LAST_DAY(sysdate)
                            - TRUNC(sysdate, 'MM')+1

        )
  )
        group by day_of_week
```

4. Finally, the provided character string is formatted to ascertain the positions of the days on the calendar, in the correct order. If the first day of the month is higher than the actual processed day, additional leading spaces are added.

The complete solution can be defined as specified in the web link given later. It is independent of the NLS_LANGUAGE settings, whereas the weekday representation is user-defined. The correlation between the numerical value of weekdays and the textual representation is expressed as 1 meaning *Monday*, 2 meaning *Tuesday*, and so on. Thus, if the NLS_TERRITORY parameter is changed, the reference list can be shifted. The complete solution is present in the following code block:

```
select case day_of_week when '1' then 'MON'
                         when '2' then 'TUE'
                         when '3' then 'WED'
                         when '4' then 'THU'
                         when '5' then 'FRI'
                         when '6' then 'SAT'
                         when '7' then 'SUN'
    end as week_day,
  case
   when TO_CHAR(TRUNC(sysdate, 'MM'), 'D') > day_of_week
     then '    '||text
       else text
   end as calendar from
  (select day_of_week, LISTAGG(lpad(day_id,4))
                          WITHIN GROUP
                                 (order by day_id) as text
   from
     (select TO_CHAR(TRUNC(sysdate, 'MM')+day_id-1, 'D')
                                              day_of_week,
             day_id
        from
          (select level as day_id
            from dual
             CONNECT BY LEVEL <=LAST_DAY(sysdate)
                                  - TRUNC(sysdate, 'MM')+1
          )
       )
     group by day_of_week
   )
   order by day_of_week;
```

The complete code for SQL calendar management can be downloaded from the GitHub repo of this chapter.

The output of the previously-defined SQL calendar is depicted here:

⬦ WEEK_DAY	⬦ CALENDAR				
MON		7	14	21	28
TUE	1	8	15	22	
WED	2	9	16	23	
THU	3	10	17	24	
FRI	4	11	18	25	
SAT	5	12	19	26	
SUN	6	13	20	27	

Figure 12.5 – Calendar – output (2)

Now that you know how to model and represent a monthly calendar in the PL/SQL and SQL languages, it's time to associate individual days with events. The next section deals with name days as a root component.

Name day management

The internet is full of people, social networking, individual names, and their characteristics. For the purposes of information systems, applications, and websites, it may be useful to display a list of names celebrating their name day today, or on any user-defined date. By browsing the net, it is easy to find a list of name days for a particular country in the form of *three* columns – the day of the month, followed by the month reference, and a list of names, which can be normalized or aggregated together to form one composite value. One way or another, it is possible to get a list based on the date, or vice versa, to get the celebration day for an individual name. In this section, the table's structure is discussed, followed by a definition of the individual methods encapsulated by the package.

A table references the day, the month, and the value itself, as can be seen in the following snippet:

```
create table nameday_tab
  (day_val integer,
   month_val char(3),
   name_list varchar(50));
```

The defined table content is loaded from an external source, usually an *Excel* sheet, *CSV*, *JSON*, or *XML*.

The proposed package has three methods for getting the name based on the specified day and month. The first method takes the month in a numerical format. The second function uses a string reference. The third function takes the name value and looks for the day and month representation provided by the output. The whole management is done by the cursor. In principle, any number of rows can be provided. Individual values provided by the parameters are bounded. When the principles are clear and the implementation is straightforward, we do not specify the implementation row by row.

The name of the package is PACK_NAMEDAY, which is formed of the *header of the package* used as the external interface and the internal implementation using the *package body*. Let's look at these two components in the subsequent sections.

The header of the package

The header of the package consists of three function definitions, which can be used anywhere outside the package body (public methods). The GET_NAME function is overloaded – the second parameter can have a string or numerical format. The GET_DAY_MONTH function takes the first name of the person (or part of it) and selects the day that pertains to that name:

```
create or replace package PACK_NAMEDAY
is
   function GET_NAME(p_day integer, p_month_int integer)
      return varchar;
   function GET_NAME(p_day integer, p_month_str varchar)
      return varchar;
   function GET_DAY_MONTH(p_name varchar)
      return varchar;
end;
/
```

The package body

The package body consists of the header and implementation of the methods themselves. The initial package body structure is in the following code block:

```
create or replace package body PACK_NAMEDAY
is
```

Now the package header is successfully compiled, we can focus on the implementation of individual functions. Please note that the package body cannot be successfully compiled without the package header being defined and accepted first. Now, let's look at the individual functions covered by the package and see how to implement them.

The GET_NAME function

The GET_NAME function has two variants, depending on the data type of the second parameter.

It takes two parameters. P_day expresses the ordinal number of the referenced day and month, which can be expressed by a numerical value (p_month_int) or in string format (p_month_str). The month_code_nt nested table is used for formatting the numerical value of the month into a textual representation, which is referenceable by the nameday_tab table consisting of the pairs *month + day* and *name list*.

In this case, cursor is built as there can be DATE values that have no associated name. The next block shows the implementation of the GET_NAME function, taking month as an integer:

```
function GET_NAME(p_day integer, p_month_int integer)
  return varchar
is
  output_string varchar(100);
  type t_month_code is table of char(3);
  month_code_nt t_month_code:=t_month_code('JAN', 'FEB',
                              'MAR','APR', 'MAY', 'JUN',
                              'JUL', 'AUG', 'SEP',
                              'OCT', 'NOV', 'DEC');
  cur_nameday_str varchar(1000);
  type t_name_day_cur is ref cursor;
  name_day_cur t_name_day_cur;
begin
  cur_nameday_str := 'select name_list
                  from nameday_tab
                  where day_val=:day_input
                      and month_val=:month_input';
  open name_day_cur for cur_nameday_str
              using p_day, month_code_nt(p_month_int);
  fetch name_day_cur into output_string;
    close name_day_cur;
      return output_string;
  end;
```

The second variant of the GET_NAME function takes the character string for the month representation, so transformation isn't required. There is, however, a limitation related to the value format. For example, let the nameday_tab table reference the English version of the month. If the language is changed for the session, the function would not return a tuple. Thus, the mapping must be done at the session level, contrary to a numerical representation, which is function embedded.

The solution for the GET_NAME function is stated in the following block, which takes the cursor and processes the output for the referenced day and month:

```
function GET_NAME(p_day integer, p_month_str varchar)
  return varchar
 is
   output_string varchar(100);
   cur_nameday_str varchar(1000);
   type t_name_day_cur is ref cursor;
   name_day_cur t_name_day_cur;
 begin
   cur_nameday_str := 'select name_list
                         from nameday_tab
                           where day_val=:day_input
                             and month_val=:month_str_input';
   open name_day_cur for cur_nameday_str
             using p_day, p_month_str;
    fetch name_day_cur into output_string;
     close name_day_cur;
      return output_string;
 end;
```

The GET_DAY_MONTH function

The GET_DAY_MONTH function uses the opposite approach to the GET_NAME method. The input parameter of the method is the name (p_name). The function returns the day and month of the name day in a character string format, obtained by the cursor. The processing is also done by the trigger. The implementation of the GET_DAY_MONTH function is depicted in the following code block:

```
function GET_DAY_MONTH(p_name varchar) return varchar
 is
   cur_val varchar(100);
   output_string varchar(4000);
   cur_nameday_str varchar(1000);
   type t_name_day_cur is ref cursor;
   name_day_cur t_name_day_cur;
 begin
   cur_nameday_str:=
```

```
    'select day_val||''.''||month_val
      from nameday_tab
        where lower(name_list)
                    like ''%''||lower(:name_input)||''%''';
  open name_day_cur for cur_nameday_str using p_name;
   loop
    fetch name_day_cur into cur_val;
     exit when name_day_cur%notfound;
       output_string:=output_string ||', ' || cur_val;
    end loop;
     return substr(output_string, 3);
   close name_day_cur;
  end;
end;
/
```

The complete code for the PACK_NAMEDAY package can be downloaded from the GitHub repo of this chapter.

An example of the method calls and the results are as follows:

```
select PACK_NAMEDAY.GET_NAME(p_day => 1,
                             p_month_str => 'MAR')
  from dual;
--> Albin, Rüdiger
select PACK_NAMEDAY.GET_NAME(p_day => 29,
                             p_month_int => 9)
  from dual;
--> Michael, Gabriel
select PACK_NAMEDAY.GET_DAY_MONTH(p_name => 'MIC')
  from dual;
--> 24.AUG, 29.SEP
```

The PACK_NAMEDAY package can provide valuable solutions when dealing with name days. Its implementation allows you to get names based on a specified day and month. Additionally, the opposite approach can be used by specifying a name as an input, resulting in getting all days that contain the specified name or part of it (where placeholders are implemented).

Summary

This chapter dealt with the monthly calendar composition through our own implementation. The principles were based on building a list of days followed by a weekday association. It was done with PL/SQL, as well as SQL. The focus was on the proper positioning of the days on the calendar. A calendar can be filled with individual events, birthdays, or name days. Therefore, a definition of a package managing name days was introduced and discussed.

So far, we have been dealing with temporality using data model extensions. However, the relational paradigm is associated with *transactions* ensuring *isolation* and *durability*. These key elements are supervised by transaction logging – *UNDO* and *REDO* structures. By extracting values from logs, historical data can be reconstructed.

The next chapter deals with the Flashback technology, providing you with images of the database, table, or query result as they existed at the defined time point or reference. Thus, if the transaction logs are available through archiving, historical data can be obtained, although it can be time- and resource-demanding.

Questions

1. The monthly calendar is limited by the *first day of the month*. Which function provides you with a particular DATE value related to sysdate?

    A. ROUND(sysdate, 'MM')

    B. TRUNC(sysdate, 'MM')

    C. FIRST_DAY(sysdate)

    D. EXTRACT(month from sysdate)

2. From the other perspective, the monthly calendar is limited by the *last day of the month*. Which statement provides you with *midnight of the last day of the month*?

    A. select LAST_DAY(sysdate) from dual;

    B. select TRUNC(sysdate) + 1 from dual;

    C. select TRUNC(ADD_MONTHS(sysdate, 1), 'MM')-1 from dual;

    D. select ROUND(sysdate) -1 from dual;

Further reading

- *Designing a Calendar Table* by Ed Pollack: In this text, the calendar is extended by applying holidays, business days, seasons, and so on. Although the solution is not implemented in Oracle Database, it still references SQL, introducing and highlighting principles. However, it is always possible to apply syntax and implement your own solution in any relational database. Text and implementation are available via this link: `https://www.sqlshack.com/designing-a-calendar-table/`.

- *Creating a date dimension or calendar table in SQL Server* by Aaron Bertrand: In this text, a calendar and a date dimension table are created using a set-based solution, making it powerful and easily customizable. Text and implemented code are available via this link: `https://www.mssqltips.com/sqlservertip/4054/creating-a-date-dimension-or-calendar-table-in-sql-server/`.

4

Part 5:
Building Robust and Secure Temporal Solutions

You already know, for sure, that integrity and overall state consistency are secured by transactions in relational systems. Each transaction creates a **transaction log**, consisting, among other critical information stored there, of **change vectors** storing original and new values for each changed data row. Thus, by extracting relevant transactional data using Flashback Technology, table content as it existed in the past can be obtained and reconstructed.

Chapter 14 highlights a significant security issue, **SQL injection**. This threat is formed of SQL code concatenation, which is then treated as a character string. You may have thought that working with date and time is safe, with the whole process secured by constructor and conversion functions, but that's not quite the case. You will come to understand the issue of complexity, as well as techniques to create secure solutions.

This part has the following chapters:

13

Flashback Management for Reconstructing the Database Image

Security, integrity, and data consistency are integral parts of relational database systems. **Transaction management** allows you to transfer the database from one consistent image to another. It must be ensured that any change in the confirmed transaction is consistent, satisfying all integrity rules and constraints. Some rules do not need to be applied immediately during the operation's execution typically related to *referential integrity*. However, to reach COMMIT (transaction approval), all data constraints must be passed.

In this chapter, we will look at how transaction logs can be used to retrieve **historical versions** of objects. We looked at the **Flashback data archive** in *Chapter 10*. Now, we will discuss **Flashback technology**, with which it is possible to reconstruct states that were valid in the past. Flashback operations can be applied at multiple levels, from the whole database to attributes defined by the queries. They extract transaction logs and apply *revert* operations to their current states.

In a nutshell, this chapter will cover the following main topics:

- Transaction as a change vector source formed by logs
- Transaction log structure and log types, including the archiving process
- Reconstructing states using Flashback technology

The source code for this chapter can be found in this book's GitHub repository at https://github.com/ PacktPublishing/Developing-Robust-Date-and-Time-Oriented-Applications- in-Oracle-Cloud/tree/960e892ada2c4ff3247f4967ca9b3d866e0f1c4f/ chapter%2013. You can also access it by scanning the following QR code:

Transaction as a change vector source

A database transaction is the main unit of work in the database system that makes the database consistent and independent of other transactions by securing a high level of parallelism. By getting any data failures or invalid operations, the transaction ensures that the changes covered by it can be removed, making the database consistent and valid again. The transaction is commonly formed of multiple data operations, which are, however, treated as atomic. Thus, all the changes are either accepted or the whole transaction is refused by rolling back the changes.

A database transaction is encapsulated by four rules – **atomicity**, **consistency**, **isolation**, and **durability** (**ACID**).

The **atomicity** of the transaction ensures that either the whole transaction is approved or it is refused completely. **Consistency** ensures that **integrity constraints** are successfully passed before transaction approval – for example, there is no character string to be stored in a numerical data type, there are no *NULL* values that we wish to store in *NOT NULL* attributes, all values pass the **check constraints**, primary key values are unique, and so on. **Isolation** means that the processes are separated. Thus, it allows access only to data that has been confirmed, which is done by the **row locks** that are applied to each changed data portion. Changes made in the current transaction are only accessible within it before approval (`COMMIT`). The spread is then performed by confirming and approving the transaction. Finally, **durability** ensures that all changes are persistent and secure in case of failure or a crash. Thus, it is always possible to reconstruct the database without any data loss, even after an instance failure. Naturally, certain data must be spread across multiple physical disks to protect against hardware failure. Transaction data changes are operated at the block level. Any data change is applied to the block located in the instance memory, which is obtained from the database. This means that the change itself is not done at the physical level but is processed by the instance memory. If an instance collapses, its memory data is lost. This would naturally cause data loss. To prevent such situations and ensure reliability and durability, each transaction is encapsulated by the transaction logs.

The importance of these logs, as well as their content, will be described in the next section. We will look at **Online** and **Archive** transaction logs next, focusing on the ability to reconstruct any historical state.

Transaction log structure and log types

Transaction logs describe the operations that are executed inside the transaction. Each log consists of the transaction position reference, the **System Change Number** (**SCN**), the individually performed operation, and the change vector. From this perspective, transaction logs can be divided into two groups – UNDO and REDO structures. An UNDO log stream stores the original data to allow the system to refuse transactions and revert individual operations to their original state. It is physically stored in the UNDO tablespace in the database. On the other hand, the REDO log stream is stored primarily in memory and then copied to the Online REDO log files, which are crucial to the Oracle database. This allows us to *replay* the approved transaction to get the state of the database just before the database instance or media failure. The **Log Writer** (**LGWR**) background process operates transaction log management.

Both streams are periodically overwritten. Successfully ended transaction data can, in principle, be removed from the instance memory to free up space for new transactions. Naturally, sufficient space must be allocated to ensure that all transactions can be securely logged. Otherwise, the whole system hangs, and new transactions cannot be operated. Inactive transaction data is then removed from the system as a natural consequence of the need to free up disk space and system resources.

The second transaction log type, REDO, is an **Archive type**. Before rewriting the Online log, a copy is created and stored in the Archive repository. This is done by the **Archiver** (**ARCn**) background process. Thanks to this, all logs are stored in the system. The user or company policy then defines the durability of the Archive logs in the filesystem and overall security policy. The primary purpose of Archive log management is security, allowing the system to be restored and recovered at any time without the risk of potential data loss.

Restore and recovery operations are used if an instance or media failure occurs. The **restore** operation involves copying backups from the secondary storage to the primary disk location. However, such files are out of date, reflecting the time point of the backup process's execution. A consecutive **recovery** operation is then used by extracting transaction logs to shift the file to the current data point.

First, Archive REDO logs are applied, then the Online REDO logs are used for that particular backup. From a reliability point of view, if the backup and all consecutive logs (Archive and Online REDO) are present, it is always possible to obtain the current state after the failure. If not, naturally, only an incomplete recovery can be made, resulting in data loss. Thus, only an image that was valid in the past is obtained, depending on the backup and logs. All subsequent changes are lost.

Note that due to consistency issues, it is impossible to restore the database only partially by ignoring the missing log files. In that case, the consistency of the data would be compromised. This operability and consistency is checked by the SCN values that are applied to each transaction and log file, as well as the Archive. Thanks to this, the relevant data can be extracted and used for image construction.

Archive log mode can be set if the database is in MOUNT mode by applying the following command:

```
ALTER DATABASE ARCHIVELOG;
```

Then, archiving is started by launching the background processes and copying the Online REDO logs to the Archive repository. This cannot be done online without an instance restart. Note that the current property can be obtained by getting log_mode from the v$database dynamic performance view:

```
SELECT log_mode FROM v$database;
```

To set up archiving, two parameters must be taken into account: DB_RECOVERY_FILE_DEST (defaults to ORACLE_BASE/flash_recovery_area) defines the location of the Archive repository – that is, FLASH RECOVERY AREA – while the DB_RECOVERY_FILE_DEST_SIZE parameter sets the maximum size that can be allocated for the Archive repository. If the requirements exceed this, database operations will hang until an extension or space cleanout.

The aim of the transaction support structures is just consistency and durability certainty. However, what does consistency mean in parallel access processing? Well, the reference point for the operation is either the operation itself or the transaction start. This means that although the data was changed, the system must always reconstruct the image to the defined time point to ensure the overall consistency of the operation. Consequently, for an individual operation, it does not need to strictly get current data, but data that was in the system at the time of the transaction or operation start. So, it relates to historical data. Thus, even in pure transaction management, temporality can be felt, can't it? Although indirectly, it is clear that by evaluating transaction data, it is possible to reconstruct states that were valid in the past. If we have all the transaction logs (Online and Archive), it is possible to obtain the state of the object at any time. However, be aware that all of them must be available; otherwise, compromised data will be produced!

With that, we have covered the core elements and principles of transaction support. Transaction logs, if all are available, can be used as input for the Flashback operation to revert the database image to the defined point. In the next section, we will look at Flashback technology in more detail by looking at various applicability levels.

Reconstructing states using Flashback technology

Flashback technology can be considered a very important rescue layer. It has surely happened to you that you forgot the Where condition while defining the Update command, or you simply changed the data the wrong way. You might have wanted to delete archive data or move it to the data warehouse and you accidentally deleted the production data that is still needed. Yes, these could cause a huge problem, even up to the level of losing your job. Surely, even just reading this gives you chills, right?

In this section, we will show you that despite finding yourself in such complicated situations, there is still a way out without causing permanent data loss. We will deal with the Flashback technology at various precision levels. FLASHBACK DATABASE reverts the whole database to the required point.

It requires the MOUNT mode of the database, so the database is not available during this operation. We will also discuss the techniques of how to get the historical image for the table, or just get the data portion as it existed in the past with no physical change to the data tuples. Let's get started!

Using FLASHBACK DATABASE to get a historical image

The FLASHBACK DATABASE command can rewind the whole database to the SCN or time point (target time) in the past. It is a variant of restore and recovery, but commonly, FLASHBACK DATABASE provides a more powerful solution.

Although it is mostly used in case of incorrect operation execution (and its transaction confirmation), it can also be applied to provide a historical image. This is done in MOUNT mode. Thus, first, shut down the database. Then, start it up in MOUNT mode. Finally, perform a flashback followed by opening the database. Once you've done this, the database will be in the previous state, as it existed at the defined SCN or time point in the past. Note that such an operation influences the Online logs, so the database opening must be preceded by the RESETLOGS operation.

The whole process is executed by **Recovery Manager** (**RMAN**) using the following steps:

```
SHUTDOWN IMMEDIATE
STARTUP MOUNT
FLASHBACK DATABASE TO SCN <scn_value>;
```

```
--> Starting flashback at 29-JUL-22
--> allocated channel: ORA_DISK_1
--> channel ORA_DISK_1: SID=104 device type=DISK
--> starting media recovery
--> media recovery complete, elapsed time: 05:12:47
--> Finished flashback at 29-JUL-22
```

```
ALTER DATABASE OPEN RESETLOGS;
```

Optionally, you can specify RESTORE POINT in SQL to which the database can be reverted:

```
CREATE RESTORE POINT <rp_name>
  GUARANTEE FLASHBACK DATABASE;
```

The FLASHBACK reference is done in RMAN, and the database must be in MOUNT mode:

```
SHUTDOWN IMMEDIATE
STARTUP MOUNT
FLASHBACK DATABASE TO RESTORE POINT '<rp_name>';
ALTER DATABASE OPEN RESETLOGS;
```

The following command allows you to reference a time point instead of SCN for processing. Let's look at the syntax:

```
FLASHBACK DATABASE TO TIME '<date_string>';
```

Let's also refer to an example:

```
FLASHBACK DATABASE TO TIME 'SYSDATE-7';
```

To enable Flashback technology, there are a few prerequisites:

- RMAN must be connected to the target database (the database to which the operation is to be applied)

- The database must be in `ARCHIVELOG` mode

- Flashback can only be processed if the database is in `MOUNT` mode during the execution itself

- The **Fast Recovery Area** (**FRA**) must be configured for maintaining `Flashback` logs, which are stored as Oracle-managed files inside the `FRA` repository

The FRA and database state can be obtained by querying the `v$database` dynamic performance view:

```
select name, log_mode, open_mode, flashback_on, current_scn
  from v$database;
--> NAME   LOG_MODE     OPEN_MODE    FLASHBACK_ON    CURRENT_SCN
--> ORCL   ARCHIVELOG   READ WRITE   NO              37787874
```

Transaction management can provide historical data images by extracting and applying change vectors; however, its user definition would be difficult and too demanding. The `DBMS_FLASHBACK` package provides robust and automated support for identifying and extracting relevant changes from transaction logs.

Understanding and referencing the DBMS_FLASHBACK package

The `DBMS_FLASHBACK` package lets you get the data version at a specified SCN or time. It proposes the interface to view the database at a particular point in the past by selectively removing individual operations across the transactions. Compared to `FLASHBACK DATABASE`, which reverts the whole database, this technology can be applied at a finer granularity level. It also provides a self-service repair availability to recover original data after approving incorrect `Update` or `Delete` statements.

The DBMS_FLASHBACK package consists of five methods:

- The DISABLE procedure: Disables Flashback mode for the entire session.

- The ENABLE_AT_TIME procedure: This method takes the time point, finds the most suitable SCN that most closely matches the specified time in the parameter, and applies the Flashback operation. The ANSI TIMESTAMP constructor can provide a time point itself, or the TO_TIMESTAMP function call can be used. Regarding the time representation, if the time elements are missing, then it defaults to the beginning point of the day. Time zone information specified in the definition is ignored.

- The ENABLE_AT_SYSTEM_CHANGE_NUMBER procedure: This takes the SCN and sets the whole session snapshot to the defined value. It enables functionality and provides a consistent data image.

- The TRANSACTION_BACKOUT procedure: This removes the impact of the specified transactions delimited either by their names or identifiers (XIDS). They are placed in array format. It analyzes the dependencies and performs **data manipulation language** (**DML**) commands by generating complex reports. It does not approve any DML and holds the locks on rows, allowing users to specify other activities. Consequently, the user must commit the data based on the preferences explicitly.

- The GET_SYSTEM_CHANGE_NUMBER function: This is the most significant function related to reversal in terms of time perspective. It provides the current SCN as a number data type. Thanks to storing such a value, it is straightforward to get the whole image precisely.

To be able to reference the DBMS_FLASHBACK package, the user must have been granted the execute privilege for the whole package:

```
grant execute on DBMS_FLASHBACK to <username>;
```

To get the current status point of the database to reference, the GET_SYSTEM_CHANGE_NUMBER function of the DBMS_FLASHBACK package can be used to provide the SCN. It can be called by using the select statement or executed by storing the output in a variable:

- The select statement:

```
select DBMS_FLASHBACK.GET_SYSTEM_CHANGE_NUMBER from dual;
```

- Using an environment variable:

```
variable scn number;
  exec :scn:=DBMS_FLASHBACK.GET_SYSTEM_CHANGE_NUMBER;
print scn;
--> 37787776900222
```

Although transaction logs are primarily used for security and recovery purposes, they can be used in a wider mode related to time management and representation. Typically, restore and recovery are used after a failure to construct the current state of the database, which is valid immediately before the failure itself. To be more general, any historical time point can be specified as a destination point. Moreover, such activity can be applied to various granularities. It can often be useful to get the historical values for a particular table, a specified query, and so on. A typical example can be related to the improper `Update` operation, but in general, it can be useful to see the table as it existed in the past. To highlight this, some data table contents are represented by the code lists and do not evolve.

Flashback technology allows you to get the data images related to the historical SCN or time point. It provides you with table content as it existed at a defined point. Flashback technology was introduced in Oracle 10g and made more complex in consecutive releases.

It has some prerequisites that must be applied before you can revert the table to the historical image:

- **ENABLE ROW MOVEMENT**: Data rows do not have to be placed on the original data block position. The address of the row can be changed during the transformation. Therefore, row movement must be enabled:

  ```
  ALTER TABLE <table_name> ENABLE ROW MOVEMENT;
  ```

- The **FLASHBACK ANY TABLE privilege**: A particular user must have been granted the `FLASHBACK ANY TABLE` privilege:

  ```
  GRANT FLASHBACK ANY TABLE TO <USERNAME>;
  ```

- **DML privileges**: A particular user must have DML privileges (`Select`, `Insert`, `Update`, and `Delete`) for the referenced table.

- **Log availability**: A sufficient amount of logs must be available to satisfy the `Flashback` operation.

Flashback technology gets the logs, extracts individual transactions, and propagates the historical image. In the following example, `tab_flash` is a reference to the particular table to be processed. The referential point for the `Flashback` operation can be either an SCN or a defined time point modeled using various techniques, as expressed in the following code:

- `FLASHBACK TABLE` to a specific SCN:

  ```
  FLASHBACK TABLE tab_flash to scn :scn;
  ```

- Specification by a logical time position (image of the table 1 minute ago):

  ```
  FLASHBACK TABLE tab_flash
      TO TIMESTAMP (SYSDATE - INTERVAL '1' minute);
  ```

- Specification by physical time position (precise time point specification by the TO_TIMESTAMP function call):

```
FLASHBACK TABLE tab_flash TO TIMESTAMP
   TO_TIMESTAMP('20.12.2021 09:30:00',
                'DD.MM.YYYY HH24:MI:SS');
```

- Specification by physical time position (precise time point specification by the ANSI constructor):

```
FLASHBACK TABLE tab_flash
   TO TIMESTAMP TIMESTAMP '2021-12-20 09:30:00';
```

Let's also deal with the finest granularity level that's applied to the query. In that case, the database's content remains the same – the operation is done at the Select statement level only and reconstructs the query image, the timing of which is defined by the AS OF clause.

Retrieving historical data using the AS OF query

FLASHBACK TABLE, as stated, physically transforms a particular table by applying the change vectors to the data blocks. Thus, after the FLASHBACK TABLE operation, that table is reverted to the defined point, and all newer states are removed. The granularity used is the table itself.

From a time management point of view, it can be even more useful to get the data as it existed at the defined point by the query. Thus, changes are not physically applied to the table. The query provides you with such a result set, which can then be applied to the table based on user preference. The Select statement can be extended by the AS OF clause, allowing you to specify the time point or SCN located as the FROM clause extension. The following bullet list summarizes individual techniques for the AS OF query referring to historical points using the SCN, the logical time reflecting the current date, or a precise time point definition:

- Using SCN:

```
select * from tab_flash AS OF SCN 37787776900222;
```

- Logical time reflecting the current date:

```
select * from tab_flash
   AS OF TIMESTAMP (SYSDATE - INTERVAL '1' minute);
```

- Precise time point definition:

```
select * from tab_flash
   AS OF TIMESTAMP(to_timestamp('20.12.2021 09:30:00',
                                'DD.MM.YYYY HH24:MI:SS'));
```

The AS OF query complements the general possibilities of getting historical data, either by reverting the table or the whole database. The main advantage of the AS OF query is that there is no impact on the table's content. Thus, it is only done dynamically for the specified query.

Summary

This chapter dealt with Flashback technology. While it wasn't intended to cover temporality and monitor object states over time, it offered techniques to get the state at a defined point in the past. The Flashback operation can be done for the whole database or even a particular table, or the historical data image can be obtained by the query itself. The Select statement provides the most powerful solution from a temporal management point of view because the stored data is not changed – just the transaction logs are applied to construct the result set. As stated, the Flashback execution is based on Online and Archive transaction logs, by which the historical state is reconstructed. The availability of the logs is crucial for the activity and reliability of the solution. Regarding performance, Flashback commonly provides a more powerful solution compared to *restore* and *recovery*. On the other hand, a precise temporal model can ensure significant performance benefits because it can be more strictly aimed at monitoring attribute changes at the level of temporal access.

The next chapter focuses on holes in security related to date and time management and covers **SQL injection**. We will show you that even the relatively trivial matters of date processing and conversion can create an unacceptable extension of the query, resulting in unauthorized data access or the possibility to modify and attack the data, structure, or infrastructure.

Questions

Answer the following questions to test your knowledge of this chapter:

1. Which transaction property ensures access just to confirmed data changes (assuming the transaction has already reached COMMIT)?

    A. Atomicity

    B. Consistency

    C. Isolation

    D. Durability

2. Which attribute signifies the transaction time reflection for the Flashback data reconstruction?

    A. UNDO

    B. REDO

    C. SCN

    D. LGWR

3. Which state of the database must be applied to perform FLASHBACK DATABASE?

   A. NOMOUNT

   B. MOUNT

   C. OPEN

   D. RESETLOGS

4. Which statement gets you the table Tab content as it existed at the SCN 37787875368953?

   A. select * from Tab AS OF SCN 37787875368953;

   B. select * from Tab AS OF TIMESTAMP 37787875368953;

   C. select * from Tab FLASHBACK to 37787875368953;

   D. select * from Tab where SCN=37787875368953;

Further reading

To learn more about the topics that were covered in this chapter, take a look at the following resources:

- *Pro Oracle Database 18c Administration (Manage and Safeguard Your Organization's Data)*, by Michelle Malcher and Darl Kuhn. It deals with database system administration. *Chapter 19* focuses on the RMAN restore and recovery operations, complete and incomplete recovery, as well as flashing back a table or whole database.

- *Oracle Database 12c – Oracle RMAN Backup and Recovery*, by Robert G. Freeman and Matthew Hart. *Part 1* deals with the base elements of RMAN, *Part 2* highlights the configuration, backup, and recovery essentials, and *Part 3* focuses on maintenance and administration tasks. RMAN as a highly-available architecture is discussed in *Part 4*. *Chapter 16* directly covers the Flashback technologies by emphasizing possible user errors. Among the stated Flashback types, it also deals with flashing back the whole transaction and flashing back the structure after dropping it using the Recycle Bin object. Specific focus is put on the Flashback Data Archive (Oracle Total Recall), which allows you to permanently archive all changes at the data table level and go back to any point in time. *Part 5* deals with RMAN media management.

14
Building Reliable Solutions to Avoid SQL Injection

Improper value management and assignment can lead to security issues related to **SQL injection**. Although it is not directly evident, it can cause significant problems if **bind variables** or other sanitization techniques are not used. Unfortunately, most people do not realize this situation and live with the incorrect assumption that the date and time values cannot be the root of the SQL injection. They put the character string sequence of individual elements representing date and time values into the command definition. However, then, the provided value is converted into a DATE or TIMESTAMP value automatically and evaluated. But there is a risk: automated conversion.

Often, date and time values are not considered security holes. They consist of individual elements, and the format is precisely specified, so where is the problem? Which parameters affect the format and representation? The session DATE and TIMESTAMP formats impact the result's representation, element meaning, and mapping. Moreover, what about the data source? It is commonly represented in a textual format that needs to be converted.

In this chapter, we will focus on the consequences of implicit conversions and techniques to limit SQL injection related to processing date and time values in dynamic SQL. We will also show you the risks, detection techniques, and methods of building reliable solutions and avoid SQL injection.

In this chapter, we're going to cover the following main topics:

- The concept of SQL injection
- Securing applications using bind variables and sanitizing the input with the DBMS_ASSERT package

For the record, this chapter just relates to date and time values. The complexity of SQL injection is much deeper. To get an overview of various streams, it is recommended that you read the additional materials and books mentioned in the *Further reading* section at the end of this chapter. Always be aware of SQL injection, as it can offer access to hackers.

The source code for this chapter can be found in this book's GitHub repository at `https://github.com/PacktPublishing/Developing-Robust-Date-and-Time-Oriented-Applications-in-Oracle-Cloud/tree/960e892ada2c4ff3247f4967ca9b3d866e0f1c4f/chapter%2014`. Alternatively, you can access the repository by scanning the following QR code:

Understanding SQL injection

Are you self-employed or part of the company? Who can access your data in the company? Well, you suppose, only your manager can access data about colleagues from the same department and not about anyone from the whole company, right? It is, therefore, important to ensure that the system inputs cannot change the query without proper authorization and thus obtain completely different values and outputs. You probably wouldn't want anyone to have access to your salary, available funds in your accounts, or other sensitive data. Consequently, it is necessary to secure and sanitize the inputs so that the user cannot influence the structure of the `select` statements or modify them to get principally different results.

The rule is simple: never rely on the user data, and always check it to ensure that suspicious code is not detected there.

SQL injection risk is a hazardous consequence of a developer prompting a user to get the value from the input. That value is then passed directly and concatenated into the query in string format. However, instead of the required values, the user places a SQL query extension by adding additional commands or conditions. The code is then compiled and executed. As a result, users can have access to improper or sensitive data. The consequences can be disastrous, such as the ability to modify the database, grant new privileges, and so on.

SQL injection is currently one of the most often used hacking techniques. It takes malicious code and places it inside the SQL query, resulting in users receiving improper or sensitive data, but it can also change the values of objects, states, or parameters. Moreover, it can destroy the database or lose its data. Thus, instead of getting a value from the user, SQL code is produced, which is dynamically executed. There are several techniques generally, and most of them are based on adding the `always true` condition (`1=1`), resulting in producing the whole dataset, even with sensitive data. Furthermore, if you perform the `union` operation to merge the result sets, you can manipulate the data from other structures and database tables.

To dig deeper, let's indicate the problem in the following example, which deals with an employment contract. The data model is shown in *Figure 14.1*. It consists of two tables dealing with the PERSONAL_ DATA and EMPLOYMENT contract characteristics, and salary management. We presume that the attribute names are self-representative and do not need special attention:

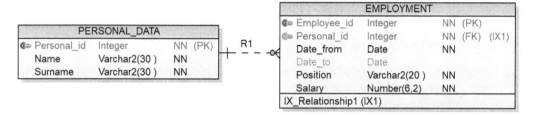

Figure 14.1 – Employee data model

All the data will be merged into one common view, V_EMP, for simplicity (*Figure 14.2*). It consists of the employee identifier (Employee_id), first name (Name), last name (Surname), job position (Position), and current salary (Salary). Moreover, two Date attributes are delimiting the contract (Date_from and Date_to). Date_from covers the first date of the employment validity, while Date_to refers to the last day:

	V_EMP		
Employee_id	Integer	NN	(PK)
Name	Varchar2(30)	NN	
Surname	Varchar2(30)	NN	
Date_from	Date	NN	
Date_to	Date		
Position	Varchar2(20)	NN	
Salary	Number(6,2)	NN	

Figure 14.2 – The V_EMP view structure dealing with the Employee structure

We will create a procedure that will list the employees that were hired on a specified date (defined as a procedure parameter). In this case, an incorrect approach will be used by concatenating the string in the v_statement local variable. This variable is then used for creating a cursor (v_cursor). By opening the cursor, it is made available for fetching until the operation gets a new data row. Finally, obtained values are stored in the local variables (v_ns for concatenating the name and surname, and v_pid for the employee identifier) and put into the console user's interface.

For evaluation, the whole statement is listed here as well:

```
create or replace procedure GET_EMPLOYEES (p_date DATE)
is
  v_statement varchar2 (10000);
```

```
v_cursor sys_refcursor;
v_ns varchar(100);
v_pid varchar(11);
begin
v_statement:='select Name || '' '' || Surname as ns,
                    employee_id
              from V_EMP
              where Date_from>='''||p_date||'''';
  open v_cursor for v_statement;
   loop
     fetch v_cursor into v_ns, v_pid;
    exit when v_cursor%notfound;
      DBMS_OUTPUT.PUT_LINE(v_ns ||': ' || v_pid);
   end loop;
  close v_cursor;
end;
/
```

The complete code of the GET_EMPLOYEES function can be downloaded from the GitHub repository for this chapter.

As stated, the default format of the date is inherited from the server parameter settings but can be overwritten by the user for the session. This notation is then used for implicit conversions and date management. Let's suppose that the format is YYYY-MM-DD and has been set by the alter session command:

```
alter session set NLS_DATE_FORMAT='YYYY-MM-DD';
```

The procedure parameter value can be specified either explicitly (using a constructor or the TO_DATE function) or by any function providing the date (for example, sysdate for getting an actual date). One way or another, any valid DATE value should be passed:

```
exec GET_EMPLOYEES(TO_DATE('1.1.2015', 'DD.MM.YYYY'));
exec GET_EMPLOYEES(sysdate);
```

Before taking a look at the result set, take note of the composed string as the input for the cursor definition:

```
select Name || ' ' || Surname as ns, Employee_Id
  from V_EMP
    where Date_from='2015-01-01';
```

The provided input value is in date format and is consecutively concatenated to the pre-prepared `select` statement, which is in string format. Thus, the referenced date value must be transformed into string format using implicit conversion. The `TO_CHAR` function call delimits such a transformation. The format is inherited from the system or session specification. In this case, the session format is `YYYY-MM-DD`. So, where can the problem occur? Is it even possible to compromise the result set? If so, how?

The problem can be deeper and more easily achievable than it may seem at first sight. The SQL injection core lies in the implicit conversion. As is evident, date-to-string transformation is done automatically without user intervention, but `NLS_DATE_FORMAT` supervises it. However, the parameter value can consist of not only the format itself but also any character string literals, as these can also, in principle, be associated with the `NLS_DATE_FORMAT` parameter. Even commands or individual statement clauses can be present. And that is just the core of the processing element and risk source. Note that a particular user does not need to have an `ALTER SESSION` privilege to change the format of the `NLS_DATE_FORMAT` parameter! Simply, it can be done by any user, irrespective of the privileges. Therefore, proper parameter treatment is crucial. Never rely on the user's behavior being correct, and use bind variables instead.

The defined `GET_EMPLOYEES` procedure should provide the list of pairs – concatenated `Name` and `Surname` as the first value – followed by the employee's identification number (`Employee_id`). By changing the `DATE` format, a query can be strongly changed. Let's consider the following command, which changes the format of the date representation:

```
alter session
  set nls_date_format = '"''' union select Surname,Salary
                                    from V_EMP --"';
```

In this case, the concatenated string forms a union with a new statement. It provides `Surname` and `Salary` instead of `Employee_id`. The two dashes at the end (`--`) express the comment, whereas one apostrophe (`'`) must be ignored from a syntax point of view. As a result, a query for the execution looks like this:

```
select Name || ' ' || Surname as ns, Employee_id
  from V_EMP
      where Date_from=''
  union
select Surname,Salary from V_EMP --';
```

Let's look deeper into the preceding code snippet. For now, the first statement is ignored – no rows are selected at all. But the first statement is unioned by the second statement, which provides the surname and current salary for each employee. Strange, isn't it? This is a severe problem, by which any table data can be provided (stolen from the system), such as access codes, personal data, PINs, account numbers, and so on. It can even reference different tables (a list of accessible tables and individual attributes can be provided by querying data dictionary views). Therefore, instead of direct value concatenation, always use bind variables. The following code snippet shows the problem. At first glance, it looks like a standard function call. However, when we look at the results, we get the salary of individual employees:

```
exec GET_EMPLOYEES(sysdate);
--> It provides the salary of the employees!!!
--> Cash:   3000
--> Smith: 1200
--> Smith: 1800
```

Thus, regarding the preceding procedure, the statement is not composed of concatenation. So, what is the right solution? How can we limit the risk of SQL injection? Well, bind variables are resistant to the threat of SQL injection. And what about other solutions? The core principles will be described in the next section.

Solutions to limit SQL injection

In the previous section, you learned about SQL injection creation. We showed what it can cause and what the consequences are. This section will drive you through the available solutions by introducing bind variables and explicit conversions. We will also discuss the DBMS_ASSERT package, which allows you to sanitize input values.

Using bind variables

Bind variables provide the relevant solution to limit SQL injection. Instead of concatenating the character strings forming the statement, the input value is applied using a bind variable, as shown in the following code block, which defines a GET_EMPLOYEES procedure. The condition in the where clause is related to the Date_from attribute and procedure parameter (p_date). This is not concatenated, but rather treated as a bind variable:

```
create or replace procedure GET_EMPLOYEES(p_date DATE)
is
 v_statement varchar2(10000);
 v_cursor sys_refcursor;
 v_ns varchar(100);
 v_pid varchar(11);
begin
 v_statement:='select Name || '' '' || Surname as ns,
                    Employee_id
              from V_EMP
```

```
               where Date_from <= :bind_date';
    DBMS_OUTPUT.PUT_LINE(v_statement);
  open v_cursor for v_statement using p_date;
   loop
     fetch v_cursor into v_ns, v_pid;
      exit when v_cursor%notfound;
     DBMS_OUTPUT.PUT_LINE(v_ns ||': ' || v_pid);
   end loop;
  close v_cursor;
 end;
 /
```

So, now, we need to check whether it is working properly and limiting the possibility of SQL injection. First, set the NLS_DATE_FORMAT value for the session:

```
alter session
  set NLS_DATE_FORMAT = '"'
   union select ''hack'',1000 from dual --"';
```

Second, execute the updated procedure:

```
exec GET_EMPLOYEES(sysdate);
```

No additional rows are listed based on the session date format specification:

```
--> Name & surname, EMPLOYEE_ID
--> John Cash: 10000
--> Simone Smith: 20000
--> Simone Smith: 20001
```

As evident, even if the date format is changed to open the possibility to extend the `select` statement definition, the additional query is not executed and is secured by the bind variable, instead of string concatenation.

To conclude this section, using bind variables is always beneficial for various reasons. First, they limit SQL injection. As stated, even Date values can be attacked by the implicit conversion delimited by the NLS_DATE_FORMAT parameter value set for the whole system or overwritten by the session. Although it works correctly for 99.99% of cases, you should always worry about the remaining 0.01% forming SQL injection. Be sure that users know such a trick. Sooner or later, they could seize sensitive data from the system.

The second beneficial aspect of bind variables is associated with the performance impact. Each statement must be parsed before execution. Bind variables propose a robust technique to evaluate the statement only once, irrespective of the changed values for referencing.

Another technique that limits the Date and Time segments of SQL injection is related to explicit conversion. The next section will cover the principles and provide examples.

Explicit date and time value conversion

Let's go back to the already created `get_employees` function, which we described in the previous section. Let's focus on the condition of the `select` statement:

```
where Date_from<=p_date;
```

The left part of the condition is the name of the attribute (`Date_from`) associated with the `DATE` data type. The right part is the name of the parameter (`p_date`), which should also take the `DATE` type. In principle, the input parameter value of the function cannot be directly influenced. As described, it can be influenced by the implicit conversion, so in principle, it can contain suspicious code. So, the task is to find a way to detect and limit it. The answer is pretty clear: just check whether the provided value is `DATE` (or `TIMESTAMP`). The easiest way to do that is based on using `TO_CHAR` as a conversion function, getting a string as the result. If the conversion cannot be done, an exception is raised. The condition extension can look like this:

```
where TO_CHAR(date_from, 'DD.MM.YYYY')
                    <= TO_CHAR(p_date, 'DD.MM.YYYY');
```

The whole statement looks like this:

```
select Name || ' ' || Surname as ns, Employee_id
  from V_EMP
   where TO_CHAR(DATE_FROM, 'DD.MM.YYYY')
                    <= TO_CHAR(p_date, 'DD.MM.YYYY');
```

What would happen if the input values were not properly formatted (potentially defining a SQL injection attack)? The `TO_CHAR` conversion cannot be done, whereas the transformed input is not formatted in terms of the `DATE` type.

Let's change the session format for the implicit `DATE` value transformation:

```
alter session set NLS_DATE_FORMAT='"hack"';
```

The original value, 15.10.2022, is converted into the `hack` value. The `TO_CHAR` conversion function raises an exception, whereas the format does not fit the second parameter value, `DD.MM.YYYY`.

The original statement looks like this code block:

```
select Name || ' ' || Surname as ns, Employee_id
  from V_EMP
    where TO_CHAR(date_from, 'DD.MM.YYYY')
            <= TO_CHAR(TO_DATE('15.10.2022'), 'DD.MM.YYYY');
```

The statement we defined previously is transformed and its physical representation is shown in the following code block. As we can see, the hack value cannot be processed as input for obtaining the DATE value. The transformed statement is as follows:

```
select Name || ' ' || Surname as ns, Employee_id
  from V_EMP
    where TO_CHAR(date_from, 'DD.MM.YYYY')
            <= TO_CHAR("hack", 'DD.MM.YYYY');
--> ORA-01861: "literal does not match format string"
--> *Cause: Literals in the input must be the same length --->
as literals in the format string (with the exception of -->
leading whitespace).
```

Using bind variables and conversion works properly for easy queries that are comparing just one value. However, there can be cases where more complex strings need to be passed. Therefore, the SQL statement must be built dynamically in a more complex manner. To protect the code, database, and application, you need to inspect and evaluate the code. Manual checking is the easiest way, but this is infeasible in the real world. So, how can we do that? We can simply use a predefined Oracle package, DBMS_ASSERT.

Sanitizing the input with the DBMS_ASSERT package

The methods in the DBMS_ASSERT package allow you to validate the input value properties. If the checking is done successfully, the original value (or the value enclosed within quote marks) is returned through the function output. Otherwise, an exception is raised. Let's look at all the methods covered by the DBMS_ASSERT package.

The ENQUOTE_LITERAL function encloses the input value in single quote marks (at the beginning and the end of the value) if they are not already there. Moreover, it checks whether existing single quote marks are properly paired. Then, the value is checked against the valid identifier list. The SCHEMA_NAME function checks whether the input value is an existing schema name or not, while SIMPLE_SQL_NAME emphasizes a simple SQL name verification. The QUALIFIED_SQL_NAME function checks the value against the qualified SQL name list. A qualified name can be composed of multiple simple SQL names expressing the schema name, object name, and database links. The `SQL_OBJECT_NAME` function evaluates whether the input value can be considered as a qualified SQL identifier of an existing SQL object. Finally, the `NOOP` (**No Operation**) function returns the original value. No checking is implemented in the body of the function.

The next section discusses the `ENQUOTE_LITERAL` function, which is the most significant function in terms of SQL injection prevention.

Implementing the ENQUOTE_LITERAL function for data enclosure

The `ENQUOTE_LITERAL` function of the `DBMS_ASSERT` package can check whether the input value can be potentially affected by the injection. It evaluates the quotation marks, as stated in the following code snippet:

```
select DBMS_ASSERT.ENQUOTE_LITERAL(''' or ''1=1')
   from dual;
```

You can see the potential query extension by placing an additional condition delimited by the `or` mark. If the processing and evaluation fail, it is denoted by raising the `ORA-06502` exception (numeric or value error). Although the raised exception type is a bit general, it can be considered an indicator of a possible SQL injection attack.

The following code snippets declare the usage and the provided output. In principle, any provided value is enclosed by single quotation marks if it has already been defined. Thus, a numerical value is extended as well. The first code block emphasizes an explicit character string declaration. However, the single quote marks that are present there are not part of the character string – they just denote the syntax. Thus, the second provided column value does not have quote marks:

```
select DBMS_ASSERT.ENQUOTE_LITERAL('Michal'),  'Michal'
   from dual;
--> 'Michal'          Michal
```

As we can see in the following code block, the `ENQUOTE_LITERAL` function encapsulates the whole expression, not a single word separately:

```
select DBMS_ASSERT.ENQUOTE_LITERAL('Michal Kvet')
   from dual;
--> 'Michal Kvet'
```

To demonstrate these principles in a bit more depth, let's process the output of the TO_CHAR function, which is transforming the DATE value into a character string. It is a string as an output, right? But not explicitly enclosed. Therefore, by calling the ENQUOTE_LITERAL function, quote marks are added. The following block gets the original value, as well as its pass to the calling ENQUOTE_LITERAL function:

```
select DBMS_ASSERT.ENQUOTE_LITERAL(TO_CHAR(sysdate,

                                        'DD.MM.YYYY')),

        TO_CHAR(sysdate, 'DD.MM.YYYY')

    from dual;

--> '12.12.2022'           12.12.2022
```

Finally, the following block shows the output if the input value is numerical. In this case, the output value is enclosed by single quote marks:

```
select DBMS_ASSERT.ENQUOTE_LITERAL(1)
    from dual;
--> '1'
```

If the NULL value is reflected as an empty string, it is extended by two single quote marks with no character inside. Hence, ensure that you emphasize the NULL value as an undefined representation. An example is shown here:

```
select DBMS_ASSERT.ENQUOTE_LITERAL(NULL)
    from dual;
--> ''
select DBMS_ASSERT.ENQUOTE_LITERAL('NULL')
    from dual;
--> 'NULL'
```

Finally, any potential risks are identified and evaluated by raising the stated exception type. In PL/SQL, this can be covered by the exception handler to reference potential SQL injection threats.

The two code blocks in the following paragraphs point to the single quote marks as part of the character string. The following code is successfully evaluated because the quote pairing is correct:

```
select DBMS_ASSERT.ENQUOTE_LITERAL
    ('character string with '''' quotes included')
    from dual;
```

The mapping stated in the following block is incorrect, resulting in raising an exception. Without the ENQUOTE_LITERAL function, it would work correctly:

```
select DBMS_ASSERT.ENQUOTE_LITERAL('''1''
                                    union select 1 from dual')
  from dual;
--> ORA-06502 exception
```

Note that the DBMS_ASSERT package is owned by the SYS user.

Although the Enquote_Literal function is commonly used for code extension detection, it can be used to identify the suspicious code delimited by the date and time values as well.

Summary

SQL injection is a technique that involves changing the original SQL statement code by extending it from the user input. Instead of getting the required value, the user provides the SQL code, which is executed. This results in access to commonly unavailable (hidden) data, the possibility of unauthorized changes to values, or dropping individual objects. The security, robustness, and correctness of the systems would be significantly impaired.

In this chapter, we focused on SQL injection problems related to Date and Time processing. We have shown that the problem can be precisely done because of the implicit character string conversion to the Date or Timestamp value. The solution is associated with the bind variables, which cannot cover additional conditions or queries.

Additionally, we covered explicit Date and Time management and looked at the DBMS_ASSERT package, which can identify identifiers and SQL names and can also limit SQL injection by using the ENQUOTE_LITERAL function. We hope that by applying the defined rules, you will be able to build a reliable solution and defend your data and structures.

The next chapter deals with Timestamp enhancements by focusing on server and client-side processing and time zone impacts.

Questions

Answer the following questions to test your knowledge of this chapter:

1. Try to figure out the SQL injection risk in your already-developed applications. Have you focused on the implicit Date and Time conversions? How can your solution be updated to eliminate it?

2. Use the data model shown in *Figure 14.1*. Assume that the following query is used to produce the name and surname values of the employee. Each employee is uniquely identified by the employee_id attribute value. The required value of the employee_id value is user specified:

```
select name, surname
  from personal_data
        join employment using (personal_id)
     where employee_id=INPUT_VAL;
```

Which input value provides SQL injection and produces all the names and surnames, irrespective of the provided employee ID?

A. NULL

B. IS NULL

C. 1=1

D. 1 or 1=1

Further reading

To learn more about the topics that were covered in this chapter, take a look at the following resources:

*   *SQL Injection Strategies: Practical techniques to secure old vulnerabilities against modern attacks*, by Ettore Galluccio, Edoardo Caselli, and Gabriele Lombari. This book serves as a theoretical as well as a practical guide to take you through the important aspects of SQL injection in web, mobile, and IoT applications. It reflects both defense and attack perspectives. You will learn how to defend systems against SQL injection attacks and about various concepts and scenarios, and the SQL injection manual attack techniques will be discovered.

*   *Real World SQL and PL/SQL: Advice from the Experts*, by Arup Nanda, Brendan Tierney, Heli Helskyaho, Martin Widlake, and Alex Nuijten. *Chapter 15* of this book discusses the threats to data security, focusing on SQL injection in a complex manner pointing to program security. It references the authorization, definer, and invoker rights models.

Part 6:
Expanding a Business
Worldwide Using Oracle Cloud

Now, you know almost everything about proper date and time management, modeling, and representation, as well as reliability, security, and so on. So, it's time to use this knowledge practically to expand your business across the globe.

This part summarizes and discusses `TIMESTAMP` enhancements to cover any region and ensure proper time zone calculation and shift across multiple clients. Date and time synchronization across multiple regions must be done. Then, we will discuss how to dynamically rewrite database queries to enable the deployment and operation of local applications worldwide, focusing on the difference between the client and server time zones by introducing the `SYSDATE_AT_DBTIMEZONE` cloud database reference.

This part includes the following chapters:

- *Chapter 15, Timestamp Enhancements*
- *Chapter 16, Oracle Cloud Time-Zone Reflection*

15
Timestamp Enhancements

One could say that life would be better and more secure if we had everything under control. In the past, businesses were entirely managed locally. Servers were placed in a local server room directly in the company, supervised by the system and database administrators. Employees knew administrators and technicians personally, mostly through filing complaints if something was not working as expected. Individual systems were locally managed and people were able to meet physically, with meetings coordinated by the locally defined time. Simply put, there was no need to deal with date and time synchronization.

With the globalization of business, multiple branch offices began to be created, opened, and managed, resulting in sharing systems and responsibilities. Customers started to spread across the world. This resulted in the necessity to coordinate activities by date and time. To ensure correctness and robustness, data needed to be managed not just locally in one branch, city, or state but on a global scale. Thus, it became necessary for data coverage to be applied to emphasize time processing, time zones, and synchronization processes.

This chapter deals with the principles of time zone management, reflecting the local and global times by the TIMESTAMP extension. It outlines the methods for time zone management, individual data type extensions, as well as transformation rules.

In this chapter, we're going to cover the following main topics:

- Setting time zones at the database and session levels
- Transforming TIMESTAMP values across time zones using the AT TIME ZONE clause and the FROM_TZ function
- Transforming DATE values across multiple time zones using the NEW_TIME function
- TIMESTAMP normalization to UTC
- Local value reflection using TIMESTAMP data type extensions
- Business growth and related problems (local and global merchantability)

The source code for this chapter can be found in the GitHub repository accessible at `https://github.com/PacktPublishing/Developing-Robust-Date-and-Time-Oriented-Applications-in-Oracle-Cloud/tree/960e892ada2c4ff3247f4967ca9b3d866e0f1c4f/chapter%2015`. Alternatively, you can access the repository by scanning the following QR code:

Applying time zones for the server and client sides

It wasn't that long ago when all applications and systems were stored locally. Offices usually included a computer that formed the central storage with a local network, often without direct access to the internet. Later, as these local devices got bigger, louder, and produced too much heat, local servers were moved to the server room of the company. Technicians and administrators were responsible for the proper functioning of individual hardware components, accessibility, and availability. Soon it became clear that one room was not enough and the data must be geographically distributed to ensure robustness and, more importantly, security. The idea was that even if any disaster or accident occurred, such as a fire, flood, or earthquake, there would always be at least one surviving repository that could handle requests and access the systems and data. However, the systems became increasingly complex, and ensuring all the features, aspects, and availability domains became more challenging.

The cloud environment brought solutions to these problems, such as system management and availability, thanks to which companies do not have to invest in their own networks, server rooms, or hardware. Instead, the entire portfolio of services is managed by a cloud vendor. Very soon, companies discovered the significant benefit of cloud services. However, if the systems are deployed in the cloud, how about moving to a higher level and expanding the business geographically? Imagine building your business in Central Europe. Why not provide your services throughout Europe? Moreover, since cloud systems are generally available, why not offer your solutions in America, Australia, or simply anywhere in the world?

To get the relevant solutions across multiple regions, it is necessary to synchronize the time values. It is important to deal with the time zones and regional settings. You don't want to get the wrong time, do you? You do not want to come to the meeting or online call two hours late, right? This is why it is important to distinguish between the server time zone and the client time zone, to make the dynamic time zone shift and recalculation from the server to the client perspective. Oracle Database offers two branches for dealing with the actual date and time value. They differ in the time zone references, which means they can be related to the server or client. It is critical to get the correct reference, mostly if the system was locally managed in the past, only across one region. And that's what this section is about: to explain sysdate and current_date functions referring to server or client region.

The DATE data type embeds the time zone representation directly in the value. There are generally two functions for getting the actual date, which differ in terms of time zone management and access. The sysdate function references the system (server) time zone, whereas current_date deals with the local session definition. The following select statement shows the difference in the output values, referring to the different time zones:

```
select sysdate, current_date from dual;
--> SYSDATE                CURRENT_DATE
--> 08.03.2022 08:27:30    08.03.2022 11:27:30
```

Based on the results in this example, it is evident that the difference between the server and session time zones is 3 hours. The currently set parameter values representing the session's time zone can be obtained by the SESSIONTIMEZONE function of the STANDARD package owned by the SYS user. However, instead of using the full definition (SYS.STANDARD.SESSIONTIMEZONE), a direct reference can be used (SESSIONTIMEZONE), where a public synonym is implemented. The return value is the time zone of the current session. It is denoted by the valid time zone offset – a character type in the format {+ | -} TZH:TZM, where TZH means time zone hour and TZM means time zone minute. The following code snippet shows the usage and output of the SESSIONTIMEZONE function call:

```
select SYS.STANDARD.SESSIONTIMEZONE from dual;
select SESSIONTIMEZONE from dual;
--> SESSIONTIMEZONE
--> +03:00
```

The time zone region name can also be referenced, as evident from the following outputs, which reference the UTC and Lisbon time zones:

- UTC:

```
    --> SESSIONTIMEZONE
    --> UTC
```

- Lisbon:

```
    --> SESSIONTIMEZONE
    --> Europe/Lisbon
```

A full list of time zone region names, user representation, and different shifts can be found in the Oracle documentation. In principle, the name of the region is typically denoted by the continent, followed by a slash and the capital city of the particular country. The following table provides a list of accessible European regions:

Name	Representation
Europe/Dublin	(00:00) Dublin
Europe/London	(00:00) London
Europe/Lisbon	(00:00) Lisbon
Europe/Oslo	(+01:00) Oslo
Europe/Stockholm	(+01:00) Stockholm
Europe/Copenhagen	(+01:00) Copenhagen
Europe/Berlin	(+01:00) Berlin
Europe/Amsterdam	(+01:00) Amsterdam
Europe/Brussels	(+01:00) Brussels
Europe/Luxembourg	(+01:00) Luxembourg
Europe/Paris	(+01:00) Paris
Europe/Zurich	(+01:00) Zurich
Europe/Madrid	(+01:00) Madrid
Europe/Rome	(+01:00) Rome
Europe/Warsaw	(+01:00) Warsaw
Europe/Prague	(+01:00) Prague Bratislava
Europe/Vienna	(+01:00) Vienna
Europe/Budapest	(+01:00) Budapest
Europe/Sofia	(+02:00) Sofia
Europe/Istanbul	(+02:00) Istanbul
Europe/Athens	(+02:00) Athens
Europe/Moscow	(+03:00) Moscow

Table 15.1 – European regions

The database (system and server) time zone can be obtained by the DBTIMEZONE function call. The principles of representation are the same, providing either the valid time zone offset – a character type in the format {+ | -} TZH:TZM – or the name representation. The following code snippet shows the usage and output of the DBTIMEZONE function call:

```
select DBTIMEZONE from dual;
--> DBTIMEZONE
--> +00:00
```

Usually, the database time zone is modeled by UTC, but there can be multiple cases for using your own implementation and referencing. The time zone can be set for the database, as well as the client. In the next section, individual time zone setting techniques are summarized.

Setting the time zone value at the database and session levels

The settings can be changed for the whole database or session itself by using the alter database or alter session command, respectively. Changing parameters at the database level cannot be done directly; it must be marked in the configuration file. Thus, to apply the change, the database has to be restarted (by shutting down and starting the instance, the configuration file is accessed). The value definition can be a character type in the format {+ | -} TZH:TZM or the *name representation* can be used as well. Now, let's look through examples of the session- and database-level time zone specification:

- The syntax for setting the session value is as follows:

```
alter session set time_zone= ' {  {+ | -} <TZH>:<TZM>
                                | <timezone_name> }';
```

- Examples of setting the session value are as follows:

```
alter session set time_zone='+3:00';
alter session set time_zone='-3:00';
alter session set time_zone='UTC';
alter session set time_zone='Europe/Lisbon';
alter session set time_zone=dbtimezone;
```

- The syntax for setting the database value is as follows:

```
alter database set time_zone= ' {  {+ | -} <TZH>:<TZM>
                                | <timezone_name> }';
SHUTDOWN IMMEDIATE;
STARTUP;
```

- Examples of setting the database value are as follows:

```
alter database set time_zone='-3:00';
alter database set time_zone='UTC';
alter database set time_zone='Europe/Vienna';
alter database set time_zone=sessiontimezone;
```

The database time zone value is defined during the database creation process, but as stated, it can be altered at any time by a user with appropriate privileges:

```
create database . . .  set time_zone='+00:00';
```

Please note that it is impossible to change the database time zone if there is at least one table column with TIMESTAMP WITH LOCAL TIME ZONE. In this case, it would be necessary to drop particular columns or alter the data types:

```
--> ORA-30079: "cannot alter database timezone when
--> database has TIMESTAMP WITH LOCAL TIME ZONE columns"
```

To deal with the DATE data type, the sysdate and current_date functions are used to provide either a session or system (database) reference. The time zone reflection is directly incorporated into the value itself. The difference between the session and database time zones can be reflected explicitly using value subtraction:

```
select (sysdate - current_date)*24 as TZH from dual;
--> TZH
--> -3
```

Now that you know how to get the session and database time zones, it's time to learn about and calculate the shift across the client's perspective.

TIMESTAMP and transformation across time zones

When dealing with higher-value precision, the TIMESTAMP data type is available to serve up to nanosecond precision. There are also two functions for getting the current date and time representation: systemtimestamp and current_timestamp. The systemtimestamp function is part of the STANDARD package owned by the SYS user. The full name (owner.package_name. function_name) does not need to be specified since the public synonym is already specified.

Thus, just using the name in the direct reference is sufficient. It provides the time zone on the server (database) side. The following code shows the `systimestamp` function call:

```
select SYS.STANDARD.systimestamp from dual;
select systimestamp from dual;
--> 08.03.22 13:53:10,324000000 +00:00
```

The reference of the local timestamp of the session can be obtained by a `current_timestamp` function call, which is part of the STANDARD package owned by the SYS user as well:

```
select SYS.STANDARD.current_timestamp from dual;
select current_timestamp from dual;
--> 18.01.22 10:35:47,943000000 +03:00
```

The output format depends on the NLS parameter called NLS_TIMESTAMP_TZ_FORMAT and can either be changed for the session or the system can be altered, as well. Note that the system parameter is `static`, referencing the initialization file, so to apply the change, the instance must be restarted. Thus, the `alter session` command can be executed directly for the session:

```
alter session
  set NLS_TIMESTAMP_TZ_FORMAT=
                        'DD.MM.RR HH24:MI:SSXFF TZH:TZM';
```

However, to apply the change at the database level, the request must be registered in the configuration file. The change itself will then be reflected after the system is restarted. The following code shows the command for the system level:

```
alter system
  set NLS_TIMESTAMP_TZ_FORMAT=
                        'DD.MM.RR HH24:MI:SSXFF TZH:TZM'
      scope=spfile;
```

If the `session` parameter value is not stated directly, the server settings are automatically applied. In the following case, the default output format, enhanced by the implicit character string transformation, is defined by the day, month, and year elements separated by dots, followed by the time representation (hour element in 24-hour format, with minutes, seconds, and fractions). Finally, the time zone is referenced. TZH denotes the time zone hour shift, while TZM denotes the minute part:

```
alter session
  set NLS_TIMESTAMP_TZ_FORMAT=
                        'DD.MM.RR HH24:MI:SSXFF TZH:TZM';
```

There can also be the definition of the region name, as well, expressed by the TZR value (meaning time zone region). The principle or value reflection and output denotation follow the rules shown in the following table. The table outlines the NLS parameter settings, time zone definition for the session/system, and output format regarding the time zone region name or hour and minute reference:

NLS parameter settings	Time zone definition for the session/system	Output format

Hour and minute components (`TZH:TZM`)
 	
`alter session set`
`NLS_TIMESTAMP_TZ_`
`FORMAT=`
`'DD.MM.RR`
`HH24:MI:SSXFF`
`TZH:TZM';`
	Hour and minute components (`TZH:TZM`)
	
	`alter session set`
	`  time_zone='3:00';`
		Hour and minute components (`TZH:TZM`)
		
		`18.03.22`
		`10:47:38,925000000`
		`+03:00`

Hour and minute components (`TZH:TZM`)
 	
`alter session set`
`NLS_TIMESTAMP_TZ_`
`FORMAT=`
`'DD.MM.RR`
`HH24:MI:SSXFF`
`TZH:TZM';`
	Region name reference (`TZR`)
	
	`alter session set`
	`  time_zone='Europe/`
	`Lisbon';`
		Hour and minute components (`TZH:TZM`)
		
		`18.03.22`
		`07:48:17,941000000`
		`+00:00`

Time zone region name (`TZR`)
 	
`alter session set`
`NLS_TIMESTAMP_TZ_`
`FORMAT=`
`'DD.MM.RR`
`HH24:MI:SSXFF TZR';`
	Hour and minute components (`TZH:TZM`)
	
	`alter session set`
	`  time_zone='3:00';`
		Hour and minute components (`TZH:TZM`)
		
		`18.03.22`
		`10:49:12,925000000`
		`+03:00`

NLS parameter settings	Time zone definition for the session/system	Output format
Time zone region name (TZR)	Region name reference (TZR)	Region name reference (TZR)

`alter session set`
 	
`NLS_TIMESTAMP_TZ_`
`FORMAT=`
 	
`'DD.MM.RR`
`HH24:MI:SSXFF TZR';`
	`alter session set`
	` time_zone='Europe/`
	`Lisbon';`
		`18.01.22`
		`07:50:14,787000000`
		`EUROPE/LISBON`

Table 15.2 – Time zone reference

When dealing with the `TIMESTAMP` format, both methods (`current_timestamp` and `systimestamp`) provide the `TIMESTAMP WITH TIME ZONE` data type. This can be seen in the following snippet:

```
create table timetab as
  select systimestamp T1, current_timestamp T2 from dual;
--> Name Null? Type
--> ---- ----- --------------------------
--> T1          TIMESTAMP(6) WITH TIME ZONE
--> T2          TIMESTAMP(6) WITH TIME ZONE
```

The `TIMESTAMP WITH TIME ZONE` data type clearly specifies the original client-side value delimited by the time zone element as part of the definition. It was introduced in Oracle 9i Release 1. Before that, values had to be modeled by the `TIMESTAMP` data type, which became obsolete over the decades and could not treat time zone reflection at all. Thus, nowadays, it is treated as non-robust, covering only local environment dimensions. The next section deals with the `FROM_TZ` function, extending the `TIMESTAMP` value with the time zone. It shows the principle of transforming pure `TIMESTAMP` into `TIMESTAMP WITH TIME ZONE`.

Extending the TIMESTAMP value with the time zone using the FROM_TZ function

Reflecting the historical evolution, original `TIMESTAMP` data values can now be extended by the time zone definition using the `FROM_TZ` function, which accepts two parameters – the `TIMESTAMP` value and the `TIME ZONE` dimension.

The syntax of the FROM_TZ function is shown in the following snippet, taking two parameters:

```
FROM_TZ( <timestamp_value>, <time_zone_value> )
```

time_zone_value is a character string in the format TZH:TZM or the character string of the valid time zone region (TZR).

Now, you are familiar with the syntax, so let's move on to the practical usage and references. They are shown in the following code block:

```
select FROM_TZ(TIMESTAMP '2022-04-28 9:40:05', '5:00')
 from dual;
--> 28.04.2022 09:40:05,000000000 +05:00
select FROM_TZ(to_timestamp ('2022-04-28 9:40:05',
                             'YYYY-MM-DD HH:MI:SS'),
            'Pacific/Honolulu')
 from dual;
--> 28.04.22 09:40:05,000000000 PACIFIC/HONOLULU
```

Using the following command, a new table is created (timetab2), consisting of two attributes (t1 and t2). They are defined based on the inner query. The t1 definition originates from the FROM_TZ function call, enclosing the TIMESTAMP constructor. It provides the TIMESTAMP WITH TIME ZONE data type. Conversely, the second attribute (t2) uses a pure TIMESTAMP constructor, resulting in obtaining the TIMESTAMP data type format:

```
create table timetab2 as
  select
    FROM_TZ(TIMESTAMP '2000-03-28 08:00:00', '3:00') t1,
    TIMESTAMP '2000-03-28 08:00:00' t2
  from dual;
```

The following code snippet shows the structure of the previously created table and the data types of the attributes:

```
desc timetab2;
--> Name Null? Type
--> ---- ----- ---------------------------
--> T1          TIMESTAMP(9) WITH TIME ZONE
--> T2          TIMESTAMP(9)
```

Thus, the TIMESTAMP constructor does not manage time zones at all. However, by using the FROM_TZ function call, a particular reference can be made. As is evident, FROM_TZ takes the TIMESTAMP value as a first parameter and extends it by the time zone:

```
--> T1:          28.03.00 08:00:00,000000000 +03:00
--> T2:          28.03.00 08:00:00,000000000
```

Another solution dealing with the TIMESTAMP value and time zone reflection can be carried out using the AT TIME ZONE clause, which is described in the next section.

Using the AT TIME ZONE clause to reflect the time zone shift

Dynamic transformation across time zones can be done with the AT TIME ZONE keyword, extending the TIMESTAMP (or TIMESTAMP WITH TIME ZONE) format:

```
select current_timestamp,
       current_timestamp AT TIME ZONE 'Australia/Sydney'
 from dual;
--> CURRENT_TIMESTAMP
-->            CURRENT_TIMESTAMPAZTIMEZONE'AUSTRALIA/SYDNEY'
--> 24.01.22 12:32:52,896000000 EUROPE/PRAGUE
-->            24.01.22 22:32:52,896000000 AUSTRALIA/SYDNEY
```

The time zone can be specified by the name, or the hour:minute format can be used:

```
select current_timestamp AT TIME ZONE 'Europe/Bratislava',
       current_timestamp AT TIME ZONE '3:00'
 from dual;
```

The original value is transformed into the value of the particular time zone. Thus, the hour and minute elements can generally be changed. The AT TIME ZONE extension provides the TIMESTAMP WITH TIME ZONE value as a result.

The AT TIME ZONE keyword can only be used for TIMESTAMP types. It is impossible to associate it with the DATE value:

```
select current_date AT TIME ZONE 'US/Eastern' from dual;
--> ORA-30084: "invalid data type for datetime primary
--> with time zone modifier"
```

The workaround is to transform the DATE format into TIMESTAMP using explicit conversion, TO_TIMESTAMP or CAST:

```
select CAST(DATE '2014-04-08' as TIMESTAMP) d1,
       CAST(DATE '2014-04-08' as TIMESTAMP)
                              AT TIME ZONE 'US/Easter' d2,
       CAST(DATE '2014-04-08' as TIMESTAMP)
                              AT TIME ZONE '3:00' d3
  from dual;
```

Note that in the preceding example, the time is not defined. We just specified the year, month, and day elements by using the *ISO DATE constructor*. Thus, time elements reflect the 00:00:00 value.

A more complex solution is shown in the following code snippet. In the first phase, the input value is cast as TIMESTAMP, followed by the time zone extension definition. As a result, TIMESTAMP WITH TIME ZONE is provided, finally transformed to the Berlin time zone using the AT TIME ZONE keyword:

```
select FROM_TZ(CAST(sysdate as TIMESTAMP), '5:00')
              AT TIME ZONE 'Europe/Brussels'
  from dual;
```

Let's process it step by step. sysdate provides DATE elements, such as 08.03.2022 15:15:17. Then, it is transformed to the TIMESTAMP format by adding second fractions: 08.03.22 15:15:17,000000000. Then, the time zone is added, calling the FROM_TZ function, resulting in the following value: 08.03.22 15:17:15,000000000 +05:00. Finally, the AT TIME ZONE clause is used, resulting in this output: 08.03.22 11:17:55,000000000 EUROPE/BRUSSELS.

For the time zone reference and calculation, FROM_TZ has already been introduced, which is associated with the TIMESTAMP value, while NEW_TIME functions are accessible that reference the DATE value. In the next section, the syntax and usage of the NEW_TIME function are presented.

Transforming DATE values across multiple time zones using the NEW_TIME function

The NEW_TIME function transforms the DATE value using input_timezone (time zone of the source) and output_timezone (time zone of the destination). It returns the DATE value, regardless of the implicit conversion of the input. All parameters need to be specified. If the original DATE value provided is NULL, then the output is NULL as well:

```
NEW_TIME(<p_date>, <input_time_zone>, <output_timezone>)
```

The limitation of the NEW_TIME method is associated with the list of available time zones that can be referenced. The following table shows a list of available values:

Time zone value reference	Meaning
AST, ADT	Atlantic Standard or Daylight Time
BST, BDT	Bering Standard or Daylight Time
CST, CDT	Central Standard or Daylight Time
EST, EDT	Eastern Standard or Daylight Time
GMT	Greenwich Mean Time
HST, HDT	Alaska-Hawaii Standard Time or Daylight Time
MST, MDT	Mountain Standard or Daylight Time
NST	Newfoundland Standard Time
PST, PDT	Pacific Standard or Daylight Time
YST, YDT	Yukon Standard or Daylight Time

Table 15.3 – List of available time zones for the NEW_TIME reference

We recommend using the 24-hour format or highlighting morning and evening references when writing code. This ensures time values can be precisely interpreted.

The following query highlights the transformation using the NEW_TIME function call:

```
select
   TO_DATE('15-12-2015 03:23:45',  'DD-MM-YYYY HH24:MI:SS')
                    "Original date and time",
   NEW_TIME(TO_DATE('15-12-2015 03:23:45    ',
                    'DD-MM-YYYY HH24:MI:SS'),
            'AST', 'PST')
                    "New date and time"
   from dual;
```

The NEW_TIME function accepts only US time zones in a textual format. It is impossible to reference TZH, TZM, or TZR without getting an exception:

```
select
   NEW_TIME(TO_DATE('15-12-2015 03:23:45',
```

```
                            'DD-MM-YYYY HH24:MI:SS'),
            '05:00', '07:00') "New date and time"
  from dual;
--> ORA-01857: not a valid time zone
```

The workaround is based on using the FROM_TZ function call dealing with the input time zone, followed by using the AT TIME ZONE clause.

Let's combine the FROM_TZ function and the AT TIME ZONE clause to provide a new function definition for converting time zones. The next section defines how to do that by introducing *function code* that takes the original and destination time zones as input values. The output is the original value with the shift applied.

Converting time zones

The following code shows the implementation of similar functionality to NEW_TIME, but in a general manner. It takes DATE as input, which is transformed to the TIMESTAMP WITH TIME ZONE value, extended by the input time zone. Then, the output time zone transformation is done using the AT TIME ZONE clause. Finally, TIMESTAMP WITH TIME ZONE is transformed to provide the DATE value:

```
create or replace function CONVERT_DATE_TIMEZONES
 (p_date DATE,
  input_timezone varchar,
  output_timezone varchar
  )
 return date
   is
     v_timestamp TIMESTAMP WITH TIME ZONE;
begin
    v_timestamp:=FROM_TZ(CAST(p_date AS TIMESTAMP),
                         input_timezone);
    return CAST(v_timestamp AT TIME ZONE output_timezone
               as DATE);
end;
/
```

The complete code of the CONVERT_DATE_TIMEZONES function can be downloaded from the GitHub repo of this chapter.

To highlight the principles, the following `select` statement gets `sysdate` as the first column. The second column extends the `sysdate` value by the input time zone. The third column takes the proposed function:

```
select sysdate "Original value",
        FROM_TZ(CAST (sysdate as TIMESTAMP), '05:00')
                "Value extended by the input time zone",
        CONVERT_DATE_TIMEZONES(sysdate, '05:00', '07:00')
                "Output"
 from dual;
--> 10.03.2022 08:38:33
--> 10.03.2022 08:38:33,000000000 +05:00
--> 10.03.2022 10:38:33
```

You know how to manage shift. However, what about the reference time? How do we normalize time? You have probably already guessed that the **Universal Coordinated Time** (**UTC**) format is the right solution. In the next section, we will provide you with a theoretical guide on how to reference time using the UTC format, and then demonstrate how to reference the `SYS_EXTRACT_UTC` function to get the proper value normalized to UTC.

TIMESTAMP normalization

In the previous section, there was an analysis of time zone management by transforming values across multiple zones or by adding a time zone definition from the pure `TIMESTAMP` value origin.

The world is divided into several time zones that differ by hour or half an hour intervals. Standard time is calculated by the number of hours offset from UTC. UTC provides the precise time secured by multiple laboratories (such as the US Naval Observatory) to serve as a referential value. UTC is a successor of GMT, in that it has taken over as the international standard time where GMT was previously used. Thus, UTC is a referential value to which individual regions apply defined time zones.

There are some examples of shift in the following table. Besides the time zone management itself, a difference related to summer and winter time can be identified as well:

Type	Representation
Eastern Daylight Time	Subtraction of 4 hours from UTC
Central Standard Time	Addition of 9:30 hours to UTC
Central Summer Time	Addition of 10:30 hours to UTC

Type	Representation
Eastern European Time	Addition of 2 hours to UTC
Eastern European Summer Time	Addition of 3 hours to UTC

Table 15.4 – UTC shifts

Let's say you live in Houston, Texas, which uses United States CST. To convert 18:00 UTC (6:00 p.m.) into your local time, subtract 6 hours to get 12 noon CST. During daylight saving (summer) time, you would only remove 5 hours, so 18:00 UTC would convert to 1:00 p.m. CDT. Note that the US uses a 12-hour format with a.m. and p.m.

Other countries use 24-hour time. So, let's say you're in Paris, France, which is in **Central European Time (CET)**. To convert 18:00 UTC into your local time, add 1 hour to get 19:00 CET. In summer, add 2 hours to get 20:00 **Central European Summer Time (CEST)**. CEST is the standard used in the summer time in European countries, which apply CET during the rest of the year. CET is expressed as UTC + 01:00.

The next section deals with the SYS_EXTRACT_UTC function, by which the UTC reference can be obtained.

Extracting UTC

UTC does not need to cover the time zone, because it works as a reference. Therefore, the TIMESTAMP data type value is provided by extracting the UTC value from the specified value, which is relevant for such usage. UTC time extraction can be done by using the SYS_EXTRACT_UTC function:

```
alter session set time_zone='3:00';
select current_timestamp,
       SYS_EXTRACT_UTC(current_timestamp)
  from dual;
--> CURRENT_TIMESTAMP
-->          SYS_EXTRACT_UTC(CURRENT_TIMESTAMP)
--> 08.03.22 16:51:05,593000000 +03:00
-->          08.03.22 13:51:05,593000000
```

Similarly, the SYS_EXTRACT_UTC function does not work for DATE data types directly. To use it, the TIMESTAMP transformation must be carried out first; otherwise, the following exception is raised:

```
ORA-30175: invalid type given for an argument
```

Thus, to limit the preceding exception, the DATE value is cast as TIMESTAMP, and then the SYS_
EXTRACT_UTC function can be used. A practical example is shown in the following snippet:

```
select SYS_EXTRACT_UTC(FROM_TZ(CAST(sysdate AS TIMESTAMP),
                         '3:00')),
   SYS_EXTRACT_UTC(FROM_TZ(CAST(current_date AS TIMESTAMP),
                         '3:00'))
 from dual;
--> 08.03.22 12:24:52,000000000
--> 08.03.22 14:24:52,000000000
```

Alternatively, the AT TIME ZONE keyword clause can be used, referencing UTC:

```
select systimestamp AT TIME ZONE 'UTC' from dual;
```

The difference between TIMESTAMP WITH TIME ZONE and UTC reflection is the time zone offset.
For the particular region, the time zone offset can be obtained by the TZ_OFFSET function, which
takes the time zone name and provides the hour and minute shift:

```
select TZ_OFFSET('Europe/Paris')
 from dual;
--> TZ_OFFSET('EUROPE/PA'IS')
--> +01:00
```

Technically, the input can be extended by the time zone definition in an hour-and-minute format.
However, there's not much sense in doing this, since the same value is received as output:

```
select TZ_OFFSET('3:00')
 from dual;
--> TZ_OFFSET('3:00')
--> +03:00
```

The syntax of the TZ_OFFSET function is as follows:

```
select TZ_OFFSET({ '<time_zone_name>'
                  | '{ + | - } <hh> : <mi>'
                  | SESSIONTIMEZONE
                  | DBTIMEZONE
                }
                )
 from dual;
```

Reflecting the syntax, the input parameter value can be specified either explicitly or by using a database or session reference.

The TZ_OFFSET function always provides the hour and minute format, not the name of the time zone region:

```
alter session set time_zone='Europe/Vienna';
select
     SESSIONTIMEZONE,
     TZ_OFFSET(SESSIONTIMEZONE)
 from dual;
--> SESSIONTIMEZONE     TZ_OFFSET(SESSIONTIMEZONE)
--> Europe/Vienna       +01:00
```

This example sets the time zone to UTC using the 00:00 value for the hour and minute shift:

```
alter session set time_zone='+00:00';
select
     SESSIONTIMEZONE,
     TZ_OFFSET(SESSIONTIMEZONE)
 from dual;
--> SESSIONTIMEZONE     TZ_OFFSET(SESSIONTIMEZONE)
--> +00:00              +00:00
```

This example sets the time zone to UTC explicitly:

```
alter session set time_zone='+UTC';
select
     SESSIONTIMEZONE,
     TZ_OFFSET(SESSIONTIMEZONE)
 from dual;
--> SESSIONTIMEZONE     TZ_OFFSET(SESSIONTIMEZONE)
--> UTC                 +00:00
```

The UTC value is typically used for the server reference. The next section walks you through the local value reflection principles by pointing out the TIMESTAMP data type extensions.

Local value reflection using TIMESTAMP data type extensions

To cover the complexity of time zone management, local timestamp reflection should also be highlighted. This section introduces data types forming the TIMESTAMP type extension. In principle, three TIMESTAMP data types are available:

- TIMESTAMP does not cover time zone elements. It is mostly used for backward compatibility or values that do not need to reference time zones at all.

- TIMESTAMP WITH TIME ZONE maps the time zone to the value by the offset extension. Using this data type, particular values can be time zone-shifted and coordinated across regions. Thus, it is normalized across the database.

- TIMESTAMP WITH LOCAL TIME ZONE is a specific type that does not state the time zone explicitly. Instead, the transformation is applied directly. Thus, the original value of the hour and minute elements is recalculated and provided to the user depending on their time zone.

Let's create a simple table combining all three data types. Then, we'll reference the same value in three different time zones by altering the session to the 3-hour offset:

```
create table timetab(t1 TIMESTAMP,
                     t2 TIMESTAMP WITH TIME ZONE,
                     t3 TIMESTAMP WITH LOCAL TIME ZONE);
alter session set time_zone='3:00';
```

Insert one row with the same timestamp value. In our case, it is provided by the character string converted using the TO_TIMESTAMP method:

```
insert into timetab
 values(TO_TIMESTAMP('11.1.2022 15:24:12.4',
                     'DD.MM.YYYY HH24:MI:SS.FF'),
        TO_TIMESTAMP('11.1.2022 15:24:12.4',
                     'DD.MM.YYYY HH24:MI:SS.FF'),
        TO_TIMESTAMP('11.1.2022 15:24:12.4',
                     'DD.MM.YYYY HH24:MI:SS.FF'));
```

Change the time zone to a different value to focus on the session reflection:

```
alter session set time_zone='5:00';
```

Obtain the content of the table with the `select` statement:

```
select * from timetab;
--> T1:     11.01.22 15:24:12,400000000
--> T2:     11.01.22 15:24:12,400000000 +03:00
--> T3:     11.01.22 17:24:12,400000000
```

As you can see, the `t1` attribute is pure `TIMESTAMP` and does not reflect the time zone shift. Simply put, there is no information about the time zone change! Consequently, this data type is not suitable for usage across multiple regions. The user must take care of time zone management on their own. This can cause problems, mostly related to improper data management.

As you can see, when changing the time zone, the original value is still provided. Therefore, the user must handle the transformation explicitly to ensure the data is correct and reliable. The `t2` attribute applies the time zone directly to the definition. The output value (respecting the session format) is `11.01.22 15:24:12,400000000 +03:00`. Thus, the original value that was inserted is obtained, but it is enclosed by the time zone definition. As already described, recalculation to the session perspective can be done through various techniques, for example, by using the `AT TIME ZONE` keyword extension clause, as shown in the following code snippet. The transformed value defines the time `16:24` instead of the original `14:24`, but the time zone shift from `+03:00` to `+05:00` is specified:

```
select t2, t2 AT TIME ZONE '5:00' from timetab;
--> T2:                    11.01.22 15:24:12,400000000 +03:00
--> T2ATTIMEZONE'5:00': 11.01.22 17:24:12,400000000 +05:00
```

The local time zone does not need to be specified explicitly by the hour and minute elements. The session time zone can be directly referenced using `SESSIONTIMEZONE` by proposing a dynamic query respecting the current settings of the time zone for the particular session:

```
select t2, t2 AT TIME ZONE sessiontimezone from timetab;
```

The last attribute (`t3`) does not use database value normalization. Instead, the client time zone is directly applied by calculating the shift. To just see the change, reference the database and session time zones. The difference is then applied directly to the value:

```
select DBTIMEZONE, SESSIONTIMEZONE from dual;
--> DBTIMEZONE     SESSIONTIMEZONE
--> +00:00         +03:00
```

Therefore, `TIMESTAMP WITH LOCAL TIMESTAMP` does not define an additional element of the time zone. Instead, it is applied to the input value:

```
select t3 from timetab;
--> 11.01.22 11:24:12,400000000
```

Note that Oracle recommends setting the database time zone to UTC (+00:00). Then, the transformation of the value to the client's perspective is straightforward, whereas the session time zone can be directly applied to the transformation to get the time for the particular client:

```
select t3, t3 AT TIME ZONE sessiontimezone from timetab;
--> 11.01.22 11:24:12,400000000
--> 11.01.22 14:24:12,400000000 +03:00
```

The importance of time zone management is strongly visible in the case of event synchronization across countries. Nowadays, online conferences and virtual meetings are widespread. Therefore, proper time zones must be applied to ensure that all the participants get the appropriate timestamp value, corresponding to their time zones and regional parameter. Simply put, it is easy to synchronize and plan meetings with the obtained values.

Let's look at a quick example of wrapping up the processing of timestamp synchronization across several sessions with different time zones. We will use a new table consisting of the row identifier (`id`) and three `TIMESTAMP` attributes with different characteristics to focus on time zone management:

- `t1` is a pure timestamp format (`TIMESTAMP`)
- `t2` covers the time zone definition as a data type extension (`TIMESTAMP WITH TIME ZONE`)
- `t3` uses local timestamp management (`TIMESTAMP WITH LOCAL TIME ZONE`)

The corresponding table definition is shown in the following code snippet dealing with three `TIMESTAMP` data type formats:

```
create table timetab5(id integer,
                      t1 TIMESTAMP,
                      t2 TIMESTAMP WITH TIME ZONE,
                      t3 TIMESTAMP WITH LOCAL TIME ZONE);
```

Then, create two sessions with different time zones and evaluate the provided data operated by the `insert` and `select` statements. The `t1` attribute value does not evolve with the time zone change, and the value always remains the same, resulting in inaccurate data. `TIMESTAMP WITH TIME ZONE` uses the time zone definition as an extension. Although the core value remains the same, there is also a time zone reflection to be covered. Thanks to that, individual offset reflection can be done easily in a reliable manner.

Finally, there is an attribute, `t3`, dealing with `TIMESTAMP WITH LOCAL TIME ZONE`. The value itself is transformed according to the session definition. As can be seen, by shifting the time zone value from `3:00` to `5:00`, a particular `t3` value is recalculated by applying output session time zone settings. So, for session 2, the time value will be transformed from `10-JAN-22 03.24.12.400000 PM` to `10-JAN-22 05.24.12.400000 PM`. Note that the output format of the timestamp representation depends on the NLS parameters. In our case, time elements are dot-separated, using the American 12-hour format. The original `15:24` value (`hour:minute`) is implicitly converted to `03.24 PM` as a character string.

Table 15.5 shows the previously described principles practically. There are two sessions with different time zones. While studying the table, focus on individual time zones, their characteristics, and the way `TIMESTAMP` value transformation works between sessions:

SESSION 1	SESSION 2
alter session set time_zone='3:00';	

insert into timetab5
 	
 values(1,
 	
        TO_TIMESTAMP
 	
('10.1.2022 15:24:12.4',
 	
'DD.MM.YYYY HH24:MI:SS.FF'),
 	
        TO_TIMESTAMP
 	
('10.1.2022 15:24:12.4',
 	
'DD.MM.YYYY HH24:MI:SS.FF'),
 	
        TO_TIMESTAMP
 	
('10.1.2022 15:24:12.4',
 	
'DD.MM.YYYY HH24:MI:SS.FF'));
 	
commit;	

SESSION 1	SESSION 2

`select * from timetab5 where id=1;`
 	
`/*`
 	
`10-JAN-22 03.24.12.400000 PM`
 	
`10-JAN-22 03.24.12.400000 PM +03:00`
 	
`10-JAN-22 03.24.12.400000 PM`
 	
`*/`	

	`alter session`
	
	`set time_zone='5:00';`

	`insert into timetab5`
	` values(2,`
	`        TO_TIMESTAMP`
	`('10.1.2022 15:24:12.4',`
	`'DD.MM.YYYY HH24:MI:SS.FF'),`
	`        TO_TIMESTAMP`
	`('10.1.2022 15:24:12.4',`
	`'DD.MM.YYYY HH24:MI:SS.FF'),`
	`        TO_TIMESTAMP`
	`('10.1.2022 15:24:12.4',`
	`'DD.MM.YYYY HH24:MI:SS.FF'));`
	`commit;`

SESSION 1	SESSION 2

	`select * from timetab5`
	
	`where id=2;`
	
	`/*`
	
	`10-JAN-22 03.24.12.400000 PM`
	
	`10-JAN-22 03.24.12.400000 PM`
	`+05:00`
	
	`10-JAN-22 03.24.12.400000 PM`
	
	`*/`

`select * from timetab5`
 	
`where id=2;`
 	
`/*`
 	
`10-JAN-22 03.24.12.400000 PM`
 	
`10-JAN-22 03.24.12.400000 PM`
`+05:00`
 	
`10-JAN-22 01.24.12.400000 PM`
 	
`*/`	

	`select * from timetab5`
	
	`where id=1;`
	
	`/*`
	
	`10-JAN-22 03.24.12.400000 PM`
	
	`10-JAN-22 03.24.12.400000 PM`
	`+03:00`
	
	`10-JAN-22 05.24.12.400000 PM`
	
	`*/`

Table 15.5 – Managing time across sessions

Be aware! There is a bug in older versions of SQL Developer, providing incorrect data of the time zone reflection for local management. However, it works well for SQL clients and SQL Developer Release 20 and newer.

Before concluding this chapter, we will provide one final section to reflect on the implications of time zone management. In this example, our goal is to get sales information for the last hour. How will the solution work for a local business? And how will it need to be changed if we need to process multiple regions and time zones? Think about the consequences.

Local versus global expansion

Let's use the example of sales management to highlight the importance of proper time zone application. A shop in a small town has just one cash desk consisting of a computer, which is also the central data repository. Trade prospers, sales increase, and the business needs to be expanded. The owner rents another space, and the number of cash desks has increased. So, the local computer is no longer enough.

The owner needs a server to be located in their office. However, the business is still thriving, and therefore the owner opens more and more shops in the city. Over time, operations have sprung up across the country. Although the server is still located in a local office room, it is clear that such a solution is not robust, and the approach must be changed. As a result, data is migrated to the local server room operated by an external company supervising the architecture, accessibility, and hardware. This decision does not bring any significant changes in terms of the business and information system. The shop just pays a significant amount to an external administrator to ensure system and data availability. It is simply too expensive and not very reliable.

After a few years, the company has operated in various parts of the world. The local server room is no longer suitable due to security reasons, robustness, and resistance. It is abundantly clear that unlimited data and information system accessibility has become a critical element. So, how do they ensure that? Sure, a cloud account has been provisioned, and the owner is happy and satisfied. Reports, statistics, and analytical modules work well.

However, when spreading the business worldwide, a significant question arises pointing to analytics. How do we get *sales in the last hour*? Is the following code suitable? Refer to the local times in *Figure 15.1*. Think of the consequences of using the sysdate function call and offer a better solution.

The following code shows a command attempting to get the orders created during the last hour, expressed by the subtraction of the 1/24 value from sysdate. Is this solution correct? If not, rewrite it:

```
select *
  from orders
    where order_date > sysdate - 1/24;
```

The core of the problem is shown in the following figure. Individual cities have different time zones, so when we look at their actual time, it is obvious that the last hour must be bound to a local value, not a database on the server (in the cloud), right? So, we have to change the preceding query to obtain valid outputs?

Figure 15.1 – Current time across the world

Well, several solutions can exist. In principle, the client-side value must be referenced, either by getting `current_date` (or `current_timestamp`) or by transforming the server (database) `Date` and `Time` values by extending and applying the time zone shift to the client's perspective.

One way or another, simply having a database value without transformation and normalization does not provide sufficient power in a multi-regional environment. Keep this in mind when designing your applications and systems. From a functionality point of view, the solution is syntactically correct, but the expected data would not be obtained.

Summary

The topics covered in this chapter are particularly important in the context of business expansion, in which it is necessary to synchronize events across multiple regions and, in general, the entire world. In this chapter, you learned about data types and synchronization techniques. You also learned about the complexity of the local and database date and time, focusing on time zone shift.

Additionally, in this chapter, the `TIMESTAMP` data type extensions were highlighted, preceded by the `FROM_TZ` and `NEW_TIME` functions. Time zone reflection and shift processing are critical. You probably wouldn't like to arrive at an online meeting an hour after it ended, right? So, now, you are prepared to trade worldwide and offer your solutions anywhere, as well as organize webinars and meetings with participants from any region.

In this chapter, session and database date and time were discussed, dealing with the `sysdate` and `current_date` differences, as well as `systimestamp` and `current_timestamp`. However, what about a situation where your application is built on a `sysdate` or `systimestamp` reference and you are unable to recompile it immediately? How do you correctly apply time zones? How do you differentiate between server- and client-side regions? And, naturally, how do you migrate such a system to the cloud? How do you use it in multiple regions? Is it even possible? Do you really need to rewrite the whole code? These are the questions you will get the answers to in the next chapter, which focuses on the Oracle Cloud environment and dynamic SQL transformation using translation profiles.

Questions

1. How do we get the client time zone? Which function can be used?

    A. The SESSIONTIMEZONE function of the STANDARD package

    B. The MYTIMEZONE function of the STANDARD package

    C. The UTC_SHIFT function of the STANDARD package

    D. The DBTIMEZONE function of the STANDARD package

2. Choose the best option specifying the database time zone:

    A. TZH only

    B. TZH:TZM only

    C. Name representation only

    D. TZH:TZM and name representation

3. What data type is used for the column of the following select statement?

    ```
    select FROM_TZ(cast(sysdate as TIMESTAMP), '5:00')
                   AT TIME ZONE 'Europe/Brussels'
      from dual;
    ```

    A. DATE

    B. TIMESTAMP

    C. TIMESTAMP WITH TIME ZONE

    D. TIMESTAMP WITH LOCAL TIME ZONE

4. Which function takes three parameters – input timestamp values and the original and destination time zone to be calculated – applying the time shift for the input value?

    A. FROM_TZ

    B. NEW_TIME

    C. TIMESTAMP

    D. AT TIME ZONE

5. Which function gets the UTC normalized value from the TIMESTAMP value?

    A. UTC

    B. FROM_TZ

    C. NEW_UTC_TIME

    D. SYS_EXTRACT_UTC

Further reading

- Refer to *DB System Time Zone* in the Oracle documentation. It focuses on the time zone options in OCI. It also provides a step-by-step guide to changing the time zone after database provisioning. It is available through this link: https://docs.oracle.com/en-us/iaas/Content/Database/References/timezones.htm.

- Check out *Date and time localization* by Shanika Wickramasinghe. This paper provides a date and time localization guide in various languages, including Python, Java, and JavaScript. It is recommended to gain insights into the principles and usage in other systems: https://lokalise.com/blog/date-time-localization/.

16
Oracle Cloud Time Zone Reflection

In the past, overall system complexity and performance were handled by the administrator. Date and time reflection was covered by the server, which ensured it was properly managed irrespective of the session definitions. As a result, DATE and TIMESTAMP values were obtained by the sysdate and systimestamp functions, respectively. With globalization, this is no longer the case. The server time zones cannot manage data because different regions and associated time zones must be taken into consideration. Therefore, session perspective reflection should be used.

Moreover, another problem can be if customers want to move their databases to a cloud environment. We often get questions related to the process of moving databases to the cloud. Time management suddenly becomes difficult because there is no proper reference to the time points. It is not enough for the server to use UTC as its reference; time zones must be properly set and evaluated. Dedicated virtual machines in the cloud can solve the problem only partially. Although the time zone can be directly set depending on the region, it removes the main benefits of the cloud by requiring a developer or administrator to look after the resources. This can consume a lot of time and resources.

In this chapter, we will first briefly point to the time zone reflection covered by the system and client perspectives, and then a transformation solution is proposed and discussed. In principle, dynamic translation profiles are used to *rewrite* statements on the fly, which brings additional demands. Furthermore, another perspective has been recently introduced at the autonomous database level – the SYSDATE_AT_DBTIMEZONE parameter, which simplifies the activity only to the correct parameter setting. Thanks to that, the migration of non-UTC databases to the cloud environment is now straightforward, reliable, and performance effective.

In this chapter, we're going to cover the following main topics:

- Summarizing client and server date and time values
- Implementing *translation profiles* to reference client perspectives on the fly without needing to rewrite and rebuild source code and complex solutions

- Using the `SYSDATE_AT_DBTIMEZONE` parameter for automated transformation in the Oracle Cloud environment

Note that the source code can be found in the GitHub repository, accessible at `https://github.com/PacktPublishing/Developing-Robust-Date-and-Time-Oriented-Applications-in-Oracle-Cloud/blob/960e892ada2c4ff3247f4967ca9b3d866e0f1c4f/chapter%2016/chapter%2016.sql`. You may also access the repository by scanning the following QR code:

Summarizing time zone management

Nowadays, data is no longer placed and processed locally. Instead, cloud technologies are used. Oracle provides many cloud locations allowing the provisioning of databases, infrastructure, and many other resources. The number of cloud locations is continuously rising, offering you a wide range of technologies and parameters. But the greatest advantage is, you do not need to hire system administrators. You do not need to manage and optimize the structure. You do not need to maintain hardware. The cloud vendors supervise all the infrastructure, management, patching, and global availability. All data is encrypted, secured, and duplicated, so robustness and reliability can be ensured. Thanks to that, more and more systems are migrated to the cloud to minimize costs and maximize performance. In terms of time management, different temporal aspects must be highlighted. Namely, there is the region where the cloud data center is located. There is also the UTC reference, which, as discussed, provides a general time reference. And on top of that, individual users (that is, the clients) can be spread across multiple regions in different time zones. To ensure the data is correct, all these date and time aspects must be taken into consideration during data retrieval and processing; otherwise, there might be errors in the synchronization and results.

Cloud-managed database services generally reference the UTC time zone in their databases, which can generally have two consequences:

- The `sysdate` and `systimestamp` functions always relate the output to UTC no matter what the database time zone is set to. So, even if the database is set to Tokyo (Asia), the provided values and overall management always come back as UTC. On the contrary, `current_date` and `current_timestamp` respect the database time zone:

  - `sysdate`: Date/time on server

  - `current_date`: Date/time on client

- Historically, applications referenced `sysdate` and `systimestamp` instead of `current_date` and `current_timestamp`. The main reason was mostly related to the output reliability, by which the time zone was ensured even if the client set an incorrect time zone value reference. When moving to the cloud and expanding the system across regions, a session reference should mostly be used. This means that `sysdate` should be replaced by `current_date`, and `current_timestamp` should replace `systimestamp`. But how? Be aware that rebuilding the application can be very demanding and cannot be done quickly. It is necessary to refactor the code and then carry out the whole approval process. This might even be impossible if the source code is not available, or you just do not have access to the sources.

Let's set the database time zone to Tokyo (Asia), reflecting the +09:00 shift:

```
alter database set time_zone = '+09:00';

select DBTIMEZONE from dual;
--> +00:00
```

As evident, the change to the database time zone is not applied straight away. Because a static parameter has been used, the instance needs to be restarted. After the restart, the time zone that we specified in the previous command is used:

```
select DBTIMEZONE from dual;
--> +09:00
```

Let's set the client time zone to the Brussels region (*Figure 16.1*). This can be done with an explicit hour reference or by using the time zone region naming convention:

```
alter session set time_zone='Europe/Brussels';
alter session set time_zone='00:00';
```

Time in Brussels (Belgium) vs. Tokyo (Japan)
Tokyo time is 9:00 hours ahead of Brussels.

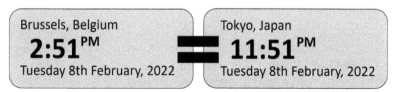

Figure 16.1 – Time zone management

Now, evaluate the difference between the server and the client by using the following function calls:

```
select sysdate, current_date from dual;
--> 08.02.2022 23:51:12
--> 08.02.2022 14:51:12
select systimestamp, current_timestamp from dual;
--> 08.02.22 23:53:37,020685000 +09:00
--> 08.02.22 14:53:37,020687000 GMT
```

However, how do we get the `current_date` value by calling `sysdate`? Why is it important? Well, existing applications may refer to the server time as a result of local processing. But if we want to move the application to the cloud environment and create an independent solution applicable in any time zone, we must take into account the time zone shift and thus reference the client through the `current_date` and `current_timestamp` functions. In the next section, we will show how to make the shift between client and server time zones automatically without needing to rewrite the code using a *SQL translation profile*. The original solution would be too demanding, complicated, and time consuming and needs to be accepted by many managers, developers, testers, and other IT employees. Moreover, it would require complex approval. Sends chills down your spine, doesn't it?

Exploring SQL translation profiles

A **SQL translation profile** can serve as the first solution for *dynamic SQL transformation*. Although it is primarily intended to transform non-Oracle SQL syntax, it can also be used internally to transform date function calls and representations. The `DBMS_SQL_TRANSLATION` package and its methods power the management of the SQL translation profile. Simply, the translation profile is created and given a name (in our case, it is called `DATE_PROF`), and then the query transformation registration is carried out. Multiple statements can be registered for the transformation by using the `REGISTER_SQL_TRANSLATION` procedure of the `DBMS_SQL_TRANSLATION` package.

The `REGISTER_SQL_TRANSLATION` procedure takes three parameters – the name of the SQL translation profile (parameter 1), the original statement (parameter 2), and the transformed statement (parameter 3). In our case, we transform two statements by referencing the `sysdate` and `systimestamp` functions:

```
begin
  -- create profile
  DBMS_SQL_TRANSLATOR.CREATE_PROFILE ('DATE_PROF');
  -- register transformation for the DATE value
  DBMS_SQL_TRANSLATOR.REGISTER_SQL_TRANSLATION
    (profile_name => 'DATE_PROF',
     sql_text=> 'select sysdate from dual',
```

```
      translated_text=>'select current_date from dual');
  -- register transformation for the TIMESTAMP value
  DBMS_SQL_TRANSLATOR.REGISTER_SQL_TRANSLATION
    (profile_name => 'DATE_PROF',
     sql_text=> 'select systimestamp from dual',
     translated_text=>'select current_timestamp from dual');
  end;
  /
```

To apply the profile, its usage must be enabled for the current session:

```
  alter session set SQL_TRANSLATION_PROFILE=DATE_PROF;
```

This solution, however, has many limitations. As evident, individual statements must be explicitly stated, which can not only require a lot of effort from the user but is also risky. Furthermore, the user must specify *all possible queries* that can be executed, referencing the mentioned functions. So, if sysdate is used in other queries (or possibly can be), all of the possible queries must be explicitly listed in the translation profile. That's a lot to code, isn't it? Although this is, in principle, feasible, it allows significant room for errors and, additionally, improper data management and evaluation.

Therefore, different principles should be used to ensure dynamic query translation. Simply, any query referencing the sysdate function should be replaced by current_date and so on. How do we do that?

Developing a package to translate SQL code

A variant of the SQL translation profile can be applied. Let's drop the existing profile and create a specific package. Its name can be specified by the user, but it must take one procedure, TRANSLATE_SQL (with the exact name), which takes two parameters – the original query (input parameter 1) and the translated query (input parameter 2). There is also another procedure for covering errors during the transformation (TRANSLATE_ERROR), but it is optional:

```
  exec DBMS_SQL_TRANSLATOR.DROP_PROFILE ('DATE_PROF');


  create or replace package DATE_TRANSLATOR_PACKAGE is
   procedure TRANSLATE_SQL(sql_text IN clob,
                             translated_text OUT clob);
   procedure TRANSLATE_ERROR(error_code IN binary_integer,
                     translated_code OUT binary_integer,
                     translated_sql_state OUT varchar);
```

```
end;
/
```

The content of the package body is responsible for the dynamic transformation by translating specific function calls. This is commonly done by using regular expressions that replace all occurrences of the function call. The translated query (`translated_text`) is composed by the REGEXP_REPLACE function:

```
create or replace package body DATE_TRANSLATOR_PACKAGE
 is
  procedure TRANSLATE_SQL(sql_text in clob,
                            translated_text out clob)
    is
     begin
       translated_text:=REGEXP_REPLACE(sql_text,
                                'sysdate',
                                'current_date',
                                1,0,'i');
       translated_text:=REGEXP_REPLACE(translated_text,
                                'systimestamp',
                                'current_timestamp',
                                1,0,'i');
     end;
    procedure TRANSLATE_ERROR(error_code IN binary_integer,
                          translated_code OUT binary_integer,
                          translated_sql_state OUT varchar)
    is
     begin
       null;
     end;
 end;
 /
```

The complete code of the DATE_TRANSLATOR_PACKAGE package can be downloaded using this QR code (https://github.com/PacktPublishing/Developing-Robust-Date-and-Time-Oriented-Applications-in-Oracle-Cloud/blob/960e892ada2c4ff3247f4967ca9b3d866e0f1c4f/chapter%2016/DATE_TRANSLATOR_PACKAGE.sql):

If the package specification and body are successfully compiled, this package can be registered as a source of the SQL translation profile. It uses three functions:

- Creating a profile using the CREATE_PROFILE function of the DBMS_SQL_TRANSLATOR package. It takes the profile name as a parameter:

```
dbms_SQL_TRANSLATOR.CREATE_PROFILE
    (profile_name => 'DATE_PROF');
```

- Setting attributes delimiting the transformation using SET_ATTRIBUTE. It requires at least three parameter values – the name of the profile (PROFILE_NAME) and the attribute name (ATTRIBUTE_NAME) to be set with a particular value (ATTRIBUTE_VALUE). As stated, translation is primarily used for non-Oracle database query conversion. In our case, however, we want to apply translation to all queries. Therefore, the ATTR_FOREIGN_SQL_SYNTAX attribute must be set to false (ATTR_VALUE_FALSE). It commonly defaults to true:

```
DBMS_SQL_TRANSLATOR.SET_ATTRIBUTE
    (profile_name => 'DATE_PROF',
     attribute_name
        => dbms_sql_translator.ATTR_FOREIGN_SQL_SYNTAX,
     attribute_value
        => dbms_sql_translator.ATTR_VALUE_FALSE);
```

- Registering the profile to the created package using SET_ATTRIBUTE. The defined package is associated with DBMS_SQL_TRANSLATOR.ATTR_TRANSLATOR:

```
DBMS_SQL_TRANSLATOR.SET_ATTRIBUTE
    (profile_name => 'DATE_PROF',
     attribute_name
            => dbms_sql_translator.attr_translator,
     attribute_value => 'DATE_TRANSLATOR');
```

The next section explains how to apply the developed translation profile practically.

Translation profile usage

The application is straightforward; we enable a particular profile specified by its name:

```
alter session set sql_translation_profile=DATE_PROF;
```

Then, the disable operation is applied by setting the translation profile to `null`:

```
alter session set sql_translation_profile=null;
```

The discussed technique using the SQL translation profile provides a universal solution applicable to any database environment. Although it is primarily intended to transform source code using different database systems highlighting individual dialects, it can be robustly used for time zone reference and migration. However, Oracle Cloud promotes a newly introduced parameter – SYSDATE_AT_DBTIMEZONE – with which the whole process can be rapidly simplified. The next section walks you through the implementation details of SYSDATE_AT_DBTIMEZONE.

Simplifying time zone management shift using SYSDATE_AT_DBTIMEZONE

As mentioned, another parameter has been introduced recently – SYSDATE_AT_DBTIMEZONE – which provides the same functionality as the previously described translation profiles. It is associated with the session and converts sysdate references to local client date and time values by specifying the precise difference between the database and client time zones. The following commands show the principles and results:

```
select dbtimezone from dual;
--> +09:00
alter session set SYSDATE_AT_DBTIMEZONE=false;
select sysdate, current_date from dual;
--> SYSDATE              CURRENT_DATE
--> 08.02.2022 14:56:00    08.02.2022 14:56:00
alter session set SYSDATE_AT_DBTIMEZONE=true;
select sysdate, current_date from dual;
--> SYSDATE              CURRENT_DATE
--> 08.02.2022 23:56:10    08.02.2022 14:56:10
```

A similar approach can be used for the TIMESTAMP format as well:

```
alter session set SYSDATE_AT_DBTIMEZONE=false;
select systimestamp, current_timestamp from dual;
--> SYSTIMESTAMP
-->          CURRENT_TIMESTAMP
--> 08.02.22 14:56:36,968599000 GMT
-->          08.02.22 14:56:36,968603000 GMT


alter session set SYSDATE_AT_DBTIMEZONE=true;
select systimestamp, current_timestamp from dual;
--> SYSTIMESTAMP
-->          CURRENT_TIMESTAMP
--> 08.02.22 23:56:48,315516000 +09:00
-->          08.02.22 14:56:48,315518000 GMT
```

In this section, we have been dealing with the Oracle Cloud-introduced SYSDATE_AT_DBTIMEZONE parameter. Compared to pre-existing solutions using SQL translation profiles, the newly provided parameter offers a straightforward solution without needing to develop code transformation functions.

Summary

Global events must take into account different time zone regions to ensure synchronization. Date and time data types can reflect time zones, delimited by the name or hour and minute offset.

Databases are commonly set to UTC using value normalization. However, many systems do not respect these principles and developers need to adjust the values to ensure synchronization. As a result, methods in the original code need to be transformed from the server perspective to the client perspective. This chapter dealt with the date and time value transformation provided by SQL translation profiles, which was explained in the first part of the chapter. This transformation of the original statement to the new one is applied automatically after the user specifies the query.

An alternative solution that provides the same results as using SQL translation profiles is using the SYSDATE_AT_DBTIMEZONE session parameter, which is, however, only concerned with the date and time perspective. Thus, the translation profile does not need to be specified. To use it, a single command simply needs to be run. The rest of the management is done autonomously by the cloud.

With that, we are now at the end of the book. You've come a long way in terms of proper date and time management. I believe that you have gained a large amount of new knowledge that you will use in practice by implementing solutions applicable worldwide to spread your business.

Questions

1. Which command correctly invalidates the translation profile called PROF1 for the session?

    A. `alter session set sql_translation_profile=null;`

    B. `set sql_translation_profile=null;`

    C. `unset sql_translation_profile=PROF1;`

    D. `alter session deregister sql_translation_profile= PROF1;`

2. What will the output of the last query in the following block be if the SYSDATE_AT_DBTIMEZONE parameter value is enabled?

    ```
    alter database set time_zone='-06:00';
      -- database system restart
    alter session set time_zone='+10:00';
    alter session set SYSDATE_AT_DBTIMEZONE=true;
    select (sysdate - current_date)*24 from dual;
    ```

    A. 0

    B. -10

    C. -16

    D. 16

Further reading

Now that we have reached the end of the book, we would like to direct you to further resources to continue your study of the development of data-driven applications. The Oracle Cloud environment offers an ideal solution through the Oracle **Application Express (APEX)** technology. It is a low-code development platform offering the ability to create scalable, robust, secure, and reliable solutions that are deployable in the cloud, making it accessible worldwide. Four books are listed here that will guide you through the world of application development in the Oracle APEX environment:

*   *Oracle APEX Cookbook* by Marcel van der Plas and Michel van Zoest. This book shows you how to create web applications by focusing on the concepts, principles, and tips to develop the solution quickly.

*   *Extending Oracle Application Express with Oracle Cloud Features: A Guide to Enhancing APEX Web Applications with Cloud-Native and Machine Learning Technologies* by Adrian Png and Heli Helskyaho. You will learn about data science, which can be supported at a high level by OCI's features, such as data anomaly detection and machine learning models. It will walk you through APEX application integration, as well as providing a general insight, so the principles are applicable to any modern web developer framework running on the OCI platform.

- *Oracle Application Express: Build Powerful Data-Centric Web Apps with APEX* by Arie Geller and Brian Spendolini. You will learn the fundamentals of development, emphasizing the individual components and techniques with real-world coding examples and best practices.

- *Mastering Apex Programming: A developer's guide to learning advanced techniques and best practices for building robust Salesforce applications* by Paul Battisson and Mike Wheeler. By reading this book, you will understand common coding mistakes in APEX, their impacts, as well as techniques to limit them. Throughout the 16 chapters, there is a focus on typical APEX development mistakes, principles of debugging, trigger management and firing, exception handling, testing, securing the code, utilizing methods, scheduling jobs, and using platform events. There is a dedicated part where the authors deal with performance – limits, profiling, scalability, and application architectures.

Assessments

This section contains the answers to the questions throughout the book. Go ahead and check whether you have got them right.

Chapter 1 – Oracle Cloud Fundamentals

1. D is correct. All the aspects are covered by OCI.

2. B is correct. Analytical databases focus on complex data retrieval, aggregations, and analytical queries. The performance of the retrieval process is ensured by the indexing, which is mostly reflected by the B+ trees and bitmap index types.

3. D is correct. ATPs are defined by small online transactions and are covered by the TP and the higher priority TPURGENT.

4. C is correct. The user is not responsible for the platform or the infrastructure. Applications are located in the cloud, and the whole administration is vendor operated and administered.

5. D is correct. The archiver process takes the online logs and copies them into the archive repository before they are rewritten. Log Writer is responsible for transactions by managing the changes and storing them in the online logs. The Process Monitor background process manages sessions and server-client interconnection. System Monitor is responsible for instance recovery in case of failure.

6. A is correct. The other architectures cannot have 1:1 mapping if we are to ensure robustness, error resilience, and optimal performance.

7. B is correct. A compartment comprises cloud resources with privileges and quotas. An image is a template covering the operating system and other software that's available for direct installation. Availability domains are used to create independent and isolated environments to ensure fault tolerance.

8. D is correct. It holds encryption keys, a list of available connection strings, and a general wallet.

Chapter 2 – Data Loading and Migration Perspectives

1. A is correct. SQL Loader is related to the third-party systems holding the data to be imported. Multiple file types and structures can be used. The essential element in this process is the instruction set defined in a control file, which specifies how to import/load and map the data from the external file, including handling undefined values and other loading instructions.

2. C is correct. Exp and Imp methods are operated on the client side.

3. A is correct. We have used this method to make the log available through Object Storage.

4. C is correct. The pre-authenticated request consists of several elements. B defines the bucket and O defines the object name.

Chapter 3 – Date and Time Standardization Principles

1. A is correct. It takes preferences to the top-level precisions, thus the year is first, followed by the month and day elements.

2. A is correct. T encloses time elements.

3. C is correct. It refers to the Gregorian calendar.

4. D is correct. Unbounded periodicity is expressed by the value -1 after the R designation.

Chapter 4 – Concepts of Temporality

1. A is correct. UTC is used for the normalization reference.

2. C is correct. The client time uses the UTC reference extended by the time zone shift definition, whereas local time applies the time zone directly to the value itself.

3. C is correct. The Gregorian calendar correctly applies the current rules for the leap year definition.

4. If he was 17 the day before yesterday, he must have had his 18th birthday yesterday.

5. In order for him to be 20 next year, he must be 19 this year. So, today must be the beginning of the year – January 1.

6. The timeline is shown in *Figure A.1*:

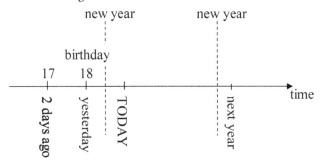

Figure A.1 – Solution: time positions

Chapter 5 – Modeling and Storage Principles

1. C is correct. Microsecond precision is used by default, consisting of six decimal places.

2. B is correct. `DATE` can operate up to second precision, whereas `TIMESTAMP` uses nanosecond precision.

3. C is correct. `DATE` and `TIMESTAMP` cannot hold time zone references. Using `TIMESTAMP` with a local time zone incorporates the shift between the server and client directly into the value itself during retrieval and processing.

4. A is correct. There is a `TIMESTAMP` keyword followed by the element value definition. There is no explicit mapping definition, so the order of elements must apply the predefined format.

5. D is correct. `SYSDATE` and `CURRENT_DATE` provide the `DATE` data type. `systimestamp` provides the database's time zone and `localtimestamp` provides the client side's time.

6. D is correct. Oracle provides the number of days elapsed when two `DATE` values are subtracted.

7. A is correct. Oracle provides `INTERVAL DAY TO SECOND` when two `TIMESTAMP` values are subtracted.

Chapter 6 – Conversion Functions and Element Extraction

1. B is correct. The format is `RM`.

2. A is correct. The `D` and `DDD` formats provide numerical output. `DY` is textual but provides only abbreviations. `DAY` provides the full name.

3. C is correct. By default, the 12-hour format is used, so `HH` and `HH12` are the same. The `HHEurope` format does not exist. The 24-hour format is referenced by `HH24`.

4. D is correct. `DATE`, `TIMESTAMP`, and `INTERVAL` types are permitted values for `EXTRACT` function operations.

Chapter 7 – Date and Time Management Functions

1. A is correct. The query takes the provided date and adds one month, resulting in the last day of February. Since 2023 is not a leap year, February 28, 2023, is the result.

2. A is correct. Any numerical value can be produced, even negative.

3. D is correct. Languages and regions influence the output of the `NEXT_DAY` function. This is why we implement the own function, which is not influenced by the NLS parameter values.

4. B is correct. The `Q` parameter value is used for quarter trimming. `CC` trims based on the whole century.

5. D is correct. The `PERSONAL_ID` value is an interesting concept of person identification; it is formed by the date of birth and gender applied for the month elements.

Chapter 8 – Delving into National Language Support Parameters

1. A is correct. NLS_DATE_FORMAT is a static parameter, and the instance must be restarted to apply the change stored in SPFILE.

2. B is correct. NLS_TERRITORY references the first day of the week, whether it is a Sunday or a Monday.

3. B is correct. NLS_DATE_LANGUAGE can be defined by the third parameter of the TO_CHAR function.

4. D is correct. For the system (database), the NLS_DATABASE_PARAMETERS data dictionary should be referenced. Session-level values can be obtained by the NLS_SESSION_PARAMETERS data dictionary. The other definitions (options B and C) are not valid data dictionary structures.

Chapter 9 – Duration Modeling and Calculations

1. A is correct. Undefined states cannot be covered; each new state automatically marks the end of the validity of the direct predecessor.

2. B is correct. The right-side open characteristics expresses the first timepoint, which is invalid for the duration spectrum.

3. C is correct. DATE value subtraction produces day granularity.

4. C is correct. INTERVAL DAY TO SECOND is produced.

Chapter 10 – Interval Representation and Type Relationships

1. C is correct. The left border remains the same. The right border is extended by one day to March 1 because 2022 is not a leap year.

2. C is correct. The not_fills relationship type does not exist. The occupies type characterizes the common start or end point or the whole coverage. The excludes category covers before and meets types, which do not overlap.

3. C is correct. If there are no undefined states, each beginning point of the validity ends the validity of the direct predecessor. Thus, no gaps can be present.

4. B is correct. The PERIOD definition takes the name specified after the FOR keyword by extending its name with _START or _END. Therefore, if it is specified as PERIOD FOR VALIDITY, then the name of the PERIOD border attributes in this case would be VALIDITY_START or VALIDITY_END.

5. A is correct. In principle, there are three options – ALL, CURRENT, and ASOF, extended by the QUERY_TIME parameter. All other options are logically and syntactically incorrect.

Chapter 11 – Temporal Database Concepts

1. D is correct. The model takes two layers. The first is formed of the primary key and references. Individual object states are temporally oriented and part of the second layer. From the temporal perspective, historical, current, and future valid data can be managed.

2. A is correct. A conventional system stores only current valid data. Any change invokes physical replacement of the original state.

3. C is correct. The aspect of relevance is related to identifying significant changes. Transaction support covers the integrity of transaction rules in a temporal environment. The aspect of correctness relates to extended temporal data integrity, like a temporal ISA hierarchy. The aspect of limited temporal usability is related to the important loss of historical states and future plan management.

4. C is correct. Three dimensions are covered – Insert, Preprocessing, and Load.

5. B is correct. In this case, the best option is to treat each attribute change separately using an attribute-oriented temporal model.

6. A is correct. `data_val` is a hierarchy root for single attributes and temporal group compositions.

Chapter 12 – Building Month Calendars Using SQL and PL/SQL

1. B is correct. It is necessary to remove all elements lower than the month itself. This can be done with a truncating operation. Values can be rounded up. The `first_day` function does not exist and should be user defined. And finally, the `EXTRACT` function gets only one element, not the `DATE` value.

2. C is correct. First, increase the original value (`sysdate`) by one month, then truncate its value. Finally, one day is subtracted, providing midnight of the last day of the month. The `last_day` function keeps the original time elements. `TRUNC(sysdate) + 1` adds one day and removes time elements. The `ROUND` function cannot provide these results.

Chapter 13 – Flashback Management for Reconstructing the Database Image

1. C is correct. The isolation property of the transaction ensures access to the confirmed changes.

2. C is correct. The **System Change Number (SCN)** is Oracle's clock, which is incremented every time a transaction is committed. SCN marks a consistent point in time in the database.

3. B is correct. The `FLASHBACK DATABASE` command can only be launched in `MOUNT` mode.

4. A is correct. Syntactically, the `AS OF SCN` clause is used, followed by the SCN value.

Chapter 14 – Building Reliable Solutions to Avoid SQL Injection

1. We are pleased that you have realized the importance of the correct access and processing of date and time and have applied the principles of reducing the possibility of SQL injection in your solutions.

2. D is correct. You have to specify valid conditions. For option D, the WHERE clause looks like this: employee_id=1 or 1=1.

Chapter 15 – Timestamp Enhancements

1. A is correct. The client time zone is obtained by SESSIONTIMEZONE. The server time zone can be obtained by the DBTIMEZONE function of the STANDARD package. MYTIMEZONE and UTC_SHIFT do not exist in that package.

2. D is correct. The time zone can be stated using the valid name representation or with numerical values using time zone hour (TZH) and minute (TZM).

3. C is correct. The DATE value is transformed into a TIMESTAMP reference and extended by the time zone reference, forming the TIMESTAMP with time zone data type.

4. B is correct. The FROM_TZ function takes two parameters. TIMESTAMP is a constructor, and At time zone is a TIMESTAMP clause extension, not a function. The only valid option is NEW_TIME.

5. D is correct. The valid UTC normalization for the TIMESTAMP value is done by the SYS_EXTRACT_UTC function.

Chapter 16 – Oracle Cloud Time-Zone Reflection

1. A is correct. Option B does not reflect the destination by using ALTER SESSION. The Unset command is not correct (option C). The Deregister clause does not exist either (option D).

2. C is correct. The difference between the database (server) and client time zone is *16 hours*. SYSDATE refers to the server time zone and CURRENT_DATE is the client time. Based on the settings, SYSDATE refers to the smaller value, therefore the output value provided by the query is -16. If the SYSDATE_AT_DBTIMEZONE parameter was set to false, the database time zone would not be applied, so the difference would be 10 hours, resulting in a negative value. Therefore, the output of the query would be -10.

Index

Packtpub.com

Subscribe to our online digital library for full access to over 7,000 books and videos, as well as industry leading tools to help you plan your personal development and advance your career. For more information, please visit our website.

Why subscribe?

- Spend less time learning and more time coding with practical eBooks and Videos from over 4,000 industry professionals

- Improve your learning with Skill Plans built especially for you

- Get a free eBook or video every month

- Fully searchable for easy access to vital information

- Copy and paste, print, and bookmark content

Did you know that Packt offers eBook versions of every book published, with PDF and ePub files available? You can upgrade to the eBook version at packtpub.com and as a print book customer, you are entitled to a discount on the eBook copy. Get in touch with us at customercare@packtpub.com for more details.

At www.packtpub.com, you can also read a collection of free technical articles, sign up for a range of free newsletters, and receive exclusive discounts and offers on Packt books and eBooks.

Other Books You May Enjoy

If you enjoyed this book, you may be interested in these other books by Packt:

Oracle Cloud Infrastructure for Solutions Architects

Prasenjit Sarkar

ISBN: 978-1-80056-646-0

- Become well-versed with the building blocks of OCI Gen 2.0 Cloud
- Control access to your cloud resources using IAM components
- Manage and operate various compute instances
- Tune and configure various storage options for your apps
- Develop applications on OCI using OCI Registry (OCIR), Cloud Shell, OCI Container Engine for Kubernetes (OKE), and Service Mesh
- Discover ways to use object-relational mapping (ORM) to create infrastructure blocks using Terraform code

Oracle Autonomous Database in Enterprise Architecture

Bal Mukund Sharma, Krishnakumar KM, Rashmi Panda

ISBN: 978-1-80107-224-3

- Explore migration methods available for Autonomous Databases, using both online and offline methods

- Create standby databases, RTO and RPO objectives, and Autonomous Data Guard operations

- Become well-versed with automatic and manual backups available in ADB

- Implement best practices relating to network, security, and IAM policies

- Manage database performance and log management in ADB

- Understand how to perform data masking and manage encryption keys in OCI's Autonomous Databases

Packt is searching for authors like you

If you're interested in becoming an author for Packt, please visit `authors.packtpub.com` and apply today. We have worked with thousands of developers and tech professionals, just like you, to help them share their insight with the global tech community. You can make a general application, apply for a specific hot topic that we are recruiting an author for, or submit your own idea.

Share Your Thoughts

Now you've finished *Developing Robust Date and Time Oriented Applications in Oracle Cloud*, we'd love to hear your thoughts! Scan the QR code below to go straight to the Amazon review page for this book and share your feedback or leave a review on the site that you purchased it from.

https://packt.link/r/1804611867

Your review is important to us and the tech community and will help us make sure we're delivering excellent quality content.

Download a free PDF copy of this book

Thanks for purchasing this book!

Do you like to read on the go but are unable to carry your print books everywhere?

Is your eBook purchase not compatible with the device of your choice?

Don't worry, now with every Packt book you get a DRM-free PDF version of that book at no cost.

Read anywhere, any place, on any device. Search, copy, and paste code from your favorite technical books directly into your application.

The perks don't stop there, you can get exclusive access to discounts, newsletters, and great free content in your inbox daily

Follow these simple steps to get the benefits:

1. Scan the QR code or visit the link below

https://packt.link/free-ebook/9781804611869

2. Submit your proof of purchase
3. That's it! We'll send your free PDF and other benefits to your email directly

www.ingramcontent.com/pod-product-compliance
Lightning Source LLC
Chambersburg PA
CBHW081457050326
40690CB00015B/2830